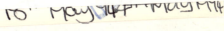

Essential Immunogenetics

A.R. WILLIAMSON
PhD, FIBiol, FRSE
Research Director, Glaxo Group Research Ltd,
Greenford, Middlesex

M.W. TURNER
PhD, FRSC, FRCPath
Reader in Immunochemistry,
Institute of Child Health,
University of London

BLACKWELL SCIENTIFIC PUBLICATIONS
OXFORD LONDON EDINBURGH
BOSTON PALO ALTO MELBOURNE

© 1987 by Blackwell Scientific
Publications
Editorial offices:
Osney Mead, Oxford, OX2 0EL
(Orders: Tel. 0865-240201)
8 John Street, London, WC1N
2ES
23 Ainslie Place, Edinburgh, EH3
6AJ
52 Beacon Street, Boston,
Massachusetts 02108, USA
667 Lytton Avenue, Palo Alto,
California 94301, USA
107 Barry Street, Carlton,
Victoria 3053, Australia

First published 1987

Set by Setrite Ltd. Hong Kong
and printed and bound in
Hong Kong by Dah Hua
Printing Press Co Ltd

DISTRIBUTORS

USA
Year Book Medical Publishers,
35 East Wacker Drive,
Chicago, Illinois 60601
(Orders: Tel. 312-726-9733)

Canada
The C.V. Mosby Company,
5240 Finch Avenue East,
Scarborough, Ontario
(Orders: Tel. 416-298-1588)

Australia
Blackwell Scientific
Publications, (Australia)
Pty Ltd, 107 Barry Street,
Carlton, Victoria 3053
(Orders: Tel. 03-347-0300)

British Library
Cataloguing in Publication Data

Williamson A R
Essential immunogenetics.
1. Immunogenetics
I. Title   II. Turner, M.W.
574.2'9   QR184

ISBN 0-632-00236-0

Contents

Acknowledgements, xi

1 **Introduction**, 1

Part 1   Proteins of the Immune System

2 **Immunoglobulins**, 11
Introduction
Cells involved in antibody production
Antibody structure: general considerations
Three-dimensional structure of antibody molecules
Immunoglobulin G
Immunoglobulin A
Immunoglobulin M
Immunoglobulin D
Immunoglobulin E
Biosynthesis of immunoglobulins
Evolution of immunoglobulins
References
Further reading

3 **Cell surface proteins of the major histocompatibility complex**, 44
Introduction
Allograft rejection: multigenic control
Plasma membrane structure
Transplantation antigens
Major transplantation antigens in man: HLA-A, -B of and -C
Major transplantation antigens in the mouse
Amino acid sequences of H-2 mutant proteins
Relationship between the major transplantation antigens of man and mouse
Major transplantation antigens of species other than man and mouse
Cell surface antigens controlled by genes linked to, but not in, the MHC
Immune associated (Ia) antigens
Structure of murine Ia antigens
Human Ia antigens: HLA-D locus products
Amino acid sequences of murine Ia antigens
Ia antigens of species other than man and mouse

Interspecies comparison of Ia antigen amino acid sequences
References
Further reading

4   **Proteins of the complement system**, 69

Introduction
Proteins of the classical pathway
Proteins of the alternative pathway
Terminal complement components
Control proteins of the complement system
Biosynthesis of complement components
Structural relationships between complement components
Complement receptors
References
Further reading

Part 2   Genes of the Immune System

5   **Classical genetics: immunoglobulins**, 97

Introduction
Isotypy in human immunoglobulins
Allotypy in human immunoglobulins
Idiotypy in human immunoglobulins
Allotypy in rabbit immunoglobulins
References
Further reading

6   **Antibody diversity**, 129

Introduction
Repertoires
Antibody response to the unexpected
Size of the antibody combining site
Three-dimensional structure of the antibody combining site
Specific antibody heterogeneity
Experimental control of antibody heterogeneity
Phenotypic markers defining individual antibody specificity
Correlation of antibody specificity, idiotypy and hypervariable
    regions
Quantitative estimates of the extent of diversity of the antibody
    phenotype
References
Further reading

7   **Molecular genetics of immunoglobulins**, 152

Introduction
Recombinant DNA technology: genetic engineering
Complementary DNA (cDNA)
Restriction endonucleases
Cloning vehicles or vectors
Tailing
Screening

Blotting: restriction endonuclease analysis
Electron microscopic analysis of cloned DNA
DNA sequencing
Genomic DNA libraries
The murine Igl locus
The murine Igk locus
V−J joining: special recombination sequences
Variability in the frame of somatic recombination
The murine Igh locus
The D_H segment
The C_H gene cluster
Each domain is encoded by an exon
The hinge region
The membrane (M) exons
Human immunoglobulin genes
Rabbit immunoglobulin genes
References
Further reading

8   **Expression of immunoglobulin genes**, 187
    Introduction
    Transcription of immunoglobulin genes
    Immunoglobulin gene expression as a function of differentiation
    Allelic exclusion
    Oncogene translocation to Ig loci
    Somatic mutations of immunoglobulin genes
    Origin of antibody diversity
    References
    Further reading

9   **Classical genetics of histocompatibility**, 215
    The mouse H-2 system
    The human HLA system
    References
    Further reading

10 **Molecular genetics of the major histocompatibility
    complex**, 256
    Introduction
    Class I genes
    Class II genes
    References

11 **Genetic aspects of the complement system**, 273
    Introduction
    C4
    C2
    C3
    Factor B
    Other complement components
    Immunogenetics of complement in other animals
    References
    Further reading

Part 3 Cells of the Immune System

12 **Differentiation antigens**, 289

Introduction
Thy-1(θ, theta): a marker for murine T cells
Thymus leukaemia antigens (TL antigens): murine thymocyte differentiation antigens
Lyt alloantigens: markers of mouse T lymphocyte functional subpopulations
Surface immunoglobulin and Ia antigens are differentiation markers of B lymphocytes
Lyb antigens: B-cell differentiation markers
Monoclonal antibodies specific for differentiation antigens
Human-lymphocyte differentiation antigens
The future
References
Further reading

13 **Genetics of cell—cell interaction**, 310

Introduction
Cell cooperation in antibody formation
Hapten—carrier-protein systems
Immune response (Ir) genes map onto cellular interactions in immune responses
Tetra-parental mice: evidence that the defect in low-responder mice is not at the B cell level
Genetic requirements for cellular interactions in immune responses: T—B cooperation
Genetic requirements for cellular interactions in immune responses: helper T cell interaction with APCs
'Soluble factors' involved in cell—cell regulation of immune responses
MHC-restricted recognition of antigen: cytotoxic T cells see self-MHC products plus foreign antigen
Allo-reactivity and the special nature of MHC antigens
Restriction specificity of T cells is acquired, not genetically predetermined
T cell antigen receptor
What determines the MHC specificity of T cell subsets?
References
Further reading

Part 4 Genetic Aspects of Immunological Diseases

14 **Genetics of the immunodeficiency disorders**, 345

Introduction
Defects of specific immunity
Defects of non-specific immunity
References
Further reading

15 Genetics of paraimmunoglobulinopathies and B lymphocyte neoplasia, 386

Introduction
Clinical and laboratory features of main syndromes
References
Further reading

16 Genetics of autoimmune disease, 402

Introduction
Diabetes
Autoimmune thyroid disease
Miscellaneous renal disorders
Myasthenia gravis (MG)
Systemic lupus erythmatosus (SLE)
Rheumatoid arthritis (RA)
Chronic active hepatitis
Coeliac disease
Dermatitis herpetiformis
References
Further reading

Glossary, 417

Index, 423

Acknowledgements

The production of this book has been more time-consuming than we could have imagined at the outset. The exponential increase in knowledge of immunogenetics that has occurred during the gestation period of this volume has necessitated much reappraisal along the way and many interim chapters became obsolete. The forbearance of the publishers has astonished us. We especially thank Per Saugman (who started the project), Peter Saugman (who revived it) and John Robson who saw it through to a conclusion. The hard work of transforming our text into print was undertaken, initially by Alison Walsh and latterly by Deborah Thompson and Jane Andrew. Oxford Illustrators skilfully converted our rough artwork into attractive figures and were most patient as we requested modification after modification. Many of our scientific colleagues have kindly offered advice and corrected some of our worst errors, although we must, of course, be held responsible for the finished product. We are particularly grateful to Roland Levinsky, James Mowbray, Marcus Pembrey, Sue Malcolm, Christine Kinnon, Ken Reid, Bob Sim and Duncan Campbell. Other colleagues have supplied illustrative material and are acknowledged in the text.

Chapter 1
Introduction

Immunogenetics can appear formidable to the new-comer. This book is intended to make the subject more approachable. We hope that *Essential Immunogenetics* will provide an entry to the field for those who may become specialists in the future, and for those already trained in many aspects of biology, for whom the dis-cipline of immunogenetics has much to offer. For the student of basic science or of medicine, *Essential Immunogenetics* provides a genetic description of immunology and an illustration of the application of immunogenetic methods to the immune system and points the way to broader applications for immuno-genetics.

Immunogenetics is much more than a collection of antisera defining antigenic variation or a list of genes. The immunogenetic method has provided a revealing view of the immune system. It also offers an approach to the description and understanding of many other biological systems. With this in mind we have tried to indicate the importance of methodology as a way of asking questions, but without dwelling on technical de-tails.

We have chosen to limit the book to an immunogene-tic introduction to immune systems because of the im-portance of those systems both in medicine and in the immunogenetic method. All aspects of the immune sys-tem are of course controlled by genes but the revelation of the working of the immune system has occurred at various levels. The division of this book into four parts reflects the way in which immunogenetics developed, some areas having been extensively studied at the level of the phenotype, either at the protein level or at the cellular level before any genetic input occurred. Other areas were studied at the level of whole animal genetics before molecular definition of phenotypes. Latterly, application of the recombinant DNA techniques of mole-cular genetics has revolutionized all molecular biology so that many areas of immunogenetics are now being revealed directly at the DNA sequence level.

1

Recombinant DNA technology was born of bacterial genetics and it is clearly a major revolution in the scientific approach to genes and gene expression. The molecular genetic method as applied to immunoglobulins is introduced in Chapter 7.

Monoclonal antibody technology is a product of the study of immunogenetics, and applications of monoclonal antibodies are making a major impact on the study of the genetics of the immune response and also on the immunogenetic method. Essentially, if a protein molecule can be defined by a monoclonal antibody then its expression and function can be studied; enough protein can be isolated by immune precipitation for a partial amino acid sequence to be determined at the micro level; from an amino acid sequence a DNA probe (albeit redundant) can be defined and chemically synthesized and the gene encoding the protein can be isolated, usually starting from complementary DNA (see Chapter 7).

Clearly the monoclonal antibody approach is a powerful way to enter any biological system. This is especially so since there is no need to have the original protein molecule in pure form at the outset.

These recent methods using recombinant DNA and monoclonal antibodies are moving our knowledge of immunogenetics forward at a rapid pace. In *Essential Immunogenetics* we have tried to cover the historical milestones so that the important findings and hypotheses illustrate how immunogenetics has progressed. Inevitably parts of the book will be out of date by the time it reaches the press. Our intention is that the book will remain valid as a frame of reference into which future findings can be slotted either as new pieces in the jigsaw or occasionally changing a piece here or there.

The book is divided into four parts that focus respectively on *proteins*, *genes*, *cells* and *diseases*. This somewhat artificial division is meant, as explained above, to reflect the development of the field.

In Part 1, 'Proteins of the immune system', we have chosen to cover immunoglobulins, major histocompatibility complex (MHC) proteins and the proteins of the complement system as being those molecules that are basic to the understanding of immunogenetics and immune responses. Also these are molecules where a biochemical description of the proteins involved preceded or paralleled an understanding of their genetics. By contrast, differentiation antigens or the T cell antigen receptor are molecules for which our understanding developed from cellular immunology and in some cases progressed directly to molecular genetics with little basic knowledge of the proteins involved. Therefore, differentiation antigens are dealt with in a separate chapter (12) later in Part 3, and the T cell antigen receptor is dealt with in Chapter 13 on the genetics of cell—cell interactions.

The second part has chapters on the classical genetics of each of the major molecules of the immune system considered in Part 1, i.e. the immunoglobulins, MHC antigens and proteins of the complement system. This is intended to give the reader a basic understanding of the approaches of classical immunogenetics to the immune system and by extrapolation to any other system. In addition to the presentation of the classical genetics of immunoglobulins in Chapter 5, a separate chapter (6) is devoted to the nature of antibody diversity, a special and fascinating aspect of immunoglobulin genetics. Following on the classical genetics of each of the major molecules is a discussion of the molecular genetics of that particular system. The immunoglobulins come first, and included in Chapter 7 on molecular genetics is a discussion of the methods of molecular genetics and the philosophy. Thereafter, the expression of immunoglobulin genes is discussed (Chapter 8), particularly because of the special controls involved in immunoglobulin gene expression and also as a vehicle for introducing the subject of gene expression in general as it will be of increasing relevance in immunogenetics. Chapter 9 on the classical genetics of the MHC, as revealed initially by transplantation of tumours and skin grafts, is followed by a chapter (10) on the molecular genetics of the MHC. A single chapter (11) covers the classical and molecular genetics of complement.

In Part 3, differentiation antigens are dealt with in Chapter 12 in the context of cells of the immune system. This is an essential base to the next chapter which deals with cell—cell interactions. The latter chapter offers a genetic description of cellular immunology. The complexities of cellular immunology have been more clearly described by the application of immunogenetics than by any other approach. The use of such simple but original immunogenetic methods as the introduction of inbred strains and the use of polyvalent alloantisera made possible important strides in our understanding of cellular immunology. The more recent use of monoclonal antibodies and the cloning of relevant genes is revealing a control system of apparent complexity, but one which may have many parallels in other differentiation systems.

Finally, Part 4 deals with genetic aspects of immunological disease. This section introduces the use of immunogenetics in the clinical situation. Immune deficiency diseases, paraimmunoglobulinopathies and autoimmune diseases are dealt with in the three separate chapters.

These areas have been selected because they highlight three different ways in which immunogenetics interacts with clinical immunology. The chapter (14) on primary immunodeficiency disorders, whilst revealing the paucity of information on the nature of the underlying defect for most of these diseases, also attempts to show how the situation may soon be transformed by the application of DNA hybridization technology. Some of the rare defects of complement proteins considered in the same chapter have given unexpected insights into the normal physiological control mechanisms regulating this complex system of proteins. The paraimmunoglobulinopathies, discussed in a separate chapter (15) are of particular historic interest because it was such studies of aberrant proteins that yielded early evidence supporting the concept of multiple genes controlling the production of a single immunoglobulin peptide chain.

The third chapter in this section (Chapter 16) discusses various autoimmune disorders, many of which show association, and in some cases linkage, with particular alleles of MHC antigens. Although the significance of these findings remains an enigma, enough is already known in some instances to guide physicians towards a more accurate assessment of a particular patient's prognosis and in other cases the choice between alternative modes of therapy may be decided by such information.

Although we have divided this book into parts and chapters for the reasons stated we would not wish the reader to miss the underlying theme of the immunogenetic method and the more fundamental genetic relationships within the immunogenetics of the immune system. It has been known for some time that immunoglobulin genes constitute a multigene family, i.e. a group

of homologous genes with similar functions. The homology link within this multigene family is a prototype
gene encoding a single immunoglobulin domain. This
single domain gene encodes about 110 amino acids;
these 110 encoded amino acids fold in a characteristic
pattern of β-pleated sheets to give the so-called immuno-
globulin- or antibody-fold that characteristically is fixed
by a covalent disulphide bond closing a loop about
60 amino acids long, centrally placed in the domain.
Historically it was the sequencing of human β_2-
microglobulin, a small protein of then (1973) unknown
function, that revealed it to be homologous to the immunoglobulin domain. Later it was found that β_2-
microglobulin is the light chain of class I MHC antigens
and that it interacts with the one domain (α_3) of the
heavy (α) chain of class I antigens, that is also an immunoglobulin homologue. This extended the evidence
deriving from the immunoglobulin molecules that immunoglobulin domains readily pair with themselves or
other immunoglobulin domains. The family of proteins
that are made up entirely or in part of homologous
immunoglobulin-like domains is now known to include,
in addition to all immunoglobulins, β_2-microglobulins,
MHC class I α chains, MHC class II α and β chains, the
antigen-specific T cell receptor chains α, β and γ, the
poly immunoglobulin receptor (originally identified as
secretory component of immunoglobulin A antibodies

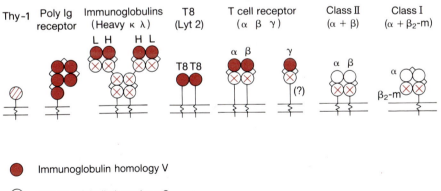

Immunoglobulin homology V

Immunoglobulin homology C

Immunoglobulin homology V/C

Immunoglobulin homology nil

Fig. 1.1 The immunoglobulin supergene family of proteins. Each is shown as a membrane
protein with a transmembrane hydrophobic peptide (〰). The domains are coded as indicated
in the key. Other members of this supergene family are known to exist.

in external secretions), Thy-1 (an alloantigenic marker on murine T cells) and T-8 (a differentiation antigen on human T cells) (Figure 1.1). Further sequences may well reveal other members of this supergene family. Determination of the exon-intron structures of genes in the immunoglobulin supergene family has shown that each immunoglobulin domain is encoded by a separate exon (Figure 1.2). This immunoglobulin domain exon has apparently been duplicated, mutated, moved to new locations and the products subjected to selective pressures to yield the supergene family as we now know it (see Table 1.1). In the special cases of immunoglobulin and T cell receptor V genes a somatic rearrangement mechanism has evolved that allows V genes to be expressed with C genes using a DNA joining mechanism. A site-specific recombination process is involved that depends on highly conserved recognition sequences

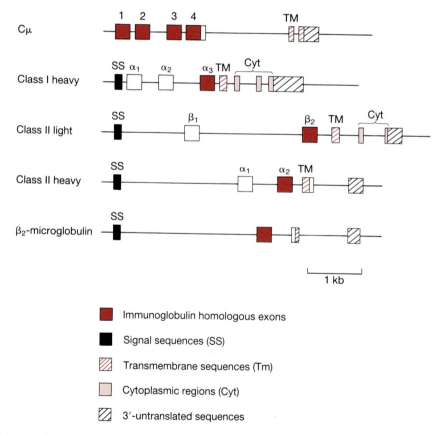

Fig. 1.2 The exon-intron structures of five genes of the immunoglobulin supergene family. Exons are designated as shown in the key.

Table 1.1 Chromosomal location of some loci of the immunoglobulin supergene family

Locus	Phenotype	Chromosome number	
		Mouse	Human
Igl	λ chain	16	22
Igk	κ chain	6	2
Igh	H chain	12	14
MHC	class I α	17	6
	class II α + β	17	6
β₂M	β₂ microglobulin	2	15
Tcra	T cell receptor α chain	14	14
Tcrb	T cell receptor β chain	6	7

common to the flanking regions of all rearrangeable elements in the immunoglobulin and T cell receptor multigene families. This mechanism allows the use of extensive repertoires of V genes and the creation of extra diversity by a lack of precision in the joining mechanism. Thus, the nature and origin of antibody diversity, one of the most fascinating challenges of immunogenetics, can now be explained in molecular genetic terms (see Chapters 7 and 8).

MHC antigens were originally recognized and defined as the major transplantation antigens (H-2), involved in rejection of tissue grafts between mice; a homologous set of MHC antigens (HLA) were identified in man. In both mice and men the extent of polymorphism of MHC antigens was unprecedentedly high. It is what are now called class I MHC antigens that are mainly responsible for rapid graft rejection and that are extremely polymorphic. Immune associated (Ia) antigen genes, now termed class II MHC antigens, were also mapped to the MHC region. The relationship between class II (Ia) and class I MHC antigens was shown by a series of experiments that demonstrated that T cells see foreign antigen presented in association with MHC antigens. Thus class I and II MHC antigens are key molecules in immune recognition. Other differentiation antigens originally identified as serological markers of T cell types are now known to function in the cell−cell interactions necessary for immune responsiveness.

A link between the MHC and complement exists in that certain complement genes map in the MHC. The complement system is an enzyme cascade mechanism activated by immune complexes and capable of lytic action on bacteria or foreign cells. In the complement cascade enzymic activity is generated by proteolytic cleavage of proenzymes but these same cleavage processes also generate many mediators of inflammation, in particular the multiple products of C3 cleavage. The twenty or so proteins constituting the complement system include at least three minifamilies of proteins showing structural and functional relationships (e.g. Clr-Cls, C2-Factor B, C3-C4-C5) and these are discussed in Chapter 4.

A hint of another multigene family has recently emerged with the publication of sequence homologies between: (1) the macrophage cell-surface glycoprotein (MAC-1 = CR3 receptor) which interacts with particles coated with C3bi; (2) the T and NK cell-surface protein LFA-1; and (3) the α-interferons. It is suggested that such molecules may constitute a very ancient group of genes encoding proteins involved in cell interactions of an antigen non-specific nature.

A new immunopharmacology is now rapidly emerging which encompasses the active peptides of the complement and kallikrein systems, polypeptide transmitters such as the lymphokines, and also low molecular weight molecules such as the prostanoids. Immunogenetics has a key part to play in the study of these lymphokines and receptor proteins for various transmitter molecules.

When the subject was in its infancy, Landsteiner's observations on blood group antigens were no doubt considered by many to be esoteric but they were, of course, destined to be of enormous practical importance. Now that the powerful resolution of the new technologies is slowly revealing more and more detail of the genome it is safe to predict that applications of immunogenetics in the clinic will continue to increase over the next few years. Much effort will be directed towards both prenatal diagnosis for a greater range of lethal conditions and reliable heterozygote identification. These tangible practical advantages for mankind may be attained much sooner than could possibly have been envisaged ten years ago and it is appropriate that the last part of this book should deal briefly with some of the clinical areas most likely to be affected.

Part 1
Proteins of the
Immune System

Chapter 2
Immunoglobulins

Introduction

The immunoglobulin are a family of structurally related proteins which mediate antibody responses and are found in most vertebrates. In higher mammals five major immunoglobulin classes are recognized (Table 2.1) and in some species subclasses have also been described. In all cases the proteins are manufactured by the plasma cells which are derived by differentiation from the B lymphocytes. Most of the immunoglobulin molecules produced by these cells are found in the serum and secretions of the body but small quantities also occur in a cell-associated form attached to the surface membranes of lymphocytes, macrophages, mast cells and basophils. The major characteristics of the recognized human immunoglobulins are shown in Table 2.2.

Cells involved in antibody production

Although antibody production is the prerogative of the differentiated B lymphocyte, the involvement of other cell types is usually essential. Only for responses to so-called T-independent polymeric antigens such as bacterial flagellin and pneumococcal polysaccharides has the presence of B cells alone been shown to be sufficient. For most of the so-called T-dependent antigens both B and T lymphocytes are required for an adequate

Table 2.1 Human immunoglobulins

Present nomenclature	Abbreviation	Previous nomenclature
Immunoglobulin G	IgG	γ-G globulin, 7S γ-globulin
Immunoglobulin A	IgA	γ-A globulin, β_2-A globulin
Immunoglobulin M	IgM	γ-M globulin, 19S γ-globulin
Immunoglobulin D	IgD	—
Immunoglobulin E	IgE	Reagin, IgND

Table 2.2 Major physicochemical, biological and metabolic characteristics of human immunoglobulins

Immuno-globulin class	Heavy chain	Molecular weight	Carbohydrate (%)	Mean serum concen-tration (mg/ml)	Complement fixation (classical pathway)	Placental transfer	Binding to mono-nuclear cells	Binding to mast cells and basophils	Half-life (days)	Distri-bution (% intra-vascular)
IgG1	γ1	146 000	2–3	9.0	++	+	+	–	21	45
IgG2	γ2	146 000		3.0	+		–	–	20	
IgG3	γ3	170 000		1.0	+++		+	–	7	
IgG4	γ4	146 000		0.5	–		–	–	21	
IgM	μ	970 000	12	1.5	+++	–	–	–	10	80
IgA1	α1			3.0	–	–	–	–		
IgA2	α2	160 000	7–11	0.5	–	–	–	–	6	42
Secretory IgA	α	385 000	7–11	0.05*	–	–	–	–	–	–
IgD	δ	184 000	9–14	0.03	–	–	–	–	3†	75
IgE	ε	188 000	12	0.00005	–	–	–	+	2†	50

*Higher concentration in external secretions.
†Half-life of cell bound molecules may be longer.

humoral response (see Figure 2.1). Such helper T cells function by focusing antigen and presenting it to the B cell in association with self-histocompatibility antigens. T−B cell cooperation is also regulated by macrophages which appear to play a role in antigen presentation, B cell activation and clonal differentiation. B cells stimulated to undergo cellular division and differentiation give rise to an expanded clone of antibody producing plasma cells (Figure 2.1). A small number of the differentiated cells are destined to become long-lived memory cells which, on subsequent contact with antigen, are able to proliferate rapidly to give the greatly amplified secondary antibody response.

Antibody structure: general considerations

The heterogeneity of antibody molecules was a major problem in early structural investigations. However, in several species, including mouse and man, pathological myeloma proteins occur which are structurally homogeneous immunoglobulins. These proteins, which

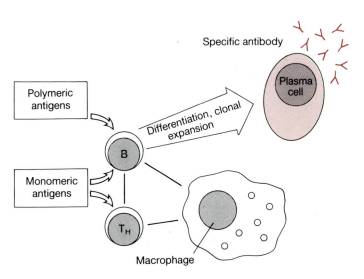

Fig. 2.1 Simplified summary of the main cell−cell interactions thought to precede commitment and differentiation of a B lymphocyte clone. Most antigens are focused onto the surface of a T helper lymphocyte (T$_H$) and are presented to the B cell in close association with self-histocompatibility antigens. Other antigens (usually polymers) are T-independent and are able to stimulate B cells directly to differentiate. The fully differentiated end-stage cell, the plasma cell, is essentially a protein factory producing large numbers of antibody molecules. See Part 3 for more details.

are the products of neoplastic plasma cells, are frequently present in serum in sufficient quantities to permit full immunochemical characterization. Without such proteins, detailed structural studies of the IgD and IgE classes would probably not have been possible.

Despite the major contribution made by myeloma proteins, our knowledge of immunoglobulin structure began with work on pooled rabbit IgG. In 1959 Porter showed that rabbit antibodies of the IgG class could be split by the plant protease papain into three large fragments which could be separated by ion-exchange chromatography. Two of these fragments (now known as Fab — *fragment antigen binding*) were identical and retained antigen binding capacity, whereas the third (Fc — *fragment crystalline*) was distinct, not able to bind antigen and associated with various effector or adjunctive functions. Subsequently, Edelman and Porter independently developed techniques for separating the peptide chains of immunoglobulins and in 1962 Porter proposed a basic four-chain structure for antibody molecules based on the two types of constituent peptide chain. This model, which is illustrated in Figure 2.2, appears to be universal throughout the vertebrate kingdom and comprises two small (light) polypeptide chains (mol. wt. 22 000) and two larger (heavy) polypeptide chains (mol. wt. 50−72 000). The heavy chains are invariably covalently joined through disulphide bridges and usually each heavy chain is linked similarly to a light chain. Overall molecular stability is maintained by non-covalent forces (see later).

The light chains are common to all classes of immunoglobulin and exist in two antigenically distinct forms called kappa (κ) and lambda (λ). In any one molecule both light chains are of the same type and, usually, both types are present in any one individual but the κ:λ ratio differs from species to species, from class to class and from subclass to subclass.

In contrast to the light chain, the heavy chain of an immunoglobulin molecule is characteristic of that class and only that class and this is indicated by the use of appropriate Greek letters for each of the distinctive chains, e.g. γ chains for IgG, α for IgA, μ for IgM, δ for IgD and ε for IgE. The chemical differences between heavy chains reside mainly in the Fc region of the molecule and are reflected in the antigenic and biological differences between the classes.

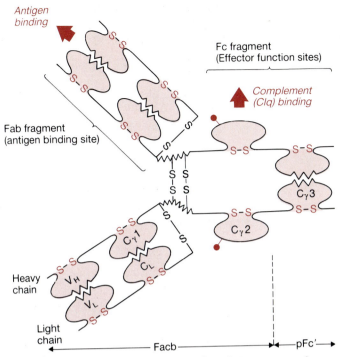

Antigen binding

Fc fragment
(Effector function sites)

Complement
(Clq) binding

Fab fragment
(antigen binding site)

$C_\gamma 3$

$C_\gamma 2$

$C_\gamma 1$

C_L

Heavy chain

V_H

V_L

Light chain

←————————— Facb —————————→ ←—pFc′—→

Fig. 2.2 Schematic representation of the four-chain structure of human IgG1 showing inter (S−S)- and intra (S−S)-chain disulphide bridges, carbohydrate side chains (●), papain-derived Fab and Fc fragments, plasmin-derived Facb fragment and the peptic pFc′ fragment. Each intra-chain disulphide bridge is associated with a structural domain (pink) and certain functions have been 'mapped' to particular domains. Antigen binding ($V_H + V_L$) and C1q binding (C_H2) sites are shown (red arrows). With the exception of $C_\gamma 2$, each domain is paired with another through strong non-covalent interactions. (Modified with permission from Roitt, 1984.)

The realization that serum myeloma proteins and urinary Bence Jones proteins represent pathological counterparts of normal, intact immunoglobulins and light chains, respectively, stimulated the application of protein-sequencing techniques to these proteins, and many such sequences are now available. Very early in the application of such techniques it became clear that both light and heavy chains are divisible into two distinct regions. In the case of light chains the carboxy-terminal half of the κ or λ chain (approximately 107 amino acid residues) does not vary except for certain minor differences which reflect either allotypic or isotypic variations. In contrast, the N-terminal region shows much sequence variability but the variability is not distributed evenly throughout the length of the region. Some positions in the sequence show exceptional variability and in both κ and λ light chains such *hyper-variable* regions are located near positions 30, 50 and 95 (see Figure 2.3). It is now generally accepted that such hypervariable residues are involved directly in the formation of the antigen binding site. For this reason they are also known as the complementarity determining regions CDR1, CDR2 and CDR3. Adjacent to these regions of hypervariability are many highly conserved amino acids called framework residues, which are thought to create the necessary rigid framework near which apparently unrestricted amino acid variability may occur. There are four framework regions (called FR1−FR4) flanking the three CDR segments and the locations of all these regions are shown in Figure 2.4.

Sequence studies of heavy chains have revealed that, in common with light chains, there are both variable (V) and constant (C) regions. In general, V_H regions are slightly longer than V_L regions (~113 and ~107 amino acid residues respectively) and have CDR regions of hypervariability between residues 31−35, 50−65 and 95−102 (see Figure 2.4). The constant regions of the

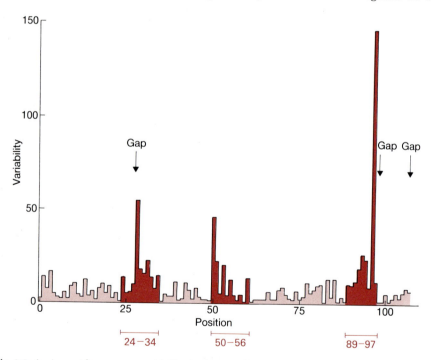

Fig. 2.3 Amino acid sequence variability in the variable region of human immunoglobulin light chains. In some sequences extra amino acids occur and in order to maximize the comparison these have been excluded as indicated by the arrows. The hypervariable regions are shown in red. (Reproduced from Wu & Kabat, 1970, with permission.)

Segment	Light chain	Heavy chain
FR1	1–23	1–30
CDR1	24–34	31–35
FR2	35–49	36–49
CDR2	50–56	50–65
FR3	57–88	66–94
CDR3	89–97	95–102
FR4	98–107	103–113

Fig. 2.4 Amino acid residue numbers of the four framework regions (FR1–FR4) and the three complementarity determining regions (CDR1–CDR3) of immunoglobulin light and heavy chain V regions. Residue numbering is from the amino-termini of the chains.

heavy chains are (depending on the class) either three or four times the length of the light chain constant region but resemble the latter in the striking level of sequence conservation within a class. Again, only minimal allotypic and isotypic variations have been observed when the amino acid sequences of heavy chains of the same class have been compared. Some of these variations are discussed in greater detail in Chapter 5.

Careful sequence comparisons of κ and λ light chain variable regions and heavy chain variable regions have shown that each is divisible into a number of subgroups as defined by the amino acids in the four framework segments. In man there are four subgroups of the κ chain V region (V_κI, V_κII, V_κIII and V_κIV) and six λ chain subgroups (V_λI–V_λVI). The amino acid residues which invariably occur at certain FR1 positions and thereby characterize these subgroups are shown in Figure 2.5. Similarly, there are characteristic residues defining these subgroups in the FR2, FR3 and FR4 regions.

The framework residues which are not invariant define a number of subsets for each FR segment of the major subgroups and such classifications have been of great value in the dissection of the genetic control of V-region diversity. It is clear, for example, that a protein having a particular FR1 set may belong to a different set when the FR2 sequence is considered, suggesting that the FR sets are independently assorted.

Human heavy chain V regions may also be classified into one of three subgroups (Figure 2.6) and various subsets as described above for κ and λ light chains.

Amino acid sequences of heavy and light chains from several animal species have been tabulated for easy comparison by Kabat et al. (1983) and the interested reader should consult this publication for further details.

Fig. 2.5

	1	2	3	4	5	6	7	8	9	10	11	12	13	14	15	16	17	18	19	20	21	22	23
Amino Acid No.																							
$V_{\kappa I}$	ASP	ILE			THR		SER	PRO	SER			SER			VAL	GLY	ARG	VAL	THR	ILE			CYS
$V_{\kappa II}$			VAL		THR		SER	PRO	LEU		LEU		VAL			GLY			ALA				CYS
$V_{\kappa III}$		ILE	VAL		THR		SER	PRO		THR	LEU	SER		SER	PRO	GLY					LEU	SER	CYS
$V_{\kappa IV}$	ASP		VAL		THR	GLN	SER	PRO			LEU	ALA	VAL	SER		GLY		ALA	THR				CYS
$V_{\lambda I}$	PCA	SER		LEU	THR		PRO	PRO	SER			SER			PRO	GLY							CYS
$V_{\lambda II}$	PCA	SER		LEU		GLN	PRO		SER			SER		SER	PRO	GLY			THR			SER	CYS
$V_{\lambda III}$	TYR			LEU			PRO	PRO	SER			SER	VAL		PRO	GLY				ILE	THR		CYS
$V_{\lambda IV}$	SER			LEU		GLN					VAL					GLY				ILE			CYS
$V_{\lambda V}$	PCA	SER	ALA	LEU	THR	GLN	PRO	PRO	SER		ALA	SER	GLY	SER		GLY	GLN	SER	VAL	THR	ILE	SER	CYS
$V_{\lambda VI}$				LEU			PRO		SER			SER		SER	PRO	GLY						SER	CYS

Fig. 2.5 Invariant residues in the FR1 segments of human κ and λ light chain V regions characteristic of the recognized subgroups. PCA indicates the 'blocked' N-terminus pyrrolid-2-one-5-carboxylic acid.

Fig. 2.6

	1	2	3	4	5	6	7	8	9	10	11	12	13	14	15	16	17	18	19	20	21	22	23	24	25	26	27	28	29	30
Amino Acid No.																														
$V_{H}I$				LEU			SER						LYS	PRO			SER					CYS				GLY		THR	PHE	
$V_{H}II$								GLY	PRO		LEU	VAL		PRO				LEU		LEU	THR	CYS			SER	GLY				
$V_{H}III$				LEU			SER	GLY	GLY					PRO	GLY		SER	LEU	ARG	LEU	SER	CYS			SER	GLY				

Fig. 2.6 Invariant residues in the FR1 segments of human heavy chain V regions characteristic of three recognized subgroups.

18   CHAPTER 2

As shown in Figure 2.2, the immunoglobulin light chain contains two *intra*-chain disulphide bridges — one in the variable and one in the constant region. Similarly, in the heavy (γ) chain (which is twice the length of the light chain) there are four such bridges. Each bridge encloses a peptide loop of between 50 and 70 amino acid residues and comparison of the amino acid sequences of these loops reveals a striking degree of sequence homology. Each loop represents the central part of what is known as a 'homology region' or domain of approximately 110 amino acid residues (see Figure 2.7). In the light chain these regions are called V_L and C_L respectively for the variable and constant parts. In the heavy chains the variable region is called V_H and there are at least three constant region domains, called C_H1, C_H2 and C_H3. Three constant region domains have been established for γ and α chains, and for human but not mouse δ chains. However, μ and ε chains are known to have four constant region domains, a feature which is consistent with the higher mol. wt. of these heavy chains (see below). The homology regions of different chains are distinguished by a specific nomenclature. Thus the constant regions of the γ chain of IgG are called $C_\gamma1$, $C_\gamma2$ and $C_\gamma3$, whereas the constant regions of the μ chain of IgM are called $C_\mu1$, $C_\mu2$, $C_\mu3$ and $C_\mu4$.

As shown in Figure 2.7, the V_H and V_L domains are larger than the constant region domains. This is consistent with the presence of an extra loop in the three-

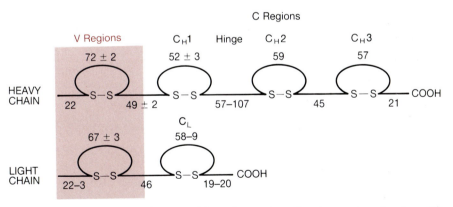

Fig. 2.7 Schematic diagram of human IgG homology regions showing number of amino acid residues enclosed by each disulphide bridge. The range for four subclasses is shown. Note that the range for the inter-C_H1–C_H2 region of γ_1, γ_2 and γ_4 chains is $57-60$ residues — the quadruplicated γ_3 hinge region is responsible for the wide range of values in this segment. (Modified from Milstein & Svasti, 1971.)

dimensional structure (see below). In general, the inter-domain distances are comparable except for that between the C_H1 and C_H2 loops which is longer. This is the so-called 'hinge' region, which is a characteristic feature of the four-domain IgG and IgA molecules, but is not shared, apparently, with the five-domain IgM and IgE molecules.

Three-dimensional structure of antibody molecules

In 1969 Edelman and Gall proposed that each homology region is folded into a compact globular structure and linked to neighbouring domains by more loosely folded sections of peptide chain. Furthermore, it was suggested that each domain had evolved to fulfil a specific function such as, for example, antigen binding (V_L + V_H domains) or complement fixation ($C_\gamma2$ domains).

X-ray crystallography of one IgA and two IgG myeloma proteins down to $0.2-0.3$ nm resolution has shown that the domains of the different proteins do indeed have the same basic folding pattern. In essence, this approximates to a cylindrical sandwich. The polypeptide chain runs back and forth in the direction of the long axis of the cylinder so that in one layer there are four stands of peptide chain and in the other layer there are three strands (Figure 2.8). In both layers adjacent segments run in opposite directions and the two layers are held together by the intra-chain disulphide bridge and hydrogen bonds in the so-called β-pleated sheet configuration. The amino acid side chains sandwiched between the layers are largely of a hydrophobic nature in all the domains, but the amino acid side chains facing outwards from the constant domains differ from those of the variable domain.

It is likely, from the gross similarities, that the variable and constant region genes are derived from a common ancestor which duplicated itself and then diverged and evolved towards the present system. In the course of this evolution the cylindrical domains of the constant regions came to interact with each other in pairs (e.g. in IgG the pairing is $C_\gamma3-C_\gamma3$, $C_\gamma2-C_\gamma2$ and $C_\gamma1-C_L$) and in each case the three-strand layers are on the outer facing surfaces while the four-strand layers interact strongly across a zone of hydrophobicity. This interaction is enhanced by an increased number of

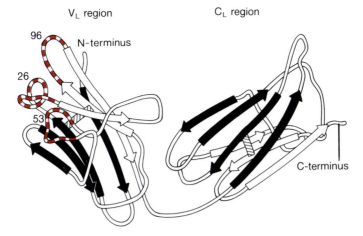

V_L region   C_L region

96
N-terminus
26
53
C-terminus

Fig. 2.8 Folding pattern of the polypeptide chain in the variable (V_L) and constant (C_L) region domains of a light chain. Note the similar sandwich-like structures with three strands in one layer and four strands in the other. The layers are stabilized by a disulphide bridge (cross-hatched) and hydrogen bonds. The hypervariable regions are indicated both by red striped segments and by certain selected residue numbers.

hydrophobic amino acid residues in the four-strand layers of the constant region domains.

In the case of the variable region domains the heavy and light chain components (V_H and V_L) interact with each other rather differently. The four-strand layers are now facing outwards and the three-strand layers interact with each other across the water-filled region which has become the antigen binding site. In areas where the structural integrity of the binding site requires it, hydrophobic residues appear to have displaced hydrophilic residues in the three-strand sections.

The antibody combining site is composed of loops of the hypervariable regions which come together to form a continuous surface interacting with the antigen through a combination of hydrogen bonding, electrostatic interactions and van der Waal's forces. This process has been studied in some detail in the case of a few antibody—hapten systems. For example, the mouse myeloma protein McPc 603 has a binding affinity for phosphorylcholine, and X-ray diffraction analysis has been used to establish the nature of some of the forces involved. The hapten is, of course, much smaller than most complex antigenic determinants and so does not fill the entire binding cleft of the antibody molecule. It is off-set towards the hypervariable regions of the heavy chain and most of the interactions that have been mapped are with heavy chain residues. The negatively charged phosphate group of the hapten interacts ionically with the positively charged NH_2 group of an arginine residue at position 52 of the heavy chain, as well as forming a strong hydrogen bond with an OH group of a neighbouring tyrosine residue (heavy chain residue number 33). The choline group has a positive charge and is close to two negatively charged glutamic acid residues (one at position 35 is in one hypervariable region of the heavy chain and the other, at position 58, occurs in the second hypervariable region). In addition, there are van der Waal's interactions between the choline group and carbon atoms in both heavy and light chains. These are shown in Figure 6.3b.

Immunoglobulin G

IgG is the major immunoglobulin in normal serum. In man it accounts for 70—75% of the total immunoglobulin pool. Isolated IgG is a monomeric protein with an apparent mol. wt. of 146 000 and a heterogeneous electrophoretic mobility ranging from α_2 to γ. In many mammalian species there are known to be subclasses of IgG. However, these appear to have arisen since speciation occurred and the subclasses of one species bear little relationship to those of another. In man four such subclasses are recognized (IgG1, IgG2, IgG3 and IgG4) and occur in the approximate proportions of 66, 23, 7 and 4% respectively. The subclasses cross-react antigenically but they also possess subclass-specific antigenic determinants in the C_H regions. The differences are, however, most dramatic in the hinge regions of the four variants (see below).

The first human γ-chain to be sequenced completely was the γ_1 chain of the Eu protein (Edelman et al., 1969) — see Figure 2.9 — and the numbering system for this chain is used frequently as a basis for comparisons between γ-chain sequences. The C_H2 and C_H3 domains of all four human G subclasses have now been sequenced (see Figure 2.10) and show a remarkably high degree of sequence homology. The sequences of the C_H3 regions show between 95 and 98% identity and C_H2 region sequences show 92—97% identity. The greatest structural differences between the subclasses seem to be found in the hinge region between the C_H1 and

```
CL   (109–129)  T V A A P S V F I F P P S D E Q – – L K S G T
CH1  (119–139)  S T K G P S V F P L A P S S K S – – T S G G T
CH2  (234–256)  L L G G P S V F L F P P K P K D T L M I S R T
CH3  (342–362)  Q P R E P Q V Y T L P P S R E E – – – M T K N Q


                              *
CL   (130–150)  A S V V C L L N N F Y P R E A K V – – Q W K V
CH1  (140–160)  A A L G C L V K D Y F P E P V T V – – S W N S
CH2  (257–279)  P E V T C V V V D V S H E D P Q V K F N W Y V
CH3  (363–383)  V S L T C L V K G F Y P S D I A V – – – E W E S


CL   (151–173)  D N A L Q S G N S Q E S V T E Q D S K D S T Y
CH1  (161–180)  – G A L T S G – V H T F P A V L Q S – S G L Y
CH2  (280–300)  D G – V Q V H N A K T K P R E Q Q Y – D S T Y
CH3  (384–404)  N D – G E P E N Y K T T P P V L D S – D G S F


                                                  *
CL   (174–196)  S L S S T L T L S K A D Y E K H K V Y A C E V
CH1  (181–202)  S L S S V V T V P S S S L G T Q – T Y I C N V
CH2  (301–323)  R V V S V L T V L H Q N W L D G K E Y K C K V
CH3  (405–427)  F L Y S K L T V D K S R W Q Q G N V F S C S V


CL   (197–214)  T H Q G L S S P V T – K S F – – N R G E C
CH1  (203–220)  N H K P S N T K V – D K R V – – E P K S C
CH2  (324–341)  S N K A L P A P I – E K T I S K A K G
CH3  (428–446)  M H E A L H N H Y T Q K S L S L S P G
```

Fig. 2.9 Amino acid sequences of C_L, C_H1, C_H2 and C_H3 homology regions of IgG1 protein Eu. Gaps, indicated by dashes, have been introduced to maximize homologies. Residues identical in three or more chains are shown in red boxes. The two stars represent the invariant cysteine residues which form the intra-domain disulphide bonds. One-letter code used for amino acid residues: A, alanine; C, cysteine; D, aspartic acid; E, glutamic acid; F, phenylalanine; G, glycine; H, histidine; I, isoleucine; K, lysine; L, leucine; M, methionine; N, asparagine; P, proline; Q, glutamine; R, arginine; S, serine; T, threonine; V, valine; W, tryptophan; Y, tyrosine. (Modified from Hahn, 1982, with permission.)

C_H2 domains. Here one finds only 50–60% sequence homology and much variation in the pattern of disulphide bonding both in the human G subclasses and those of other species (Figures 2.11 and 2.12). In the human subclasses there are two inter-chain disulphide bonds between both γ_1 and γ_4 chains and four between γ_2 chains. The number of such bonds in IgG3 is still controversial but may be as high as fifteen. The IgG3 hinge has approximately 47 extra amino acids compared to the other subclasses and this gives the γ_3 chain its higher molecular weight compared to γ_1, γ_2 and γ_4 chains.

The peptide structure of the hinge region is susceptible to the attack of proteolytic enzymes such as pepsin, papain and trypsin and over the past two decades each

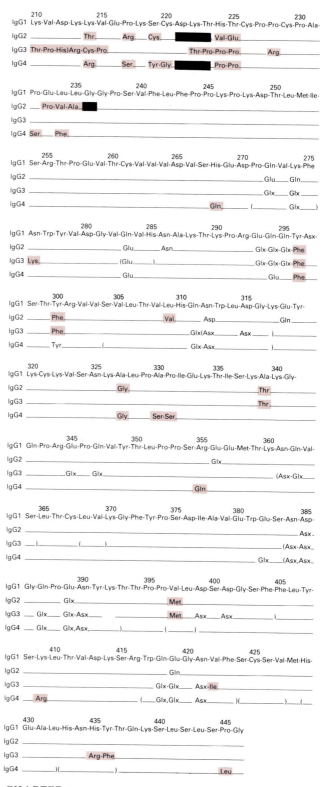

Fig. 2.10 Amino acid sequences of the C$_H$2 and C$_H$3 regions of human IgG1, IgG2, IgG3 and IgG4 subclass proteins. The IgG2, 3 and 4 sequences differ from that of IgG1 as indicated by the substitutions shown in pink. Boxed sections in black indicate gaps introduced to maximize homologies.

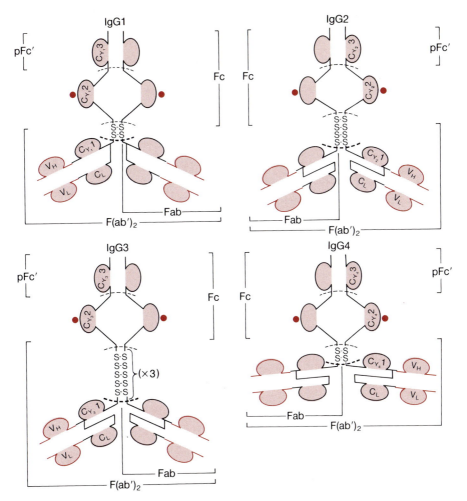

Fig. 2.11 Polypeptide structure of four human IgG subclasses showing constant and variable regions (black and red lines respectively) of light and heavy chains with the subclass-specific nomenclature for each domain. The major cleavage points of pepsin (heavy dashed line) and papain (light dashed line) are indicated, together with the nomenclature of the various fragments generated. Oligosaccharide side chains are represented (●). (Modified from a figure originally used in *Immunology Today* (1980) **1**, 1.)

of these has been exploited in structural studies of immunoglobulins. Using papain and trypsin Fc and Fab fragments are generated from all four subclasses although there are marked differences in susceptibility. IgG3 is very rapidly cleaved and the other subclasses have decreasing sensitivity in the order IgG1 > IgG4 > IgG2. In each subclass the enzymes cleave the γ-chains on the N-terminal side of the inter-heavy chain disulphide bonds but cleavage at other points (particularly between

25 IMMUNOGLOBULINS

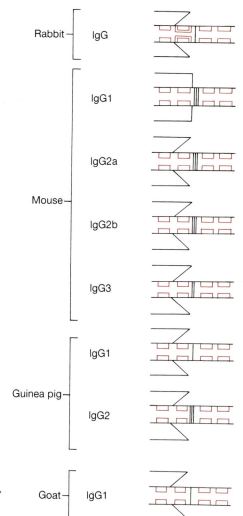

Rabbit — IgG

Mouse — IgG1
IgG2a
IgG2b
IgG3

Guinea pig — IgG1
IgG2

Goat — IgG1

Fig. 2.12 Gross structure of IgG subclasses from rabbit, mouse, guinea pig and goat. Heavy and light polypeptide chains are indicated by the long and short black lines, respectively, and inter- and intra-chain disulphide bridges by red lines.

domains) also occurs. A major 'secondary' fragment released from the C_H3 domain after prolonged papain digestion is the Fc' fragment. The ease with which the enzymes papain and trypsin cleave the IgG molecule into two Fab and one Fc fragment suggests that the molecule has a natural tripartite structure. Electron microscopy of IgG antibodies supports this view and indicates that IgG is extremely flexible in the hinge region. Detailed crystallographic analyses of IgG Fc and Fab fragments have shown the manner in which domains interact with each other in the intact molecule. The $C_\gamma3$ domains appear to pair in a similar manner to the C_L and $C_\gamma1$ domains whereas the $C_\gamma2$ domains are

Fig. 2.13 The three-dimensional structure of IgG as revealed by X-ray crystallographic analysis. The molecule comprises four peptide chains: two identical heavy chains (shown in dark grey) and two identical light chains (shown in light grey). Carbohydrate bound to the Fc portion of the molecule is also shown (marked with a cross). The structure of the immunoglobulin was determined by David R. Davies and his colleagues (Silverton *et al.*, 1977). The figure was generated by computer graphics using the system developed by Richard J. Feldmann at the National Institutes of Health, USA.

not in contact with each other (see Figure 2.13). The oligosaccharide side chain attached to the asparagine residue at position 297 may contribute to this lack of interaction by preventing appropriate residues from making contact (see Feinstein & Beale, 1977, for more detailed discussion). The conformation of the $C_\gamma 2$ domains of IgG appears to be critical for the expression of several effector functions of the molecule. Much future effort will no doubt be directed towards a more complete understanding of the topology and flexibility of this region.

Immunoglobulin A

In man IgA accounts for about 15–20% of the total serum immunoglobulin pool, where it exists mainly in the form of four-chain monomers ($\alpha_2 L_2$). However, higher mol. wt. polymers also occur and in most mammals are the predominant form — especially 10S dimers. Monomeric IgA has a mol. wt. of 160 000 and the α heavy chains are of similar size to γ chains. In man two subclasses of IgA have been identified, IgA1 and IgA2. The IgA1 subclass predominates in serum (80% of total IgA) but in seromucous secretions the two subclasses occur in approximately equal proportions. The complete amino acid sequence of a human IgA1 protein (see Figure 2.14) confirmed that the molecule did indeed have three constant homology regions. In two of these domains ($C_\alpha1$ and $C_\alpha2$) there was found to be an additional intra-chain disulphide bridge. Another major structural difference between IgA and IgG is the presence in the former of an additional C-terminal octadecapeptide with a penultimate cysteine. However, this feature is shared with IgM and in the polymeric forms of both proteins provides a covalent link to a low mol. wt. peptide called the J chain (see Figure 2.15).

As in the case of the human IgG subclasses the major differences between IgA1 and IgA2 molecules appear to be located in the hinge region. The α_2 chains lack 12 amino acid residues and two galactosamine-rich oligo-saccharide units which are present in the hinge of α_1 chains. In addition, there is a duplicated sequence of 7 amino acid residues in the α_1 hinge which is not found in the α_2 hinge.

It is now generally believed that the major role of IgA in the body is to provide antibody activity at or near mucosal surfaces. Seromucous secretions such as saliva, tracheobronchial secretions, bile, colostrum, milk and genitourinary secretions are a particularly rich source of a special form of IgA called secretory IgA. This molecule exists mainly as a dimer of two IgA four-chain units, covalently linked by J chain and incorporating another peptide chain called secretory component (SC). The J chain is a product of the plasma cell secreting the IgA molecule and is added just before secretion. In contrast, the secretory component is not synthesized by plasma cells but probably by epithelial cells. It is believed that SC actively assists in the process of transporting the IgA dimer across epithelial cell layers and into secretions.

Recent studies have shown that the secretory component is not synthesized as such but is part of a larger

Fig. 2.14 — Primary amino acid sequences of immunoglobulin constant regions (sequence alignment chart).

Band 1 (residue positions 1 – 30)

```
           1               10                  20                  30
κ          Thr Val Ala Ala Pro Ser Val Phe Ile Phe Pro Pro Ser — — Asp Glu Gln Leu Lys Ser Gly — Thr
λ          — — — Gln Pro Lys — — Ala Ala Pro Ser Val Thr Leu Phe Pro Pro Ser Ser Glu Glu Leu Gln Ala Asn — Lys
γ1 C1      — — Ala Ser Thr Lys — — Gly Pro Ser Val Phe Pro Leu Ala Pro Ser Ser Lys Ser Thr — Thr
   C2      — — — Gln Leu — — — Leu Gly Glu Pro Ser Val Tyr Thr Leu Pro Pro Ser — Arg Asp — Glu Leu Thr Lys Asn — Gln
   C3      — — — — Gln Pro Arg — — Glu Pro Gln Val Tyr Thr Leu Pro Pro Ser — Arg Asp — Glu Leu Thr Lys Asn — Gln
μ  C1      — — — — — Ala Pro Thr Leu Phe Pro Leu Val — Ser [Cys] Glu Asn Ser Pro — Asp — Thr
   C2      Val Ile Ala Glx Leu — — Pro Pro Lys Val Ser Val Phe Ala Ile — Pro Arg Asp Gly Phe Phe Gly Asx Pro Arg Lys
   C3      Val Pro Asp Glx — — Asx Thr — Ala Ile Arg Val Phe Ala Ile Pro Pro Ser Phe Ala Ser Ile Phe Leu Thr — Lys Ser
   C4      Val — Ala — — — — — Leu His Arg Pro Asp Val Tyr Leu Leu Pro Pro Ala — Arg — Glu Gln Leu Asn — Leu Arg Glu
α1 C1      — — Ala Ser Pro — — Thr Ser Pro Lys Val Phe Pro Leu Ser Leu — Leu [Cys] — Ser Thr Glx Pro Asx Gly Asx — 
   C2      — — — — His Pro Arg Leu Ser Leu His Arg Pro Ala Leu — — Glu Pro — — Ser Gln Glu Leu Ala Leu Asn Gln Leu
   C3      — — — — Asn Thr Phe His — Pro Glu Leu — Pro Pro Val Val Leu — — Asn — Ser — Asn Ala Thr
ε  C1      Gly Ser — — — — Thr — Thr Gly Pro Thr Val Lys Ile Leu Glx — Ser Ser [Cys] Asp — Leu Gly — His Phe Pro
   C2      — — — Arg Asx Phe Thr Pro Pro Thr Val Lys Ile Leu Glx Ser Ser [Cys] Asx — Pro Ser — Leu Gly — His Phe Pro
   C3      Asp Ser Asp — Pro Arg — — Gly — Val Ser Ala Tyr Leu Ser Arg Pro Ser Pro Phe Asp — Leu Phe Ile — Arg Lys Ser
   C4      — — — Pro Arg — — Ala Ala Pro Glu Val Tyr Ala Phe Ala — Thr Pro — Glu Trp — Gly Ser Arg Asp Lys
```

Band 2 (residue positions ~37 – 64; conserved Cys boxed at ~38 and ~57)

```
                40                       50                      60
κ          Ala Ser Val — Val [Cys] Leu Leu Asn Asn Phe Tyr Pro — — Arg Glu Ala Lys Val Gln — — Trp Lys Val — Asp — Asn Ala Leu
λ          Ala Thr Leu — Val [Cys] Leu Ile Ser Asp Phe Tyr Pro — — Gly Ala Val Thr Val Ala — — Trp Lys Ala — Asp — — Ala Leu
γ1 C1      Ala Ala Leu — Gly [Cys] Leu Val Lys Asp Tyr Phe Pro — — Glu Pro Val Thr Val Ser — Trp Asn Ser — — Gly Ala Leu
   C2      Val Ser Leu — Thr [Cys] Leu Val Lys Gly Phe Tyr Pro — — Ser Asp Ile Ala Val Glu — Trp Glu Ser — Asx Gly Glu Pro —
   C3      Ala Ala Leu — Gly [Cys] Leu Val Lys Gly Phe Tyr Pro — — Ser Asp Ile Ala Val Glu — Trp Glu Ser — Asx Gly Glu Pro —
μ  C1      Val Ala Val — Gly [Cys] Leu Ala Gln Asp Phe Leu Pro — — Arg Gln Ile — Gln Val Ser — — Trp Leu Arg — Glu Gly Lys Gln Val
   C2      — Lys Leu — Ile [Cys] Gln Ala Thr Gly Phe Ser Pro — — Asp Gln Ile Ser Val Gln — — Trp Leu Arg — — Gln Gly Lys Gln Val
   C3      Thr Lys Leu — Thr [Cys] Leu Val Thr Asp Leu Thr Tyr — — Ala Asp Ser Val Phe Gln — — Trp Gln Met — Arg Gly Gln Pro Leu
   C4      Ala Thr Ile — Thr [Cys] Leu Ala Thr Gly Phe Ser Pro — — Gln — — Gly Val Thr Val Ser Trp Ser — — Glx Pro Asx Gly Glx —
α1 C1      — Thr — — Ala [Cys] Thr Leu Thr Gly Leu Arg Asp — — Ala Ser — Gly Val Thr Phe Thr Trp Thr Pro — — Ser Gly —
   C2      Ala Asx Leu — Thr [Cys] Thr Leu Thr Gly Leu Arg Asp — Lys — Asp — Val Leu Val Arg — — Trp Leu Gln — — Gly Ser Glx —
   C3      Val Thr Leu — Thr [Cys] Leu Ala Arg Gly Phe Ser Pro — — Glu Asp Val Leu Val Arg — — Trp Leu Gln — — Gly Ser Glx —
ε  C1      Val Thr Leu — Gly [Cys] Leu Ala Thr Gly Tyr Phe Pro — — Glu Pro Val Met Val — — — Trp Asx — — — Thr — Gly Ser Leu
   C2      Pro Thr Ile — Glx [Cys] Leu Val Ser Gly Tyr Thr Pro — — Gly — Thr Ile Asn Ile Thr — — Trp Leu — — — Glx Asx Gly Glx —
   C3      Pro Thr Ile — Thr [Cys] Leu Val Val Asx Leu Ala Pro Ser Lys Gly Thr Val Asn Leu Thr — — Trp Ser Arg Ala Ser Gly —
   C4      Arg Thr Leu — Ala [Cys] Leu Ile Gln Asn Phe Met Pro — — Glu Asp Ile Ser Val Gln — — Trp Leu His Asn Glu Val Gln Leu Pro
```

Band 3 (residue positions ~65 – 95)

```
           65          70                   80                      90        95
κ          Gln Ser Gly Asn — — Ser Gln — — — — Glu Ser Val Thr — Gln Gln Asp Ser Lys Asp Ser — — Thr Tyr Ser Leu Ser
λ          Lys Ala Gly Val — — — — — — Glu Thr Thr Thr Pro Ser Lys Gln Ser — Asn Asn — Lys Tyr Ala Ala Ser
γ1 C1      Thr Ser Gly Val — — — — His — Thr Phe Pro Ala Val Leu Gln Ser — Ser Gly — Leu Tyr Ser Leu Ser
   C2      — — — — — His — — Asn Ala Lys Thr Lys Pro Arg Glu Glu Gln Tyr — Asx Ser Thr Tyr Arg Val Val
   C3      Glx Asp — Asn Tyr — — — His — Asn Ala Lys Thr Lys Pro Arg Glu Glu Gln Tyr Asx Ser Thr Tyr Arg Val Val
μ  C1      Lys Asx Asn Tyr Lys Asp Ile Ser — Thr — — Arg Gly Phe Pro Ser Val Leu Arg Gly Gly Lys Tyr Ala Ala Thr Ser
   C2      — — — — — — — — Thr — Thr Asx — Gln Val Met Lys Thr — Ala Lys Thr Gln Pro Arg Trp Met — — Gly Glx Arg Ala — [Cys]
   C3      Lys Thr His Thr Asx Ile Ser — Glx — — Thr Ser Pro Met Pro Glu Pro Gln Ala Pro Gly Arg — Tyr Phe Ala His
   C4      — — Gly Val — Lys Tyr Val — — Thr — Ala Leu Glu Gln Pro Glu Gln Glx Ser — Ala Ala Pro Gly — Arg Tyr Thr Thr Ser
α1 C1      — Gly — Val — — — — Thr Phe — Thr Val Thr Ser Thr Leu Pro Val — — Gly — His [Cys] Tyr Thr Thr Ser
   C2      Arg — — Glx Lys Tyr Leu — — Thr — Trp Ala Ser Arg Gln Glu Pro Ser Gln Gly Thr Glu Leu Pro His Pro Ala
   C3      — Asx Gly — — Thr — — Thr Leu Pro — Ala Thr Pro Glu Asp Asx Gly Thr — Tyr Thr Ala Tyr Ser Thr Ile
ε  C1      Met Asp — Val — Ala Ser — — Leu Ser — — — Ala Ser Thr Glu Gln Gly Leu Glu Glu Thr Glu
   C2      Lys Pro — — Val Asx — — His — — — Thr Ala Ser Arg Thr Leu Glu Lys Gln Arg Asn Gly Thr Leu Thr Val
   C3      — — — — Asp — Ala Arg — — His Ser — Thr Thr Gln Pro Arg Lys Thr Lys Gly Ser — Gly — Phe Phe Val Phe
   C4      — — — — Asp Ala Arg — His Ser — Thr Thr Gln Pro Arg Lys Thr Lys Gly — Ser Gly — Phe Phe Val Phe
```

Band 4 (residue positions ~98 – 125; conserved Cys boxed at ~115 and ~125)

```
                100                     110                      120
κ          Ser Thr Leu Thr Leu Ser — Lys Ala — Asp Tyr — Glu Lys — His Lys Val Tyr Ala — [Cys] Glu Val Thr His Gln Gly — — Leu Ser Ser
λ          Ser Tyr Leu Ser Leu Thr Pro — Glu — Gln Trp Lys Ser — His Arg Ser Tyr Ser — [Cys] Gln Val Thr His Glu Gly — — Ser Thr
γ1 C1      Ser Val Val Thr Val Pro Ser Ser — Ser Leu — Gly Thr Gln Thr Tyr Ile — — [Cys] Asn Val Asn His Lys Pro — — Ser Asn Thr
   C2      Ser Val Leu Thr Val Leu — — His Gln Asp Trp Leu Asn Gly — Lys Glu Tyr Lys — [Cys] Lys Val Ser Asn Lys Ala — — Leu Pro Ala
   C3      Ser Val Leu Thr Val Asp — Lys Ser — Arg Trp Gln Gln Gly Asn Val Phe Ser — [Cys] Ser Val Met His Glu Ala — — Leu His Asn
μ  C1      Ser Gln Val Leu Leu Pro Ser Lys — — Asp Val Met Gln — Gly Thr Asp Glu His Val Val [Cys] Lys Val Glx His Pro — — Asx Gly Asx
   C2      Ser Thr Leu Thr Ile Lys Glu — Ser — Asp Trp Leu Ser Gln Ser — Met Phe Thr — [Cys] Arg Val Asp His Arg Gly — — Leu — Thr
   C3      — — — — — — — — Glu Asx Trp Asn Thr Gly Glu — Thr Tyr Thr — [Cys] Val Ala His Glu Ala — — Leu Pro Asn
   C4      Ser Ile Leu Thr Val Ser — Glu Glu — Glu Trp Asn Thr Gly Glu Thr Tyr Thr [Cys] Val Val Ala His Glu Ala Leu His Asx His — Thr Asx Lys Ser
α1 C1      Ser Ile Leu Thr Val Ser Gly Ala — Leu [Cys] Leu Ala Ala — Lys Gln Val Thr — [Cys] His Val Lys His Tyr Thr — — Asx Pro Ser
   C2      Ser Val Leu Ser Gly Leu Pro — Ala Thr Glx [Cys] Leu Lys Ser — Trp Asp His Gly — [Cys] Thr Ala Gln His Pro Glu — — Ser Lys Thr
   C3      Ser Ile Leu Arg Val Ala — Ala — — Glx Asx Trp Leu Lys — Gly — Gln Met Phe Thr — [Cys] Arg Val Ala His Thr Pro — — Ser Ser Thr
ε  C1      Ser Ile Leu Thr Val Ser — Gln — — Ala — Asp Arg Thr Tyr — — Thr — [Cys] Glu Val Thr Tyr Glx Gly — — His Thr Phe
   C2      Ser Glu Leu Thr Trp Gly — Arg Thr — Val — — Arg Thr Ser Thr — Ser Glx — [Cys] Thr Val Thr His Pro — — His Leu Pro Arg
   C3      Ser Thr Leu Thr Val Ser Gly — Ala Trp Arg — Glu Thr Thr Tyr Glx — [Cys] Arg Val Thr His Pro His — — Leu Pro Arg
   C4      Ser Arg Leu Glu Val — — Thr Arg Ala Glu — Trp Gln Glu Lys — Asp Glu Phe Ile — [Cys] Arg Ala Val His Glu Ala Ala Ser Pro Ser
```

Band 5 (residue positions ~128 – 159; interchain Cys residues boxed)

```
               130                       140           150          159
κ          Pro Val Thr — Lys Ser Phe Asn Arg Gly Glu [Cys]
λ          — Val — — Glu Lys Thr Val Ala Pro Thr Glu [Cys] Ser
γ1 C1      Lys Val — Asp Lys Lys Val Glu Pro Lys Ser [Cys] Asp Lys Thr His Thr [Cys] Pro Pro [Cys] Pro Ala Pro
   C2      Pro Ile — Glu Lys Thr Ile Ser Lys Ala Lys Gly
   C3      His Tyr Thr Gln Lys Ser Leu Ser Leu Ser Pro Gly
μ  C1      Lys — — Glu Lys Leu Asp Asp — Val Lys Met [Cys]
   C2      Phe Gln — — — Asx — — Ala Ser Ser Met [Cys] — — Pro Thr Leu Tyr Asx Val Ser Leu Val Met Ser Asx Thr Ala Gly Thr [Cys] Tyr
   C3      Arg Val Thr Gln Arg Thr Val Asp Lys Ser Thr Gly Lys Pro Thr
α1 C1      Asx Asn — Val Lys Thr Phe Ser Val [Cys] Ser — — Ser Cys Cys
   C2      Pro Leu Thr Ala — Thr Leu Ser Lys — Ser Gly His Val Asx Val Ser Val Val Met Ala Glu Val Asp Gly Thr [Cys] Tyr
   C3      Ala Phe Thr Gln Lys Thr Ile Asp Arg Leu Ala Gly Lys Pro Thr
ε  C1      Asx Asn — Val Lys Thr Phe Ser Val — — [Cys] Ser
   C2      Pro — — Glx Asx Ser — Thr Leu — Ser [Cys] Ala
   C3      Ala Leu — — Met Arg Ser Thr Thr Lys Thr Ser Gly
   C4      Gln Thr Val Gln Arg Ala Val Ser Val Asn Pro Gly Lys
```

Fig. 2.14 Primary amino acid sequence of the constant regions of human κ, λ, γ₁, μ, α₁ and ε chains. Gaps have been introduced to maximize homology. (Adapted from Kratzin *et al.*, 1975, with permission).

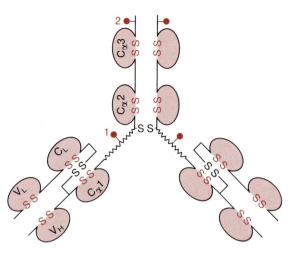

Fig. 2.15 Polypeptide structure of human IgA1 molecule. The approximate location of galactosamine and glucosamine oligosaccharides are indicated by red circles numbered 1 and 2 respectively. For the sake of clarity additional intra-chain S−S bridges between the hinge region and the $C_\alpha 1$ domain and between the hinge region and the $C_\alpha 2$ domain are omitted.

precursor receptor molecule which is present at the basolateral surface of many glandular epithelial cells. Here it binds to polymeric IgA (and to a lesser extent IgM) by both covalent and non-covalent interactions. The complex then undergoes endocytosis and is transported across the cell in vesicles before release across the apical cell surface. During transport the receptor molecule is cleaved by a proteolytic enzyme (hence the term 'sacrificial receptor') and the portion remaining attached to the immunoglobulin is that which was previously called the secretory component. Mostov *et al.* (1984) have cloned and sequenced cDNA for the rabbit receptor and shown that it consists of a short signal peptide (18 amino acid residues), an extracellular immunoglobulin binding region (629 residues), a transmembrane hydrophobic region (23 residues) and a carboxy-terminal cytoplasmic region (103 residues). Structurally, the immunoglobulin binding region was found to comprise five disulphide-bridged domains of approximately 100−115 amino acid residues and, interestingly, each showed a significant degree of homology with the immunoglobulin superfamily. It is possible that the interaction between the receptor domains and the domains of the immunoglobulin is similar to that between the Ig domains themselves. It is likely that the fifth domain of the receptor is disulphide linked to one of the IgA subunits and that the remaining domains of the receptor interact non-covalently with both IgA monomers. A possible structure for the secretary IgA molecule is shown in Figure 2.16.

Immunoglobulin M

IgM occurs predominantly in man as a pentamer $(\mu_2 L_2)_5$ with a mol. wt. of 970 000. Significant amounts of a low molecular weight monomeric form occur in the serum

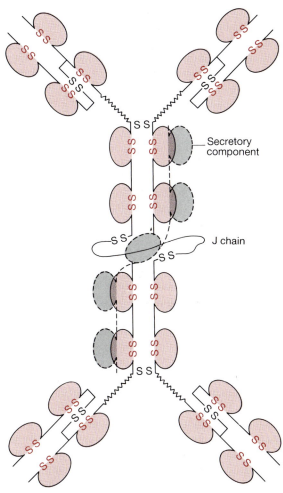

Fig. 2.16 Schematic diagram of human secretory IgA showing possible arrangement of IgA monomers, J chain and secretory component. Details of the pairing between the domains of the SC and the α chain have yet to be established.

of some patients with immunological disorders (in particular Waldenström's macroglobulinaemia and ataxia telangiectasia).

Structurally, IgM consists of five subunits based on the usual two heavy (μ) chains and two light chains polymerized in association with one J chain (Figure 2.17). Each μ chain has five intrachain domains, each of similar size to the γ-chain domains (see Figure 2.14). The μ chain is rich in carbohydrate, having five oligosaccharide groups per chain. Each is attached to an asparagine residue in the obligatory sequence Asn-X-Ser/Thr and both simple and complex side chains occur. Their role (if any) is not clear but they are known to increase the solubility of the molecule and to influence its conformation.

An IgM-like protein may be found in the serum of most vertebrates but its structure differs as one moves up the evolutionary tree. In most teleost fish the molecule is tetrameric, in amphibia it is usually hexameric and in reptiles, birds and mammals it is pentameric. Primitive immunoglobulins are structurally more closely related to human IgM than to human IgG or IgA.

Immunoglobulin D

IgD is a trace serum protein and accounts for less than 1% of the serum immunoglobulin pool. It should be regarded as essentially a lymphocyte membrane protein and its true function is probably to be found in the poorly understood processes of lymphocyte activation and differentiation.

Until recently, structural studies on IgD were less complete than was the case with the other immunoglobulins. However, two groups have now independently sequenced the Fc region of the human δ chain and it is established that the relatively high mol. wt. of IgD (184 000) is not due to the presence of a fifth domain as occurs in IgE (mol. wt. 188 000). The probable domain structure is shown in Figure 2.18.

The protein has a single disulphide bridge between the δ chains. During isolation IgD is liable to undergo spontaneous proteolysis by plasmin if ε-amino caproic acid is not added as an enzyme inhibitor. Structural studies have shown that the hinge region of IgD is similar to the hinge of IgA1. Indeed, the recent sequence data suggests that the δ chain gene emerged very early in the evolution of Ig heavy chain genes, branching off

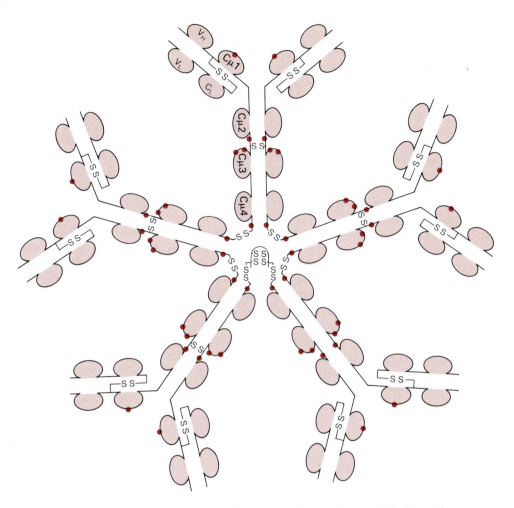

Fig. 2.17 Schematic diagram of human IgM showing homology regions, carbohydrate side chains (●) and possible location of J chain (centre).

the α-chain gene shortly after the divergence of the μ- and α-chain genes.

The amino acid sequence of the mouse δ chain has also been deduced following DNA sequence analysis of the gene coding for the δ heavy chain. This provided the unexpected information that mouse IgD possessed only two constant region domains per δ chain (see Figure 7.29). From sequence homologies with other classes these are probably C_H1 and C_H3 domains. These regions are separated by a long hinge segment which lacks the cysteine bridge found in human δ chains. There is, however, a disulphide bridge between the chains in the carboxyterminal 'tail piece' of the secreted molecule.

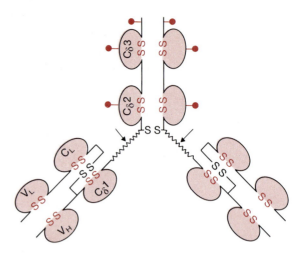

Fig. 2.18 Polypeptide structure of human IgD showing the single inter-heavy chain disulphide bond and probable sites of enzyme cleavage (arrows).

The possible functional role of IgD continues to engage the attention of immunologists. The identification of the protein in at least seven species to date supports the view that it has such a role, but the loss of an entire structural domain from the protein in rodents appears not to have compromised the biological viability of these organisms.

Immunoglobulin E

The amounts of IgE in the circulation of healthy individuals are vanishingly small. As with IgD, it is likely that the vast bulk of the protein is carried in a cell-bound form. However, unlike IgD, where the cell is the lymphocyte, IgE is found strongly associated with a surface receptor on the membrane of tissue mast cells and circulating basophils.

IgE has a mol. wt. of 188 000 and the protein exists as a four-chain monomer with two light chains and two ε chains. Structural analysis has shown that the ε chain has four constant region domains (see Figure 2.14) and thus resembles IgM. However, the sequence data has shown that it is probably more closely related to IgG (33% sequence homology).

When IgE is digested with papain a large Fc fragment (comprising the $C_\varepsilon 2$, $C_\varepsilon 3$ and $C_\varepsilon 4$ domains and having a mol. wt. of 98 000) is released. This fragment still retains the ability to bind to the mast cell surface and appears to overlap structurally with the peptic F(ab')$_2$ fragment (see Figure 2.19). IgE has one unusual structural feature: two inter-heavy chain disulphide bridges separated by a complete (C$_\varepsilon$2) domain. Like IgM, the protein is heavily glycosylated.

Some functional studies suggest the existence of two subpopulations of IgE with different mast cell binding characteristics. However, only one ε-chain gene has been found.

Biosynthesis of immunoglobulins

Lymphoid cells actively secreting immunoglobulins have a network of rough endoplasmic reticulum and synthesize light and heavy chains on polyribosomes of different sizes. L-chain synthesis occurs on polyribosomes sedimenting at about 200S, whereas H chains are synthesized on larger polyribosomes sedimenting at about 300S. Various intermediates are generated during the assembly of four-chain molecules and these play a role in the control of synthesis. Freshly synthesized light chains are released from the polyribosomes and

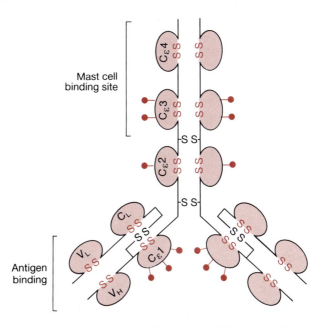

Fig. 2.19 Polypeptide structure of human IgE. Oligosaccharides are indicated by red circles. For the sake of clarity an additional intrachain S−S bridge within the C$_\varepsilon$1 domain is not shown. (Data from Bennich & von Bahr-Lindstrom, 1974.)

contribute to an intra-cisternal pool of free L chains which are then used to assemble H_2L_2 molecules. Various covalent intermediates have been identified during this process, notably H_2, H_2L and HL. The intermediates found from class to class depend on the order of disulphide bond formation between H chains and between H and L chains. After the H_2L_2 molecules are released into the cisternae of the endoplasmic reticulum they pass to the exterior via the Golgi apparatus. Oligosaccharides are added, via the lipid carrier dolichol on which they are synthesized, by membrane-bound glycosyl transferases. In the case of IgM, but not IgG, there is no evidence that the molecule must be glycosylated before secretion can occur. Before polymerization of IgA and IgM occurs there appears to be a requirement that J chain be covalently bonded to two monomer subunits of the immunoglobulin. A disulphide interchange enzyme present at high concentrations in the Golgi apparatus is probably essential for this stage. In the case of IgM, further polymerization to give a pentamer is achieved by non-covalent interactions of monomer subunits with the dimer, whereas with IgA such interactions are thought to be infrequent and thus the formation of higher polymers is relatively rare. Completely assembled immunoglobulin molecules are either secreted from the cell or become incorporated into the plasma membrane. A molecule that is to remain membrane bound to the B cell has an additional stretch of hydrophobic amino acids at the carboxy-terminal end of the heavy chains which becomes inserted into the lipid bilayer of the surface membrane of the cell and serves to anchor the molecule. This additional stretch of peptide comprises three distinct regions: a cytoplasmic segment which remains within the cytosol of the B cell, a transmembranal segment which spans the lipid bilayer of the membrane, and a spacer segment outside the lipid bilayer (see Figure 7.27). Although the transmembranal segment uniformly consists of 26 amino acid residues, the lengths and composition of the spacer region and the cytoplasmic segment differ between the immunoglobulin classes (see Table 2.3). In the case of the spacer region these differences may be relevant to the functional role of particular classes in lymphocyte development.

In contrast to the membrane bound variety, those molecules which are destined to be secreted are synthesized with a segment of $2-21$ hydrophilic amino acids

Table 2.3 Comparison of secreted and membrane-bound forms of mouse immunoglobulins

	IgM	IgD	IgG1	IgG2a	IgG2b	IgG3	IgE	IgA
Number of constant region domains	4	2	3	3	3	3	4	3
Size of hinge region	0	35	13	16	22	17	0	14
Size of C-terminus in secreted form	20	21	2	2	2	2	8	20
Membrane form								
Spacer	12	26	17	17	17	17	18	25
Transmembrane segment	26	26	26	26	26	26	26	26
Cytoplasmic segment	3	3	28	28	28	28	28	14

With the exception of the data on domain structure all numbers in the table refer to the amino acids associated with particular molecular regions. This information is largely derived from genomic DNA sequencing. (Adapted from Blattner & Tucker, 1984.)

at the carboxy-terminus (see Figure 2.20). It has been established that different messenger RNA molecules exist for the two types of μ chain, and that these arise by differential transcription (and/or processing) of a single gene (see Chapter 7).

Evolution of immunoglobulins

Immunoglobulin-like molecules are confined to vertebrates and are found even in primitive species such as the hagfish and the lamprey. Structural studies of the latter have revealed a molecule with two heavy chains (~70 000) and two light chains (~25 000). Amino acid sequence analysis of this protein suggests that the ancestral four-chain Ig molecule may have emerged in primitive fishes 450 million years ago. Higher cartilaginous fishes, which arose about 350 million years ago, have both monomeric immunoglobulin and high mol. wt. IgM-like molecules. In the case of the shark the pentameric 'IgM', but not the monomeric Ig, incorporates a J chain-like molecule. More significantly, sequencing of the variable regions of shark light and heavy

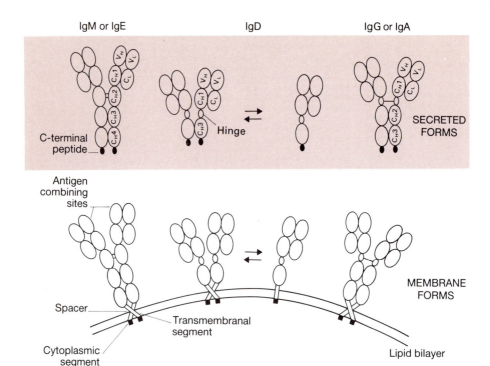

Fig. 2.20 Gross structural characteristics of secreted and membrane-bound mouse immunoglobulins. Ellipsoids represent domains and small ellipsoids are used to denote 'hinge' regions. H—H inter-chain disulphide bridges are shown by short bars but the subclass variation found in IgG is not indicated. IgD on B cell membranes is an equilibrium between conventional H_2L_2 molecules and half-molecules of H—L. It is not clear whether this is also the case for secreted IgD. (Modified from Blattner & Tucker, 1984.)

chains reveals that these are more closely related to their mammalian equivalents than they are to each other and therefore that the V_L and V_H domains diverged more than 400 million years ago.

Amphibians and reptiles also have both monomeric immunoglobulin and high mol. wt. IgM-like molecules, and it is tempting to interpret the presence of a monomeric 7S immunoglobulin as evidence of early divergence of the γ chain in the lower vertebrates, but not all investigators are prepared to accept this conclusion. Atwell and Marchalonis (1976) found such marked differences between amphibian 7S immunoglobulin and mammalian IgG that they proposed the term 'IgRAA' for this protein (immunoglobulin of reptiles, amphibians and aves).

There is also evidence of a low mol. wt. (5.7S) immunoglobulin (called IgN) in reptiles and birds, but there are large species variations in its occurrence — it is the major immunoglobulin in the duck, but a minor protein in the chicken.

A molecule resembling IgA has been described in birds and most mammals. Similarly, an IgD-like molecule has been readily demonstrated in birds and mammals and asserted in reptiles. This would suggest an evolutionary origin for IgD before the divergence of birds and mammals from reptiles (~180 million years ago). There is evidence for the existence of an avian IgE class but much further work is required in this area.

Since the advent of amino acid sequencing it has been possible to compare much more precisely the extent to which two proteins are related. One of the techniques which has been particularly useful in this respect is the calculation of the minimum number of mutations required to convert one amino acid sequence to another (the so-called minimum mutational distance or MMD) and then contrasting that number with the MMD between randomized sequences from the two proteins. Such calculations permit estimates of relatedness between the same protein in different species as well as shedding light on possible relationships between apparently diverse proteins. More than a decade ago the repeating nature of the internal disulphide bridges of heavy and light chains led to the proposal by Hill *et al.* (1966) that the immunoglobulin domain of some 110 amino acid residues represented a primitive unit from which all immunoglobulins had arisen by processes of gene duplication. More recently, Shinoda *et al.* (1981) have shown that immunoglobulin domains may, for comparative purposes, be divided into segments of 50–60 amino acid residues having a half-cysteine residue at the centre. Comparison of the sequences in such segments gives 18–28% sequence homology, similar to that observed for the intact domains. This has prompted the suggestion that the immunoglobulin constant region genes may have evolved from a gene coding for one half of a domain. If a unit of this size is considered as a possible precursor, it may explain some of the recent observations of Ig-like sequences in such diverse molecules as Thy-1 and the 49 residue thymic hormone thymopoietin, in addition to the more well known relationships with β_2-microglobulin and the HLA $\alpha3$ domain.

A possible series of genetic events leading to the evolution of the five known immunoglobulin classes is outlined in Figure 2.21. This represents a modification by Lin and Putnam (1981) of an earlier computer generated model published by Barker *et al.* (1980). This model envisages a series of duplications commencing with a gene coding for a half-domain. This would have undergone successive duplications to give precursors of the H and L chain constant regions. The heavy chain precursor gene then gave rise to two ancestral genes, one coding for a μ-δ-α chain precursor molecule and one for a γ-ε chain precursor. The μ-δ-α chain precursor gene at some stage acquired an additional nucleotide segment coding for the hydrophilic carboxy-terminal tail piece before undergoing duplication into ancestral genes for the μ chains and a δ-α precursor (Figure 2.21). Finally the δ-α precursor and γ-ε precursor genes underwent complete duplication to give the genes coding for the known heavy chain classes.

Further genetic 'processing' has given rise to the hinge regions of δ, γ and α chains and to the subclasses of γ and α chains.

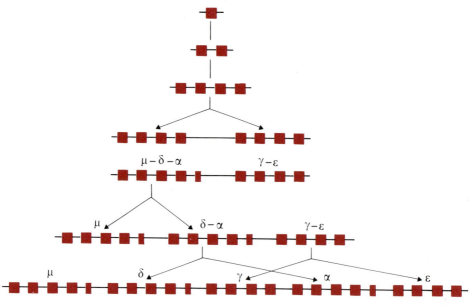

Fig. 2.21 Possible sequence of evolutionary events leading to five heavy chain classes. Each square red box represents a gene coding for an immunoglobulin domain and the intervening short black lines represent intronic DNA. The single primordial ancestral DNA gene is shown undergoing a series of equal and unequal duplications. This would have given rise to two genes coding for proteins ancestral to μ, δ and α chains and λ and ε chains respectively. Subsequently, duplication would have given ancestral genes for all the known heavy chain classes. Further duplication and mutation has occurred to produce subclasses of γ and α chains and the hinge regions of δ, γ and α chains.

Putnam *et al.* (1973) suggested that the hinge region of immunoglobulins might represent a 'collapsed' domain. Both amino acid and nucleotide sequencing studies have confirmed that there are indeed strong sequence homologies between the hinge regions and intact domains. A possible explanation for the origin of the hinge would be the moving of an RNA splice site from an inter-domain position to an intra-domain location so that only a portion of the domain would subsequently be translated (see Chapter 7, p. 181).

Although most hinge regions have remained in the

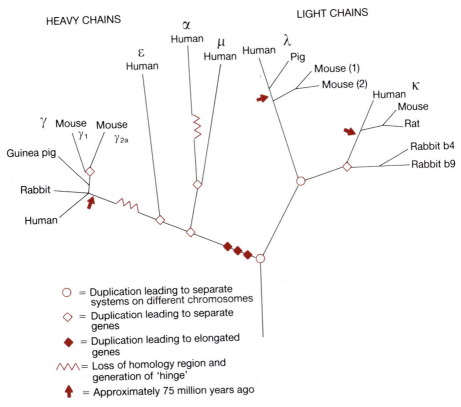

Fig. 2.22 Possible evolutionary tree for immunoglobulin constant region genes. It seems likely that V and C genes and joining mechanisms had been established by the time light and heavy chain genes diverged early in vertebrate evolution. A series of internal duplications (solid red diamonds) produced a precursor heavy chain constant region gene four times the length of the light chain C gene. Further duplications of the heavy chain gene (open red diamonds) gave rise to all of the major classes and had occurred well before the spread of the mammals some 75 million years ago. The precursor γ, δ and α chain genes appear to have undergone similar but independent contractions before the mammalian radiation. This may have been the result of unequal crossing-over events and led ultimately to the loss of the second domain and the generation of the hinge regions of these three classes. The branch lengths are expressed in accepted point mutations per 100 residues. (Adapted from Barker *et al.*, 1980.)

collapsed state the IgG3 subclass appears to have under-gone further modification. Unlike the other three IgG subclasses which have hinge regions of 12−15 residues the IgG3 hinge contains 62 amino acid residues. Structural analysis has revealed that this comprises a 17 residue segment followed by three identical 15 residue segments. Recently Huck et al. (1986) have confirmed that the IgG3 hinge coding region comprises four distinct exons compared to only one exon for the hinge coding regions of the other IgG subclasses.

The Barker computer model may be used to construct an evolutionary tree of immunoglobulin constant region gene diversification (see Figure 2.22). This suggests that heavy and light chain genes diverged very early in vertebrate evolution and both types of light chain were present in the genome before mammalian speciation began some 75 million years ago.

References

Atwell J.L. & Marchalonis J.J. (1976) Immunoglobulin classes of lower vertebrates distinct from IgM immunoglobulin. In *Comparative Immunology*, (ed. Marchalonis J.J.), p. 276. Blackwell Scientific Publications, Oxford.

Barker W.C., Ketcham L.K. & Dayhoff M.O. (1980) Origins of immuno-globulin heavy chain domains. *J. Mol. Evol.* **15**, 113.

Bennich H. & von Bahr-Lindstrom H. (1974) Structure of immunoglo-bulin E (IgE). In *Progress in Immunology*, Vol. II, (eds. Brent L. & Holborow J.), p. 49. North-Holland Publishing Co., Amsterdam.

Blattner F.R. & Tucker P.W. (1984) The molecular biology of immuno-globulin D. *Nature* **307**, 417.

Edelman G.M., Cunningham B.A., Gall W.E., Gottlieb P.D., Rutishau-ser U. & Waxdal M.J. (1969) The covalent structure of an entire γG immunoglobulin molecule. *Proc. natn. Acad. Sci. USA* **63**, 78.

Edelman G.M. & Gall W.E. (1969) The antibody problem. *Ann. Rev. Biochem.* **38**, 415.

Feinstein A. & Beale D. (1977) Models of immunoglobulins and antigen-antibody complexes. In *Immunochemistry: An Advanced Textbook*, (eds. Glynn L.E. & Steward M.W.), p. 263. John Wiley & Sons, Chichester.

Hahn G.S. (1982) Antibody structure, function and active sites. In *Physiology of Immunoglobulins: Diagnostic and Clinical Aspects*, (ed. Ritzmann S.), p. 193. Alan Liss Inc., New York.

Hill R.L., Delaney R., Fellows R.E. jr & Lebovitz H.E. (1966) The evolutionary origins of immunoglobulins. *Biochemistry* **56**, 1762.

Huck S., Fort P., Crawford D.H., Lefranc M.P. & Lefranc G. (1986) Sequence of a human immunoglobulin γ3 heavy chain constant region gene: comparison with the other human Cγ genes. *Nucleic Acid Research* **14**, 1779.

Kabat E.A., Wu T.T., Bilofsky H., Reid-Miller M. & Perry H. (1983) *Sequences of Proteins of Immunological Interest*. US Dept of Health and Human Services, Public Health Service, National Institutes of Health.

Kratzin H., Altevogt P., Ruban E., Kortt A., Staroscik K. & Hilschmann N. (1975) Die primärsstruktur eines monoklanalen IgA-immuno-globulin (IgA Tro) II. Die amino saure sequenz der H-Kette, α Typ: Subgruppe III. Struktur des gesamten IgA-molekuls. *Hoppe-Seyler's Z. physiol. Chem.* **356**, 1337.

Lin L.C. & Putnam F.W. (1981) Primary structure of the Fc region of human immunoglobulin D: implications for evolutionary origin and biological function. *Proc. natn. Acad. Sci. USA* **78**, 504.

Milstein C. & Svasti J. (1971) Expansion and contraction in the evolution of immunoglobulin gene pools. In *Progress in Immunology*, Vol. I, (ed. Amos B.), p. 35. Academic Press, New York.

Mostov K.E., Friedlander M. & Blobel G. (1984) The receptor for transepithelial transport of IgA and IgM contains multiple immunoglobulin-like domains. *Nature* **308**, 37.

Porter R.R. (1959) The hydrolysis of rabbit γ-globulin and antibodies with crystalline papain. *Biochem. J.* **73**, 119.

Porter R.R. (1962) The structure of γ-globulin and antibodies. In *Symposium on Basic Problems in Neoplastic Disease*, (eds. Gelhorn A. & Hirschberg E.), p. 177. Columbia University Press.

Putnam F.W., Florent G., Paul C., Shinoda T. & Shimizu A. (1973) Complete amino acid sequence of the mu heavy chain of human IgM immunoglobulin. *Science* **182**, 287.

Roitt I. (1984) *Essential Immunology*, 5th edn. Blackwell Scientific Publications, Oxford.

Shinoda T., Takahashi N., Takayasu T., Okuyama T. & Shimizu A. (1981) Complete amino acid sequence of the Fc region of a human δ chain. *Proc. natn. Acad. Sci. USA* **78**, 785.

Silverton E.W., Navia M.A. & Davies D.R. (1977) Three-dimensional structure of an intact human immunoglobulin. *Proc. natn. Acad. Sci. USA* **74**, 5140.

Wu T.T. & Kabat E.A. (1970) An analysis of the sequences of the variable regions of Bence Jones proteins and myeloma light chains and their implications for antibody complementarity. *J. exp. Med.* **132**, 211.

Further reading

Capra D. & Edmundson A.B. (1977) The antibody combining site. *Sci. Am.* **236**, 50.

Davies D.R. & Metzger H. (1983) Structural basis of antibody function. *Ann. Rev. Immunol.* **1**, 87.

Hahn G.S. (1982) Antibody structure, function and active sites. In *Physiology of Immunoglobulins: Diagnostic and Clinical Aspects*, (ed. Ritzmann S.E.), p. 193. Alan R. Liss Inc., New York.

Kabat E.A., Wu T.T., Bilofsky H., Reid-Miller M. & Perry H. (1983) *Sequences of Proteins of Immunological Interest*. US Dept of Health and Human Services, Public Health Service, National Institutes of Health.

Nisonoff A. (1984) *Introduction to Molecular Immunology*, 2nd edn. Sinauer Associates Inc., Baltimore.

Putnam F.W. (1974) Comparative structural study of human IgM, IgA and IgG immunoglobulins. In *Progress in Immunology*, Vol. II, (eds. Brent L. & Holborow J.), p. 25. North-Holland Publishing Co., Amsterdam.

Turner M.W. (1977) Structure and function of immunoglobulins. In *Immunochemistry: An Advanced Textbook*, (eds. Glynn L.E. & Steward M.W.), p. 1, John Wiley & Sons, Chichester.

Chapter 3
Cell surface proteins of the major histocompatibility complex

Introduction

All immune responses are initiated by and regulated by antigens. A universal aspect of immunogenetics is, therefore, the genetics of each antigen. This aspect is only essential to an understanding of the genetics of immune responsiveness when the antigens are determinants on molecules involved intrinsically in the mechanism of immune responsiveness. That this is true of the immunoglobulins (discussed in the previous chapter) is self-evident. More recently the importance of the cell surface proteins of the major histocompatibility complex (MHC) in immune responsiveness has been revealed.

The histocompatibility antigens of the MHC were recognized initially by the rapidity of allograft rejection when the donor and recipient of the graft differ antigenically at the MHC.

Allograft rejection: multigenic control

Tissue grafts between two outbred members of the same species (allografts) are usually rejected. This rejection was recognized as an immune response and the antigenic differences between donor and recipient have been mapped genetically.

The ABO blood group antigens were the first transplantation antigens to be recognized. Landsteiner defined erythrocyte ABO antigens by the use of antibodies specific for each alloantigen. The use of typing antisera as reagents to characterize the erythrocyte phenotype illustrates the basic methodology of immunogenetics (Figure 3.1). The inheritance of phenotypic traits is followed by using antisera directed against alloantigenic differences and the genes controlling each trait are mapped.

The genetics of rejection of grafts of tissue other than blood have proved to be complex. Transplantation of tumour tissue in mice provided the first demonstration that many genes are involved in controlling the acceptance of a tumour graft, i.e. susceptibility to being killed by the tumour. The genetics of transplantation are described in Part 2.

Many different tissue antigens show alloantigenic differences and a single alloantigenic incompatibility between the donor of the graft and the host is sufficient for graft rejection. The rapidity of rejection varies according to the nature of the alloantigenic difference. In mice, alloantigens mapping genetically at a locus termed *Histocompatibility-2* or *H-2* account for the most rapid rejection of grafts. The H-2 alloantigens involved in rejection reactions have been defined serologically as cell-surface molecules.

Haemagglutination

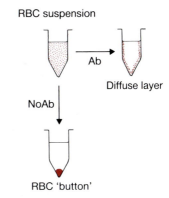

Fig. 3.1 *Haemagglutination*: Antibodies to erythrocyte antigens were first recognised as agglutinins. Cross-linking of red cells by antibodies is termed haemagglutination; it is a sensitive visual test for specific antibodies. It can be adapted to detect antibodies to any antigen capable of being linked to the red cell surface: *Haemolysis*: The sensitivity of the agglutination test is increased by adding complement to effect red cell lysis or haemolysis. Fresh rabbit or guinea pig serum is a usual source of complement. The extent of lysis is proportional to both antibody and complement concentration. A variation on this assay is complement fixation: antigen-antibody concentration and is measured by complement is added; the amount of complement fixed is a measure of the original antibody concentration, and is measured by using a standard haemolysis assay to determine the level of free complement.

Haemolysis

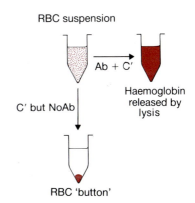

Plasma membrane structure

The plasma membrane provides a self-contained identity for the cell. The membrane is a bilayer consisting in the main of phospholipid and glycolipid with some cholesterol. This lipid bilayer is a hydrophobic envelope for the cell contents and acts as a selective permeability barrier. The plasma membrane also contains proteins; these are either integral proteins contained wholly or partly within the membrane bilayer, or peripheral proteins bound to the surface of the bilayer or to integral proteins by electrostatic attraction and hydrogen bonding.

Integral membrane proteins are usually glycoproteins. The carbohydrate portions of both glycolipids and glycoproteins appear to be located exclusively on the external surface of mammalian cells. The membrane is, therefore, structurally asymmetric.

Plasma membrane structure is best described by the fluid mosaic model of Singer and Nicholson (1972) (Figure 3.2). This model depicts membrane as a two-dimensional solution. The lipid bilayer behaves like a sea in which integral proteins drift like icebergs, though in many cases much of the protein is above the membrane sea. The observed freedom of membrane proteins and lipids to diffuse rapidly in the plane of the membrane is incorporated into the fluid mosaic model.

The pattern of peptide and carbohydrate determinants displayed on the outer surface of the plasma membrane provides the antigenic identity of the cell.

Transplantation antigens

Polymorphic forms of surface antigens are recognized as foreign determinants on allografts. Antibodies specific for these alloantigenic differences define either protein or carbohydrate determinants. The molecules bearing the antigenic determinants can be purified by biochemical techniques using specific antisera to assay the product at each stage. If the antiserum is specific for a peptide antigen then the protein product of the polymorphic gene should be isolated by this procedure. If the antiserum is specific for a carbohydrate antigen then the polymorphic gene will probably code for a glycosyl transferase, one of the enzymes involved in the step-wise synthesis of the oligosaccharides of glycolipids and glycoproteins. The blood group antigens are carbohydrate in nature.

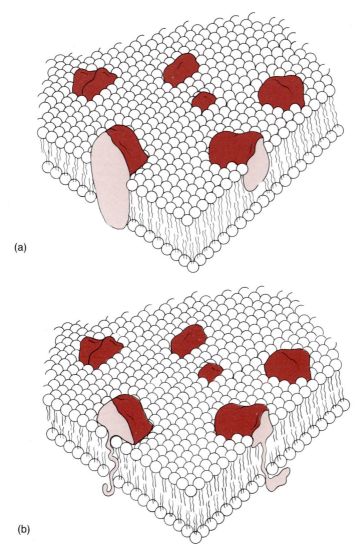

(a)

(b)

Fig. 3.2 (a) The fluid-mosaic model of membrane structure showing integral proteins immersed in the phospholipid bilayer. This model, proposed by Singer & Nicolson (1972), can be likened to icebergs floating in the sea. (b) The fluid-mosaic model redrawn to indicate the finding that the transmembrane section of membrane proteins is usually a short hydrophobic helix.

Major transplantation antigens in man: HLA-A, -B and -C

Isolation

The major transplantation antigens (HLA-A, -B and -C) of human tissue have been isolated by biochemical fractionation of the membrane proteins of lymphoblastoid cell lines. These cell lines are established in culture by Epstein—Barr Virus (EBV) transformation of human B lymphocytes. The membranes of such cells are chosen for biochemical study because they are rich in HLA antigens. However, almost all cells have HLA (-A, -B and -C) antigens, now termed class I antigens, on the surface (Figure 3.3).

Structure

The class I antigens are on glycoproteins, each consisting of two polypeptide chains in non-covalent association (Figure 3.4). The heavier chain in each case is the product of one of the MHC loci (A, B or C), defined by transplantation and serology. Each heavy chain is glycosylated and has an approximate mol. wt. of 45 000. The lighter polypeptide chains of each protein are indistinguishable from one another and have been shown to be identical to β_2-microglobulin, a polypeptide of mol. wt. 12 000, the human counterpart to which was first isolated from urine. There is no glycosylation of β_2-microglobulin, either in its free state or when it is

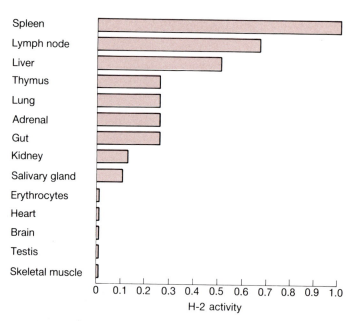

Fig. 3.3 The relative quantity of H-2 antigens (class I) detected on various mouse tissues (from Klein, 1975). Class I HLA antigens are similarly distributed on human tissue with the important exception that human erythrocytes lack HLA antigens.

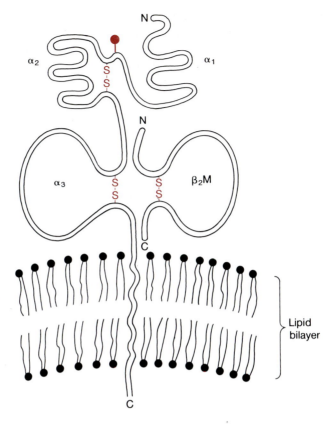

Fig 3.4 A diagrammatic representation of the domain structure of a class I MHC antigen. Domain α_3 is homologous with immunoglobulin domains as is β_2-microglobulin (see Fig. 3.6). The transmembrane region consists of a hydrophilic helix about 20 residues long. ● represents an oligosaccharide unit.

the light chain of a transplantation antigen on the cell surface.

HLA antigens are integral transmembrane proteins

The glycoproteins carrying HLA-A, -B or -C antigens are bound to the plasma membrane by a hydrophobic peptide. This peptide (17 residues long) is located about 50 residues from the carboxy-terminal end of the HLA heavy chain (Figure 3.4). The antigenic protein can be released from the membrane by treatment with detergents. This treatment yields the glycoproteins complexed with detergent. Alternatively, the proteolytic enzyme papain can cleave transplantation antigens from the cell surface. The papain-released molecules are fully allo-antigenic; they lack the carboxy-terminal hydrophilic peptide and the penultimate hydrophobic membrane-binding peptide.

The transmembrane orientation of the heavy chain of HLA transplantation antigens has been confirmed by direct vectorial iodination (Figure 3.5). Plasma membrane is isolated in the form of closed vesicles, having either the original orientation of the cell membrane or the reversed orientation, i.e. inside out vesicles. Separation of these opposite orientations is possible and each form is labelled with radioactive iodine using the enzyme lactoperoxidase to catalyse the reaction uniquely with the available outer surface of the vesicles. Isolation of the HLA molecules then shows that the heavy chain can be iodinated from either side of the membrane and is, therefore, transmembrane, but the light chain is located exclusively on the outer surface of the cell membrane.

Amino acid sequences

The amino acid sequence of human β_2-microglobulin shows clear sequence homology with an immunoglobulin domain (Figure 3.6). β_2-microglobulin has a single intra-chain disulphide bond in a position analogous to that in an immunoglobulin domain. This loop and the extent of homology suggest that β_2-microglobulin may adopt the β-pleated sheet structure characteristic of an immunoglobulin domain.

The amino acid sequence of one papain-solubilized HLA heavy chain, HLA-B7, was completed in 1979. The amino-terminal, extracellular portion of the heavy chain is organized into three domains, α_1, α_2 and α_3; the α_2 and α_3 domains are disulphide bonded loops (Figure 3.4). Each of these loops is similar in size to an immunoglobulin domain loop. For the α_2 domain there is no clear sequence homology with an immunoglobulin domain except for a few residues around the amino-terminal cysteine. By contrast, the α_3 domain containing a 55 residue disulphide loop has significant sequence homology with an immunoglobulin domain (Figure 3.7). As with β_2-microglobulin, the sequence of the second domain of HLA-B7 heavy chain suggests that it may adopt the immunoglobulin fold. This leads to the hypothesis that β_2-microglobulin interacts non-covalently

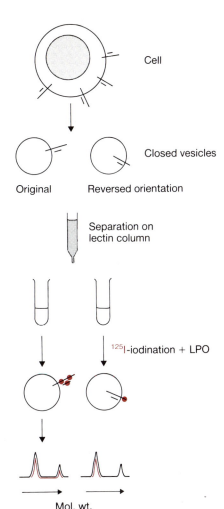

Cell

Closed vesicles

Original   Reversed orientation

Separation on
lectin column

^{125}I-iodination + LPO

Mol. wt.

Fig. 3.5 The transmembrane orientation of a class I MHC antigen (as shown in Fig. 3.4) was demonstrated experimentally by preparing membrane vesicles and separating them by concanavalin-A affinity chromatography according to whether they were right-side out or inside-out (reversed). Enzymic radio-iodination of exposed membrane proteins followed by electrophoretic analysis shows that both α and β_2-microglobulin chains are available on the outside of the cell, while only the α chain crosses the membrane and can be labelled on the inside surface of the cell membrane. Experiment performed by Walsh & Crumpton (1977).

with the second heavy chain domain in a manner analogous to the interaction of two immunoglobulin domains.

Extensive amino acid sequence data for the HLA-A2 heavy chain allowed a comparison to be drawn with the B7 sequence and striking homology between the two sequences was revealed (Figure 3.8). In the amino-terminal 167 residues, homology is at least 80%. This homology points to a probable ancestral gene common to the HLA-A and -B genes. Homology is not uniform. For the amino-terminal 60 residues, B7 and A2 are 93% homologous but in the peptides 66−80 and 105−114 homology is only 40% and 50% respectively.

51 CELL SURFACE PROTEINS OF THE MHC

β₂-MICROGLOBULIN
EU C$_L$ (RESIDUES 109-214)
EU C$_H$1 (RESIDUES 119-220)
EU C$_H$2 (RESIDUES 234-341)
EU C$_H$3 (RESIDUES 342-446)

```
                                               1                               10
                            ILE GLN ARG THR PRO LYS ILE GLN VAL TYR SER
                            THR VAL ALA ALA PRO  -   -  SER VAL PHE ILE
                            SER THR LYS GLY PRO  -   -  SER VAL PHE PRO
                            LEU LEU GLY GLY PRO  -   -  SER VAL PHE LEU
                            GLN PRO ARG GLU PRO  -   -  GLN VAL TYR THR


                                                            20
ARG HIS PRO ALA  -  GLU  -   -   -   -  ASX GLY LYS SER ASX PHE LEU ASN CYS TYR VAL
PHE PRO PRO SER ASP GLU GLN  -   -  LEU LYS SER GLY THR ALA SER VAL VAL CYS LEU LEU
LEU ALA PRO SER SER LYS SER  -   -  THR SER GLY GLY THR ALA ALA LEU GLY CYS LEU VAL
PHE PRO PRO LYS PRO LYS ASP THR LEU MET ILE SER ARG THR PRO GLU VAL THR CYS VAL VAL
LEU PRO PRO SER ARG GLU GLU  -   -  MET THR LYS ASN GLN VAL SER LEU THR CYS LEU VAL


                  30                                            40
SER GLY PHE HIS PRO SER ASP ILE GLU VAL  -   -  ASP LEU LEU LYS ASP GLY GLU ARG ILE
ASN ASN PHE TYR PRO ARG GLU ALA LYS VAL  -   -  GLN TRP LYS VAL ASP ASN  -  ALA LEU
LYS ASP TYR PHE PRO GLU PRO VAL THR VAL  -   -  SER TRP ASN SER  -  GLY  -  ALA LEU
VAL ASP VAL SER HIS GLU ASP PRO GLN VAL LYS PHE ASN TRP TYR VAL ASP GLY  -   -  VAL
LYS GLY PHE TYR PRO SER ASP ILE ALA VAL  -   -  GLU TRP GLU SER ASN ASP  -   -  GLY


            50                                          60
GLX LYS VAL ASX  -  HIS SER GLX LEU SER PHE SER LYS ASN  -  SER TRP PHE TYR LEU LEU
GLN SER GLY ASN SER GLN GLU SER VAL THR GLU GLN ASP SER LYS ASP SER THR TYR SER LEU
THR SER GLY  -  VAL HIS THR PHE PRO ALA VAL LEU GLN SER  -  SER GLY LEU TYR SER LEU
GLN VAL HIS ASN ALA LYS THR LYS PRO ARG GLU GLN GLN TYR  -  ASP SER THR TYR ARG VAL
GLU PRO GLU ASN TYR LYS THR THR PRO PRO VAL LEU ASP SER  -  ASP GLY SER PHE PHE LEU


          70                                              80
TYR SER TYR  -  THR GLU PHE THR PRO THR  -  GLU LYS  -  ASP GLU TYR ALA CYS ARG VAL
SER SER THR LEU THR LEU SER LYS ALA ASP TYR GLU LYS HIS LYS VAL TYR ALA CYS GLU VAL
SER SER VAL VAL THR VAL PRO SER SER SER LEU GLY THR GLN  -  THR TYR ILE CYS ASN VAL
VAL SER VAL LEU THR VAL LEU HIS GLN ASN TRP LEU ASP GLY LYS GLU TYR LYS CYS LYS VAL
TYR SER LYS LEU THR VAL ASP LYS SER ARG TRP GLN GLN GLY ASN VAL PHE SER CYS SER VAL


              90                                           100
ASX HIS VAL THR LEU SER GLX PRO  -   -   -  LYS ILE VAL  -  LYS TRP ASP ARG ASP MET
THR HIS GLN GLY LEU SER SER PRO VAL THR  -  LYS SER PHE  -   -  ASN ARG GLY GLU CYS
ASN HIS LYS PRO SER ASN THR LYS VAL  -  ASP LYS ARG VAL  -   -  GLU PRO LYS SER CYS
SER ASN LYS ALA LEU PRO ALA PRO ILE  -  GLU LYS THR ILE SER LYS ALA LYS GLY
MET HIS GLU ALA LEU HIS ASN HIS TYR THR GLN LYS SER LEU SER LEU SER PRO GLY
```

Fig. 3.6 Comparison of the amino acid sequence of β₂-microglobulin with the sequences of the homology regions C$_L$, C$_H$1, C$_H$2 and C$_H$3 of the IgG1 human immunoglobulin Eu. Deletions, indicated by dashes, have been inserted to maximize homologies. Identical residues are enclosed in boxes. Numbering is for β₂-microglobulin.

Major transplantation antigens in the mouse

Isolation and structural characterization of murine MHC antigens (H-2 K and H-2 D) has depended mainly

```
        • DPPK THVTHHPISDHEAT LRCWALGFYPAEITLTWQR DG EDQTQDTELVETRPA
       • RTPKI QVYSRHPAENGKSNFLNCYVSGFHPSDIEVDLLK DG ER IEKVEHSDLSFS
          EPQVYTLPPSREEMTKNQVSLTCLVKGFYPSDIAVEWESNDG E  PENYKTTPPVLD

B7α3     GDRTFEKWAAVVV PS G EEQRYTCHVQHEGLP KPLTLRW •
β2M      KDWSFYLLYYTEFTPT   EKDEYACRVNHVTLS QPKIVKW •
Cγ3      SDGSFFLYSKLTVDKSRWQQGNVFSCSVMHEALHNHYTQKSL
```

Fig. 3.7 Comparison of the amino acid sequences of class I MHC antigen B7 α_3 domains with immunoglobulin Cγ3 domain and β_2-microglobulin sequences. Exon boundaries (see Chapter 7) are indicated by •. Gaps are introduced to maximize homology alignments shown in red.

on radiochemical and immunochemical techniques (Figure 3.9). The H-2 K and D antigens of murine spleen cells are either catalytically radio-iodinated or biosynthetically labelled with radioactive amino acids, released from the cell membrane by detergent lysis and immunologically precipitated with specific alloantisera. The radioactive H-2 glycoproteins are characterized by polyacrylamide gel electrophoresis.

The isolated H-2 K and D glycoprotein antigens are structurally analogous to the HLA-A, -B and -C antigens, each consisting of a heavy chain (~45 000 mol. wt.) encoded by the K or D gene and a light chain (~12 000 mol. wt.) identical with murine β_2-microglobulin.

Amino acid sequences of major transplantation antigens

Sequence determination, of necessity, required the application of radiochemical microtechniques, but complete sequences are difficult to obtain by these methods. The advent of DNA sequencing afforded the quickest route to complete amino acid sequences (see Chapters 7 and 10).

Extensive stretches of amino acid sequence were determined for the H-2 K^b polypeptide (see Figure 3.8). Partial amino-terminal sequences of the K^k, K^q, K^d, D^b and D^d products were also obtained. Comparison of these sequences (Figure 3.10) showed the following relationships.

1 Allelic products of the H-2 K locus and of the H-2 D locus show sequence homology ranging from 63 to 85%.

2 Allelic products of the H-2 K locus differ one from another by multiple amino acid residues; similarly, multiple differences are seen when comparing the two partial sequences of H-2 D locus products. Over the limited amino-terminal regions compared allelic products differ at approximately one-third of their amino acid residues.

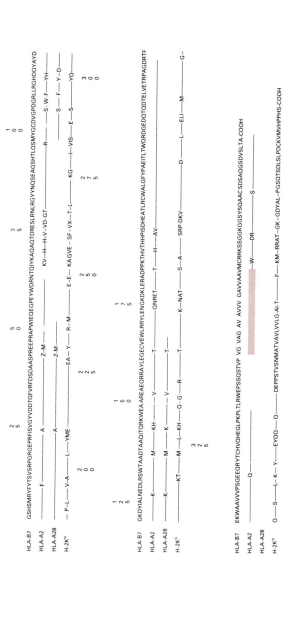

Fig. 3.8 Comparison of the amino acid sequence of the α chain of HLA-B7 with that of HLA-A2. A partial amino terminal sequence of HLA-A28 and the complete sequence of the murine antigen H-2k^b are also shown. Extensive homologies are seen between all of these proteins, though the two allelic products A2 and A28 are most closely similar in sequence.

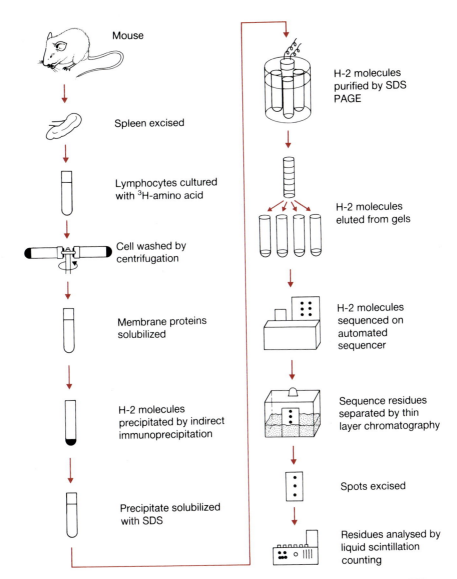

Mouse

Spleen excised

Lymphocytes cultured with ^{3}H-amino acid

Cell washed by centrifugation

Membrane proteins solubilized

H-2 molecules precipitated by indirect immunoprecipitation

Precipitate solubilized with SDS

H-2 molecules purified by SDS PAGE

H-2 molecules eluted from gels

H-2 molecules sequenced on automated sequencer

Sequence residues separated by thin layer chromatography

Spots excised

Residues analysed by liquid scintillation counting

Fig. 3.9 The experimental design used in amino acid sequence analysis of murine MHC antigens. (From Silver et al., 1976.)

3 The degree of sequence homology between K and D products is similar to that seen when comparing two allelic products of either locus. This observation, based on limited amino-terminal sequences, must be viewed cautiously since it implies that allelic products at one locus are as closely related to the products of the second locus as they are to each other. This conclusion contradicts the serological finding that K and D molecules each exhibit distinct antigenic specificities; these

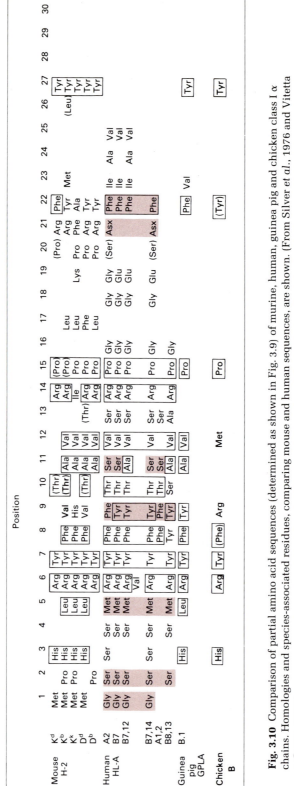

Fig. 3.10 Comparison of partial amino acid sequences (determined as shown in Fig. 3.9) of murine, human, guinea pig and chicken class I α chains. Homologies and species-associated residues, comparing mouse and human sequences, are shown. (From Silver *et al.*, 1976 and Vitetta & Capra, 1978.)

are public specificities common to the allelic products of one locus but not found on the allelic products of the second locus (see Part 2). This paradox is now resolved through knowledge of DNA sequences of K and D genes (Chapter 10).

Amino acid sequences of H-2 mutant proteins

The complex nature of H-2 allelism with many amino acid differences between allelic products means that correlation of structure with function is difficult. A more promising system is the collection of H-2 mutants mapped at K or D(L) loci. A series of mutants mapping at the K locus of the H-2^b haplotype have been detected by reciprocal skin grafting (see Chapter 9). The K locus products from wild type H-2^b and from mutants H-2^{ba} (H-2^{bm1}) and H-2^{bd} (H-2^{bm3}) can each be isolated by specific immune precipitation using an antiserum directed against the major private specificity of the K^b product.

The mutant K^b glycopolypeptides differ from each other and from the wild type K^b by a small number of amino acid residues (Figure 3.11). The alterations in K^{bm1} are at positions 152, 155 and 156 (Figure 3.12).

It is important to note that while some mutant products (e.g. bm1, bm3, bm8, bm10) have distinct and unique amino acid sequences, other mutant sequences appear to have been isolated twice (bm5, bm16) or even thrice (bm6, bm7, bm9) as independent events. Correlating structure and function, a single amino acid residue difference between gene products (e.g. bm8, bm10 or bm11 versus wild type) is sufficient to determine rejection of skin grafts exchanged between the two mouse strains. Recognition of the single amino acid difference can also be demonstrated by the graft-versus-host reaction, the mixed lymphocyte reaction (MLR) and cell-mediated cytotoxicity. These reactions involve T lymphocytes. It is clear, therefore, that the T lymphocyte receptor (see Chapter 13) can recognize an antigenic difference as small as a single amino acid side chain. This compares directly with the discrimination shown by antibodies in recognizing similar single side chain differences between allelic forms of immunoglobulin polypeptide chains (see Figure 5.6, p. 111).

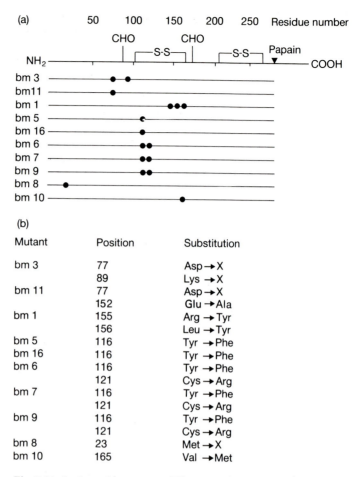

Fig. 3.11 Amino acid sequence differences of mutant H-2k^b α chains (bm1 *et seq.*) compared to the parental k^b sequence. (a) The map position of each amino acid residue difference. (b) The nature of the substitution where known. Note that identical single and double changes have been found in independently isolated mutants.

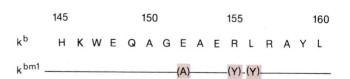

Fig. 3.12 Comparison of the amino acid sequences of k^b and k^{bm1} α chains through the only region in which a sequence difference was found. (NB residue 152 has been shown by DNA sequencing to be alanine (A) in k^{bm1}.)

Relationship between the major transplantation antigens of man and mouse

Comparison of the amino acid sequences of H-2 K and D products with HLA-A and -B products shows considerable homology (see Figures 3.8 and 3.10). Short, partial amino-terminal sequences of various K, D, A and B polypeptides illustrate the homology and show that the two murine products are equally closely related in sequence to either of the two human products.

A few longer stretches of continuous amino acid sequence of $H-2K^b$, HLA-B7 and HLA-A2 can be compared and they show about $70-80\%$ homology. This finding confirms the impression given by the comparison of short stretches of sequence. The striking sequence homologies point to the probability of the genes coding for the major transplantation antigens of mouse and man having evolved from a common ancestral gene.

The comparison of the amino-terminal sequences shows, in addition to the interspecies homology, purely intraspecies homology. Positions 1, 2, 5, 9, 11 and 22 have been described as species-associated amino acid residues, i.e. the K and D products can have the same residue at these positions and the A and B products can also be homologous with each other but have an amino acid residue different from that common to the K and D products at the corresponding position.

Species-associated residues have considerable genetic implication which will be noted here and discussed further in Part 2. Intraspecies homology is taken to indicate the evolution of multiple transplantation antigen loci by gene duplication. Interspecies homology points to a possible common ancestral gene. Therefore, species-associated residues suggest that gene duplication occurred after speciation. Alternatively, it is possible (but not probable) that species-associated residues arose by independent mutation and selection; this alternative can be contemplated only because of the extensive polymorphism of the MHC loci, the complex nature of the allelism and our ignorance of the selective pressures operating on the phenotypes. Including in the comparison the limited sequence information available for the GPLA heavy chain (Figure 3.10) shows sequence correlation with H-2 heavy chains at positions 5 and 11 and with HLA heavy chains at 9 and 22. More sequence data are needed to allow clearer interpretation of these comparisons.

Major transplantation antigens of species other than man and mouse

A major transplantation complex has been identified in representatives of the mammals, birds and amphibia. In all species so far examined (including man, Rhesus monkey, dog, guinea pig, rat, mouse, chicken and frog), at least two loci coding for major transplantation antigens have been detected serologically.

It is reasonable to assume that these antigens are analogues of the H-2 K, D and L or HLA-A, -B and -C products. Preliminary data add weight to this assumption. In several species the major antigens have been characterized as alloantigenic heavy chains, mol. wt. approximately 45 000, associated on the cell surface with a species specific β_2-microglobulin. A very limited amount of sequence data from the heavy chains of guinea pig and chicken show that these polypeptides are homologous with the H-2 and HLA major transplantation antigen heavy chains (Figure 3.10). It is anticipated on the basis of this evidence, and on the functional criteria described in Part 3, that all vertebrates will be found to have analogous major transplantation antigens.

Cell surface antigens controlled by genes linked to, but not in, the MHC

Linked to H-2 on chromosome 17 of the mouse are loci coding for other cell surface antigens. The two most studied are the T/t complex locus and the Tla locus (see Part 2). The complex and pleiotropic nature of the T/t locus evokes comparisons with the $H-2$ locus, thus leading to speculation on possible structural, functional or evolutionary relationship between the two loci. Genetic interaction between T/t and $H-2$ has led to the concept of an extended haplotype.

The Tla locus codes for an antigen expressed on T lymphocytes and related leukaemiac cells. The product of the Tla locus has been investigated because of its relationship to the immune system as a differentiation antigen marker for lymphocyte development (see Chapter 12). However, the structure of the product shows a striking resemblance to the structure of the class I major transplantation antigens. The Tla antigen consists of a glycoprotein chain of approximate mol. wt. 45 000 present on the cell surface in non-covalent linkage with β_2-microglobulin. The homology of the glycoprotein chain with class I heavy chain is confirmed by DNA probe analysis (see Chapter 10).

Immune associated (Ia) antigens

Immune associated (Ia) alloantigenic differences are controlled by the I region of the H-2 complex locus. The I region was defined initially by mapping of genes controlling immune responsiveness (i.e. Ir genes). Products of I region genes were sought by raising alloantisera using donor and responder mice differing only at the I region (see Chapter 9). These antisera precipitate glycoproteins termed Ia antigens (class II MHC antigens).

Ia glycoproteins are cell surface antigens and, like the H-2 K and D products, can also function as transplantation antigens. Ia antigens have a more limited tissue distribution by comparison with the ubiquitous H-2 K and D products. B lymphocytes carry Ia antigens but T lymphocytes have little or no Ia antigens. Antigen presenting cells should have Ia antigens on their surface (see Chapter 13). In line with this concept, a subset of macrophages express Ia antigens. Epithelial cells of the thymus, and possibly other epithelial cells, express Ia; in that context Ia antigens, together with H-2 antigens, act as the fundamental demonstration of the self-histocompatability type (see Chapter 13). Alloantigenic differences of Ia molecules are detectable by a sensitive cell culture assay which is analogous to graft rejection *in vivo*. This assay is termed a mixed lymphocyte reaction (MLR) or mixed leucocyte culture (MLC); it is described in more detail in Chapter 9. Antisera specific for Ia antigens can block an appropriate MLR.

Structure of murine Ia antigens

Ia antigens comprise a family of glycoproteins related to one another in general structure. Radiochemical and immunochemical methods similar to those used in studying H-2 K and D products have been employed to characterize murine Ia antigens. Each Ia antigen consists of two glycosylated polypeptide chains, termed α and β, of approximate mol. wt. 34 000 and 28 000 respectively. These two chains are associated non-covalently (Figure 3.13).

Peptide maps and partial amino-terminal amino acid sequences show the α and β chains of Ia antigens to be distinct and separate gene products. The family of Ia

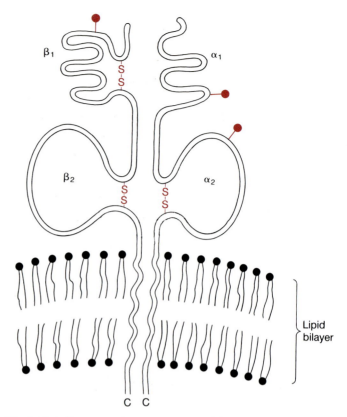

Fig. 3.13 A diagrammatic representation of the domain structure of a class II (immune associated, Ia) MHC antigen. Domains β_2 and α_2 are homologous with immunoglobulin domains. ♦ represents an oligosaccharide unit.

proteins appears to be controlled by a cluster of I region genes. The structural genes for both α and β chains of each Ia antigen map within the I region (see Part 2).

Human Ia antigens: HLA-D locus products

Immune response genes are not easily recognized in the outbred human population. Consequently, neither an I region of the human MHC nor the related Ia antigens have been defined by similar methods to those used for mice. The D locus of HLA can be equated with the H-2 I region on the evidence that alloantigenic differences at the HLA-D locus evoke the strongest MLRs involving human leucocytes. The DR locus (originally defined as being either identical to, or in tight linkage disequilibrium with, HLA-D, see Part 2) codes for a serologically detected set of alloantigens present on the surface of B lymphocytes, but not in easily detectable amounts on T lymphocytes (cf. H-2 Ia antigen expression). HLA-D region is now known to encode DR, DQ and DP molecules — each structurally distinct and comprising in *toto* a minimum of seven proteins (see Chapter 9, Figure 9.11).

The structure of DR antigens confirms the hypothesis that they are the human analogues of mouse Ia antigens. The DR antigens have been isolated by biochemical fractionation of the membrane proteins of lymphoblastoid cells and are obtained in the glycoprotein fraction, together with the A and B antigens. Gel electrophoresis of samples not exposed to dissociating conditions such as heat or urea indicates an apparent mol. wt. of 55 000 for DR antigens. This molecule can be dissociated, without prior reduction of disulphide bonds, into two polypeptide chains, of apparent mol. wt. 33 000 and 28 000.

Antisera raised in rabbits against isolated DR glycoprotein act like DR-specific alloantisera in that they are potent inhibitors of the MLR. Also, Fab fragments of the rabbit antibodies block cytotoxic lysis of DR antigen-bearing B lymphocytes elicited by specific alloantisera plus complement.

Amino acid sequences of murine Ia antigens

Sequence analysis, prior to DNA sequencing, relied upon radiochemical microsequence techniques. The data obtained were limited to amino-terminal partial sequences. They did, however, reveal a number of features of Ia structure.

Products of the I-A and I-E subregions have been isolated by specific immune precipitation from appropriate spleen cells cultured in the presence of radioactive amino acids. The α and β chains of the Ia antigens are separated by polyacrylamide gel electrophoresis and subjected to automated sequencing procedures (see Figure 3.9). The following conclusions were drawn from the partial sequences thus obtained (Figure 3.14).

1 The α and β chains are not closely related. This postulate is supported by the comparison of peptide maps of α and β chains.

2 The α and β chains of Ia antigen controlled by the I-A subregion are distinct from the respective α and β chains of Ia antigen controlled by the I-E subregion. DNA sequences have revealed that the α chains are 50% homologous and the β chains are 65% homologous.

Fig. 3.14 Comparison of partial amino acid sequences of murine class II (Ia) antigens. This type of limited sequence reveals the major differences between the α and β chains at the A and E loci and the minor differences between allelic products of these loci.

3 Allelic products controlled by the I-E subregion have extensively homologous α or β chain sequences.

4 The α chains of I-E^d and I-E^k have indistinguishable partial sequences. By contrast, the β chains of these two allelic antigens differ at several positions. The extrapolation of these limited comparisons leads to the suggestion that polymorphism of I-E subregion antigens is determined by structural differences in the β chain. DNA sequences confirm that E$_\alpha$ alleles of k and d show very few nucleotide, or predicted amino acid, differences. Polymorphism at the protein level is about 1% but allelic differences are clustered in short sequences rather than being spread through the gene. The protein data indicate that the structural gene for the β chain of the I-E antigen is located in the I-E subregion. No conclusion can be drawn from these data concerning the location of the structural gene for the α chain but this is now known from gene cloning (see Chapter 10). Peptide maps of the α and β chains of allelic Ia antigens also showed that it is the β chains and not the α chains that show sequence differences.

Ia antigens of species other than man and mouse

The guinea pig was the first species in which an Ir gene was discovered. Many guinea pig Ir genes are now recognized and have been mapped to the MHC, termed the GPLA complex. Progress in understanding the genetics of guinea pig Ir genes and of the GPLA complex has been limited by the paucity of inbred strains and the lack of recombinant guinea pigs.

Recognition of products of guinea pig Ir genes came from the use of alloantisera to block proliferative responses *in vitro* of T lymphocytes to antigens that evoke Ir gene control *in vivo* (see Part 3). These alloantisera define guinea pig Ia antigens that are structural homologues of murine and human Ia antigens and that show a similar distribution amongst cells of the lymphoid series. The guinea pig I region can be divided into subregions by serological immunogenetic methods, as has been done for the murine H-2 I region (see Part 2 for details). In contrast to the mouse, the guinea pig Ia antigens can be divided into three subtypes on gross structural evidence. Ultimately amino acid sequences define the array of structural genes within the I region of any species.

Three types of Ia molecules were independently pre-cipitated and identified by applying radiochemical methods (see Figure 3.9) to the guinea pig system. The type of Ia molecule and the serological specificity of the antiserum used to precipitate the antigen are as follows.

1 Ia13.1,6: a glycoprotein with polypeptide chains of mol. wt. $26-27\,000$, possibly existing as dimers. No corresponding molecule has yet been identified in strain 2 guinea pigs.

2 Ia13.3,5, Ia2.2: a glycoprotein with two, non-covalently associated, polypeptide chains of mol. wt. $33\,000$ (α) and $25\,000$ (β).

3 Ia13.7, Ia2.4,5: a glycoprotein with two, disulphide bond linked, polypeptide chains of mol. wt. $33\,000$ (α) and $25\,000$ (β).

On the reasonable assumption that the alloantisera recognize peptide antigens, a minimum of three struc-tural genes for Ia antigens are needed to code for the products described (assuming that antigenic differences are restricted to one polypeptide chain).

Interspecies comparison of Ia antigen amino acid sequences

Alignment of partial amino acid sequences of human murine and guinea pig Ia antigen polypeptide chains reveals homologies that suggest evolutionary relation-ships between the structural genes (Figure 3.15).

1 The α chain of the murine I-E locus antigen is strik-ingly homologous to the human DR α chain sequence. The p34 polypeptide chain can be seen from the sequences to be homologous to the DRα chain but the partial sequence of the latter does not show whether there are allelic differences in the human chains. In the outbred human population these two α chains are most pro-bably from individuals with different alleles at the DR locus. Therefore, these data suggest that for both the murine and human Ia antigens allelic sequence differ-ences have not been found on the α chains. These data are confirmed by gene sequencing. I-Eα and DRα are homologous (see Chapter 10).

2 The β chains of the Ia antigens of man, mouse and guinea pig show homologies, although these chains show allelic differences and the mouse and guinea pig have two types of β chains.

(a) The obvious homology is between the murine I-E β chain sequence and the DRβ chain sequence. This

	1	2	3	4	5	6	7	8	9	10	11	12	13	14	15	16	17	18	19	20	21	22	23	24	25	26	27	28	29	30	31	32
$E_\alpha^{d,k}$	I	K	K	E	H	V	I	I	N	A	E	F	Y	L	L	P	D	K	R	G	G	F	M	F	D	F	D	G	D	E	I	F
DR_α	I	K	E	E	H	V	I	I	Q	A	E	F	Y	L	N	P	D	Q	N	G	E	F	M	F	D	F	D	G	D	E	I	F
p34		K	E	E	H	V	I	I	Q	A	E	F	Y	L	N	P	D	Q	N	G	E	F	M	F	D	F	D	G	D	E	I	F
E_β^k	V	A	S	F	R	P	W	F	L	E	Y	C	K	S	E	C	H	F	Y	N	G	T	Q	R	V	R	L	L	V	R	Y	F
DR_β	[ ]							F	L	Y	Y	V						F	F								F	L			Y	F
E_β^d	V	R	D	T	R	P	R	F	L	E	Y	V	T	S	E	C	H	F	Y	N	G	T	Q	H	V	R	F	L	E	R	F	L
p29	[ ]	G	D	V	P			F	L	E	Q	V																				
13.7	[ ]	I	I		Y			F	L	F								Y														
2.4.5	[ ]	I	Y			Y	R	F							Y	Y																
13.3,5	[ ]			L	P	R	R	F		F	Y	F			Y	Y																
A_β^b	[ ]	G	D	S	E	R	H	F	V	Y	Q	F	M	G	E	C	Y	F	T	N	G	T	Q	R	I	R	Y	V	T	R	Y	I

Fig. 3.15 Comparison of partial amino acid sequences of murine (Eα: Eβ; Aβ), human (DRα; p34: DRβ: p29) and guinea pig (13.7: 2.4.5; 13.3,5) class II antigens. A clear pattern of homology is visible.

is also confirmed by gene sequences.

(b) The sequence of p29 (putatively a β chain of a DR allele) shows homology with DRβ chain consistent either with allelic differences residing in the β chains of human Ia antigens or with two homologous, non-allelic β chain genes.

(c) A more tenuous homology exists between the guinea pig β chains and those of mouse and man. The three guinea pig β chain sequences show these chains to be the products of distinct but homologous genes. In contrast, the β chains of the I-A and I-E antigens do not show significant homology to each other. The following tentative conclusions can be drawn: (1) man, mouse and guinea pig each have at least two distinct Ia β chain structural genes; (2) these gene pairs represent duplicates of a common ancestor; and (3) β chain gene duplication preceded speciation.

The studies of cell-surface transplantation antigens described here have laid a basis for the understanding of the products of the MHC. That base has been developed through extensive DNA sequences (see Chapter 10) that confirm many of the conclusions drawn from partial amino acid sequences. Amino acid sequencing, either using radioactive or modern microsequencing methods, remains an important starting point for entry into recombinant DNA studies via synthetic oligonucleotide probes (see Chapter 7).

References

Klein J. (1975) *Mouse Major Histocompatibility Complex.* Springer Verlag, Berlin.

Silver J., Cecka J.M., McMillan M. & Hood L. (1976) Chemical characterization of products of the H-2 complex. *Cold Spring Harb. Symp. Quant. Biol.* **41**, 369.

Singer S.J. & Nicholson G.L. (1972) The fluid mosaic model of the structure of cell membrane. *Science* **175**, 720.

Vitetta E.S. & Capra J.D. (1978) The protein products of the murine 17th chromosome: genetics and structure. *Adv. Immunol.* **26**, 147.

Walsh F.S. & Crumpton M.J. (1977) Orientation of cell-surface antigens in the lipid bilayer of lymphocyte plasma membrane. *Nature* **269**, 307.

Further reading

Bodmer W.F. (Ed.) (1978) The HLA System. *Brit. Med. Bull.* **34**, 324.

Nathenson S.G. (1981) Primary structural analysis of the transplantation antigens of the murine H-2 major histocompatibility complex. *Ann. Rev. Biochem.* **50**, 1025.

Ploegh H.L., Orr H.T. & Strominger J.L. (1981) Major histocompatibility antigens: the human (HLA-A, -B, -C) and murine (H-2K, H-2D) class I molecules. *Cell* **24**, 287.

Chapter 4
Proteins of the complement system

Introduction

The complement system, which comprises at least twenty plasma proteins (see Tables 4.1−4.4), is one of the major activation systems present in the blood of all vertebrate animals. The activation of this cascade system is arguably the most important effector function of antibodies. Immune elimination of foreign micro-organisms proceeds much less efficiently in individuals with deficiencies of certain key complement components and such individuals are particularly vulnerable during early life (see Chapter 14). Numerous peptide fragments are generated during activation and are known to have potent biological activity. Some act as chemotactic factors and encourage the migration of the phagocytic cells; some possess a specific receptor for the phagocytes and thereby promote immune adherence; other products bring about changes in capillary permeability; and yet others damage the lipid layer of cell membranes. In addition to this effector role, it is also possible that complement has a more central function in the immune response.

The so-called classical pathway of complement activation was first recognized functionally 80 years ago and is still the subject of intense investigation. An antibody independent mechanism of complement activation — the alternative pathway — was originally described by Pillemer in the 1950s and then 'rediscovered' in the early 1970s. Both pathways yield enzymes (convertases) which cleave the quantitatively predominant complement protein C3. The differences between the pathways are to be found in both the nature of the early acting components required to generate the respective convertases and in the distinctive mechanisms of activation.

Most complement components circulate as inactive molecules (often as proenzymes). Activation of any one component is frequently achieved by the proteolytic attack of the preceding component revealing an enzymatically active site which will in turn act on another component. This process led early workers to describe the classical pathway as the complement 'cascade'. Activated components usually have very short biological half-lives and will decay to an inactive form if a substrate molecule is not encountered. Moreover, there are several regulator and control proteins which play a critical role in protecting the tissues of the host against the potentially damaging effects of uncontrolled activation.

The classical and alternative pathways converge on the relatively abundant C3 component and cleavage of this component is probably the most important single event of the activation process. The largest cleavage product (C3b) is also an integral part of the alternative pathway convertase and so activation of the classical pathway inevitably leads to recruitment of the alternative pathway and the generation of further convertase. Thus a marked amplification effect occurs at this stage as a result of positive feedback. This is illustrated in Figure 4.1, which summarizes present knowledge of the interaction of the two pathways.

The proteins of the complement system may be conveniently considered in four sections. Firstly, there are those proteins specific to the classical pathway, together with C3 which is shared with the alternative pathway. Secondly, there are the proteins of the alternative pathway. Thirdly, there are four late acting proteins which constitute the so-called 'attack complex'. Finally, there are the regulatory proteins. Each of these groups will now be discussed in greater depth.

Proteins of the classical pathway

Including C3, there are altogether six classical pathway proteins. The major physicochemical characteristics of the proteins are shown in Table 4.1.

The C1 complex

The C1 complex comprises three subcomponents C1q, C1r and C1s, in the ratio 1:2:2, held together in the presence of calcium ions by non-covalent interactions. Although binding and activation of C1 usually occur with aggregated immunoglobulin (either in immune complexes or otherwise aggregated), alternative mechan-

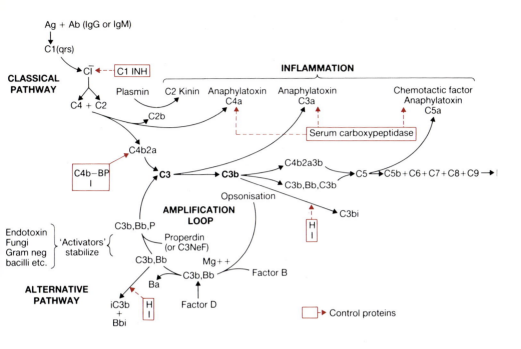

Fig. 4.1 The human complement system showing the classical and alternative pathways of activation. The key role of C3 is emphasized. C3NeF is found in the sera of certain patients with mesangiocapillary glomerulonephritis and partial lipodystrophy. It is an IgG antibody to the C3b,Bb convertase and stabilizes it against decay-dissociation. Properdin is a non-immunoglobulin with similar properties and is present in all normal sera. (Adapted from Turner, 1983.)

isms of C1 activation are now recognized. For example, C reactive protein complexed with C polysaccharide, heparin—protamine complexes, polynucleotides or lipid A of lipopolysaccharides all activate C1 in the absence of antibody. In each case it seems possible that binding

Table 4.1 Classical pathway components of complement

Component	Serum concentration (μg/ml)	Mol. wt.	Mol. wt. of peptide chains after reduction (unactivated)	Substrate cleaved
	150	460 000	$(6 \times 22\,000) + (6 \times 23\,000) + (6 \times 24\,000)$	
	50	160 000	$2 \times 83\,000$	C1s
	50	83 000	83 000	C4, C2
	400	206 000	$93\,000 + 78\,000 + 33\,000$	
	15	110 000	110 000	C3, C5
	1200	190 000	$115\,000 + 75\,000$	

and initiation of the activation mechanism is via the C1q subcomponent which has the ability to bind to immune complexes.

C1q

It is the C1q subcomponent which binds to the immunoglobulin portion of complexes. The protein is structurally unique, being half-globular and half-collagenous. It has a mol. wt. of 460 000 and is assembled from eighteen peptide chains, six A chains, six B chains and six C chains. Each of these chains is about 226 amino acid residues in length and each is glycosylated (12, 8 and 4% carbohydrate respectively). Comparison of the three types of chain has revealed striking sequence homologies. For instance, each has a stretch of about 81 residues near the N-terminal end which is characteristically collagen-like, i.e. contains the repeating triplet Gly-X-Y, where the Y residue is either hydroxylysine or hydroxyproline. The sequences of the 110 C-terminal residues are very similar in the three chains and each has one intra-chain disulphide bridge in this region. The eighteen peptide chains of C1q are associated through the collagenous regions to give six fibrils, each with three chains (A + B + C) interacting in the form of a minor helix. These fibrils further associate through their collagenous regions to give a characteristic hexameric structure with six interacting N-terminal fibrillar regions and six C-terminal globular regions (see Figures 4.2 and 4.3). In electron micrographs the protein is generally agreed to resemble a bunch of flowers with six globular 'flower heads', from which six individual stems fuse into a thickened collagenous region. The immunoglobulin binding sites of the C1q molecules are thought to be located in the globular regions.

C1r and C1s

It is convenient to discuss these proteins together since they are structurally very similar. The association between C1q and these two components is relatively weak in the fluid phase but very strong when C1q has become bound to antigen−antibody aggregates. The C1r−C1s tetramer binds to the collagenous regions of the C1q molecule (see Figure 4.4). The precise C1 activation

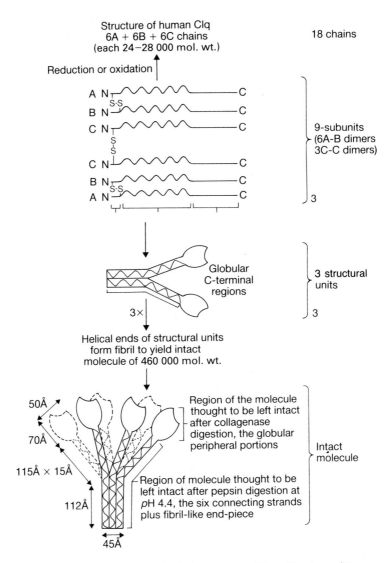

Structure of human Clq
6A + 6B + 6C chains
(each 24–28 000 mol. wt.)

18 chains

Reduction or oxidation

9-subunits
(6A-B dimers
3C-C dimers)

3

Globular
C-terminal
regions

3 structural
units

3

Helical ends of structural units
form fibril to yield intact
molecule of 460 000 mol. wt.

50Å

70Å

115Å × 15Å

112Å

45Å

Region of the molecule
thought to be left intact
after collagenase
digestion, the globular
peripheral portions

Region of molecule thought to be
left intact after pepsin digestion at
pH 4.4, the six connecting strands
plus fibril-like end-piece

Intact
molecule

Fig. 4.2 Proposed peptide chain structure of C1q. (Courtesy of Dr K.B.M. Reid.)

sequence is still unclear but it has been suggested that on binding of C1q to immunoglobulin an active site becomes exposed in the proenzyme C1r molecule. This activated proenzyme then either undergoes cleavage itself or acts on the adjacent C1r molecule to yield C1r̄ which in turn activates C1s to yield C1s̄ (C1 esterase). (NB The superscript bars denote activated enzymes.)

Both C1r and C1s consist of a single polypeptide chain having a mol. wt. of approximately 83 000, but in solution C1r occurs as a dimer, whereas C1s remains

73 PROTEINS OF THE COMPLEMENT SYSTEM

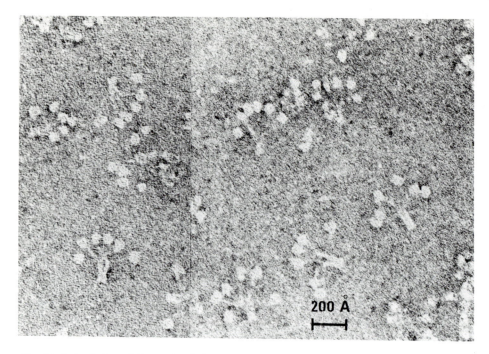

Fig. 4.3 Lateral view of C1q molecule by electron microscopy. Final magnification ×500 000. (Reproduced with permission from Knobel *et al.*, 1975.)

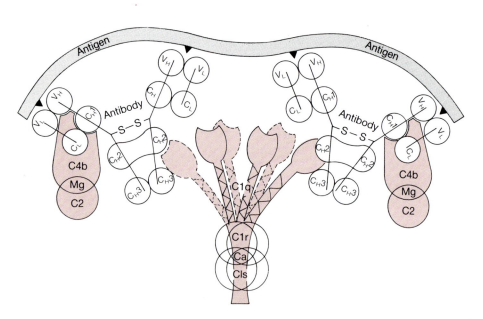

Fig. 4.4 Suggested assembly of the early components of the classical pathway of complement on antibody bound to a cell surface. The C1q molecule binds through its globular regions to the C_H2 domain of the antibody and the C1r-Ca^{2+}-C1s complex binds to the collagenous region of the C1q. (Modified from an original kindly supplied by Dr K.B.M. Reid.)

C1r   Ile - Ile - Gly - Gly - Gln - Lys - Ala - Lys - Met - Gly - Asn - Phe - Pro - Try - Gln - Val - Phe - Thr - Asn - Gln

C1s   Ile - Ile - Gly - Gly - Ser - Asp - Ala - Asp - Ile - Lys - Asn - Phe - Pro - Try - Gln - Val - Phe - Thr - Asp - Asn -

Factor D   Ile - Leu - Gly - Gly - Arg - Gln - Ala - Gln - Ala - His - Ala - Arg - Pro - Tyr - Met - Ala - Ser - Val - Gln - Leu -

Fig. 4.5 N-terminal sequences of the catalytic chains of C1r, C1s and Factor D. Identities among the sequences are shown in pink.

monomeric. The amino acid compositions and partial sequences (Figure 4.5) of these proteins show them to be remarkably similar serine-esterase proteases. It seems probable (see p. 87) that they arose as a result of a gene duplication. Upon activation, both proteins are cleaved into a large and small peptide chain still linked by a disulphide bridge (Figure 4.6). In both cases the exposed active site appears to be on the small peptide chain.

C4

C4 is one of the two substrates for the enzyme C1 esterase. It is a glycoprotein of β-electrophoretic mobil-

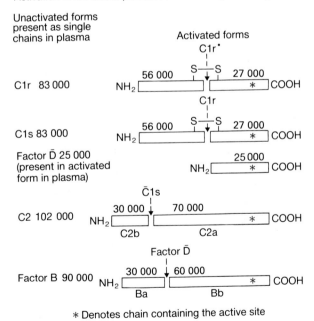

Activation of the serine proteases of the complement system

* Denotes chain containing the active site

Fig. 4.6 Activation of the serine proteases of the complement system. C1r• denotes the single-chain proenzyme form of C1r which has enzymatic activity. (Modified with permission from Reid & Porter, 1981.)

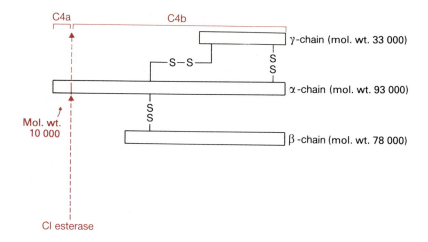

Fig. 4.7 Polypeptide chain structure of C4 showing cleavage by the C1 esterase enzyme.

ity and has a mol. wt. of 206 000. It comprises three covalently linked polypeptide chains designated α (93 000), β (78 000) and γ (33 000) (Figure 4.7). The three constituent peptide chains are held together by non-covalent and disulphide bonds. During activation the molecule is cleaved by C1 esterase only in the α chain (see Figure 4.7). A small fragment of the α chain (C4a: mol. wt. 10 000) is released into the fluid phase and its loss is believed to expose a hydrophobic binding site in the larger C4b fragment which permits approximately 10% of such fragments to bind to either adjacent cell membranes or the antigen–antibody aggregate itself. Recent work suggests that a small proportion of C4b covalently binds to the antibody molecule (probably to the C_H1 domain) and that this is then able to form a complex with C2 which is the second substrate of C1 esterase.

The α chain of the C4b fragment is itself susceptible to further cleavage by the C3b inactivator protein (now called Factor I). This splits C4b into a large fragment called C4c (mol. wt. 157 000) and a smaller peptide called C4d (mol. wt. 35 000).

C2

C2 is a single chain glycoprotein of mol. wt. 110 000 which is cleaved by C1 esterase to give two fragments, C2a (mol. wt. 70 000) and C2b (mol. wt. 30 000) (see Figure 4.6). The C2a fragment has a serine-protease

76 CHAPTER 4

enzymic site which specifically cleaves C3 and C5 molecules. (The complex C4b2a is a C3 convertase but the incorporation of C3b into the complex changes the specificity of the enzyme to that of a C5 convertase.) C2 is the product of a gene in the major histocompatibility complex (MHC) and is closely related to Factor B of the alternative pathway (see later).

C3

C3 is by far the most important component of the complement cascade. Its concentration in serum (1.0–1.5 mg/ml) is similar to that of IgM and it may rightly be regarded as the pivotal protein in the system. C3 has a mol. wt. of approximately 190 000 and consists of two covalently linked polypeptide chains; the α chain with a mol. wt. of 115 000 and the β chain with a mol. wt. of 75 000.

It is the α chain which is cleaved by the C3 convertase enzymes generated by the classical and alternative pathways. This cleavage point is known to lie between the arginine and serine residues at positions 77 and 78 of the α chain (see Figure 4.8). The smaller fragment which is generated (C3a) has a mol. wt. of 9000 and possesses anaphylatoxin activity. The major fragment (C3b) has a mol. wt. of 180 000 and four major functions: (1) it participates in the formation of the amplification C3 convertase; (2) it complexes with C4b2ab or C3bBb to produce a C5 convertase; (3) it binds to specific immune adherence receptors on phagocytic cells; and (4) it binds covalently through another site to a wide range of cell membranes and thereby acts as an opsonin. The latter function appears to resemble the binding of C4b fragments in that in both cases a shortlived nascent site is revealed which is able to form a covalent (probably ester) bond with appropriate structures (see

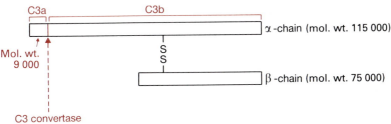

Fig. 4.8 Polypeptide chain structure of C3 showing cleavage point of the C3 convertase enzymes C4b2a and C3bBbP.

Table 4.2. Alternative pathway components of complement

Component	Synonym	Serum concentration (μg/ml)	Mol. wt.	Mol. wt. of peptide chains after reduction (unactivated)	Su cle
D	(GBGase, C3 proactivator convertase)	5	24 000	24 000	B
B	(GBG, C3 proactivator)	200	90 000	90 000	C3
P	Properdin	25	220 000	4 × 55 000	
C3		1200	190 000	115 000 + 75 000	

p. 89). Another feature common to both C4 and C3 is that the regulator protein, Factor I (previously known as C3b inactivator and C4b inactivator), cleaves a bond in the α chain of both proteins without overt fragmentation but leading in both cases to loss of biological function. Subsequent cleavage into C3c and C3d fragments is the result of further proteolytic action.

Proteins of the alternative pathway

Three proteins are required for complement activation by the alternative pathway, namely C3, Factor B and Factor D. A fourth protein P (or properdin) plays a stabilizing role and will also be considered in this section.* Table 4.2 shows the main physicochemical characteristics of these four proteins.

The available evidence suggests that there is a continuous low-grade interaction between these proteins to generate the alternative pathway convertases (see Figure 4.9). This may begin with an initial interaction of C3 with B to give a bimolecular complex (C3B). In the presence of Factor D the convertase C3Bb is generated by cleavage of B to reveal an enzymically active site on Bb with specificity for fluid-phase C3. This convertase is able to generate C3b fragments which can then complex with further Factor B to give the more efficient amplification convertase C3bBb. The half-life of this convertase may be considerably prolonged if the control protein properdin binds to it and such properdin-stabilized amplification convertase is thought to be present in the plasma at low levels at all times. Its continuous production is balanced by continuous decay under the action of two control proteins, Factors H and

*The nomenclature of the alternative pathway components used is that proposed by the I.U.I.S. and published in full in *J. Immunol.* (1981) **127**, 1261.

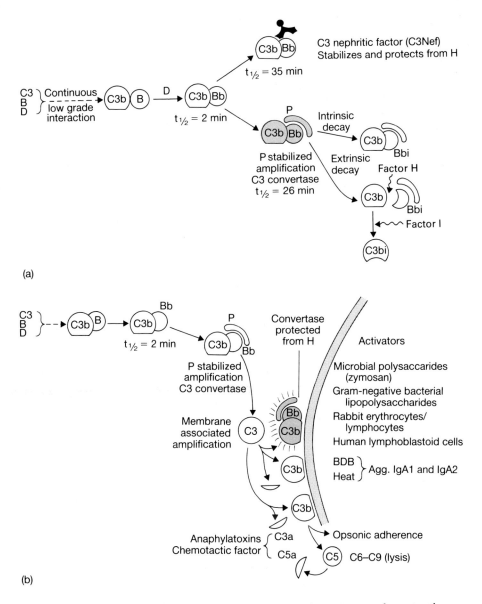

(a)

(b)

Fig. 4.9 (a) The fluid-phase interaction of the proteins of the alternative complement pathway. The C3 convertase generated by continuous low-grade interaction (C3bBb) has a short biological half-life but can be protected from decay by complexing with a stabilizing protein. Physiologically, this role is filled by properdin but although the convertase C3bBbP has a greatly increased half-life it is ultimately degraded by the concerted action of the control proteins Factors H and I. The intrinsic decay pathway is probably of minor importance. (b) Activation of the alternative pathway by surfaces. A wide variety of substances activate the alternative pathway. They probably do so by providing a microenvironment which protects the C3bBbP convertase from the actions of the control proteins Factors H and I. Thus, the convertase is permitted to act on the C3 in the fluid phase and fragment it to C3a and C3b. This provides the so-called amplification effect. (Adapted from Turner, 1983.)

I, but this balance is deregulated by the presence of 'activating' substances such as microbial polysaccharides and endotoxin. It is believed that a characteristic of all such activating substances is their ability to provide protective surfaces to which the convertase binds. The control proteins are thereby less able to inactivate the enzyme complex which acts on further fluid-phase C3 generating more C3b to function as either further convertase or as C3b opsonin.

Factor D

Factor D is a trace serum protease of α mobility and a mol. wt. of about 23 500. It consists of a single peptide chain, and enzymically active D can be derived from precursor D by brief treatment with trypsin. However, no natural plasma protein activator of D has yet been identified and the invariable presence of active D in plasma and serum eliminates any critical requirement for a specific activator. Partial amino acid sequence data tend to support this view since the first 16 N-terminal residues of human D showed 65% identity with the 'group specific' protease isolated from rat intestine. The latter is thought to be synthesized and secreted in its active form.

Factor B

Factor B is a single-chain β-globulin of mol. wt. 90 000. In the presence of C3b the protein is cleaved by D into a small fragment, Ba (mol. wt. 30 000) and a large fragment Bb (mol. wt. 60 000). B has many structural and functional properties in common with C2. Both proteins form Mg^{2+}-dependent reversible protein–protein complexes (C2 with C4b and B with C3b). Both are cleaved by a serine esterase to give unequal fragments, the larger of which forms complexes having esterase activity (i.e. C4b2a and C3bBb). Both of these enzyme complexes cleave the α chain of C3 at an arginyl–X peptide bond and are able to bind additional C3b and so modify their substrate specificity from C3 to C5. Furthermore, Factor B, like C2, is coded for by a gene in the MHC and it seems very probable that both proteins arose by a process of gene duplication.

The gene structure of human Factor B was determined

by Campbell *et al.* (1984) and the intron-exon arrangements suggest the existence of three homologous regions of about 60 amino acid residues in the Ba fragment. Similar homology regions are also a feature of C2 and C4-binding protein. The common function of all three proteins is interaction with C3b/C4b fragments and the homology regions may therefore be involved in these processes.

cDNA-derived amino acid sequences from human and mouse Factor B have revealed 83% sequence homology and a high level of evolutionary conservation for the protein.

Properdin (P)

The protein properdin for many years gave its name to the alternative method of complement activation ('properdin pathway'). It was first isolated in 1954 but its precise role in the activation process only became clear relatively recently. Properdin is a γ-globulin with a mol. wt. of approximately 220 000. It functions as a stabilizing protein in the alternative pathway where it binds with high affinity to the C3b portion of the C3bBb convertase and presumably sterically blocks access of the control protein H.

Properdin is a glycoprotein containing 9.8% carbohydrate. It appears to consist of four identical noncovalently linked subunits of mol. wt. 55 000. Ultracentrifugation studies indicate a highly asymmetric molecule.

Terminal complement components

The proteins C5, C6, C7, C8 and C9 constitute what is sometimes referred to as the 'membrane attack sequence'. These proteins function as a unit capable of inflicting membrane damage which usually culminates in cell lysis. With the exception of C5, it is not yet clear whether proteolytic cleavage is required for their activation. Some of the characteristics of these components are given in Table 4.3.

C5

C5 is a β-globulin having a mol. wt. of approximately

Table 4.3 Terminal complement components

Component	Serum concentration (μg/ml)	Mol. wt.	Mol. wt. of peptide chains after reduction (unactivated)
C5	80	190 000	115 000 + 75 000
C6	70	95 000	95 000
C7	65	120 000	120 000
C8	80	163 000	83 000 + 70 000 + 10 000
C9	200	79 000	79 000

190 000. It resembles C3 in many respects and is composed of a long α chain (mol. wt. 115 000) and a β chain (mol. wt. 75 000) (see Figure 4.10). Activation of C5 involves proteolytic cleavage by one of the C5 convertases (C4b2a3b or C3bBbC3b) adjacent to the arginine residue at position 75 of the α chain to yield a small peptide C5a (mol. wt. 12 000) and a large residual C5b fragment.

The C5a fragment has some structural homology to C3a (see Figure 4.11) and both function as potent anaphylatoxins, able to induce smooth muscle contraction and increase vascular permeability. In addition, C5a is a potent neutrophil chemotactic factor.

C6, C7, C8 and C9

C6 is a single-chain β-globulin with a mol. wt. of 95 000. C7 is also a single-chain β-globulin with a slightly higher mol. wt. of 120 000. C8 is a γ-globulin of mol. wt. 163 000 and apparently consisting of three chains: α (83 000), β (70 000) and γ (10 000). The γ chain is thought to be covalently bound to the α chain, whereas the α and β chains appear to be non-covalently linked. C9 is a single-chain α-globulin of mol. wt. 79 000.

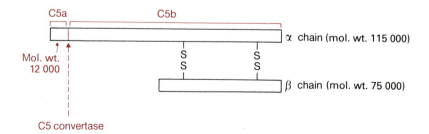

Fig. 4.10 Polypeptide chain structure of C5 showing the cleavage point of the C5 convertases C4b2a3b and C3bBbC3b.

I.   NH$_2$ • SER - VAL - GLN - LEU - Met - GLU - LYS - ARG - MET - ASN - LYS - Leu - GLY - Gln - TYR - Ser - LYS - GLU - LEU -

II. NH$_2$ • SER - VAL - GLN - LEU - Thr - GLU - LYS - ARG - MET - ASN - LYS - Val - GLY - LYS - TYR - Pro - LYS - GLU - LEU -

III. NH$_2$ • Thr - LEU - Gln - Lys - LYS - Ile - Glu - Glu - Ala - Ala - LYS - TYR - Lys - His - Ser - Val -

1   10

- ARG - Arg - CYS - CYS - GLU - His - GLY - MET - ARG - Asn - ASN - PRO - MET - Lys - PHE - SER - CYS - GLN - ARG -

- ARG - LYS - CYS - CYS - GLU - ASP - GLY - MET - ARG - Gln - ASN - PRO - MET - Arg - PHE - SER - CYS - GLN - ARG -

- Val - Lys - LYS - CYS - CYS - Tyr - ASP - GLY - Ala - Cys - Val - Asn - Asp - Glu - Thr - CYS - Glu - Gln -

20   30

- ARG - ALA - Gln - PHE - ILE - His - Gln - GLY - Asn - ALA - CYS - Val - LYS - ALA - PHE - LEU - Asn - CYS - CYS - Glu - TYR -

- ARG - Thr - Arg - PHE - ILE - SER - LEU - GLY - Glu - ALA - CYS - Lys - LYS - VAL - PHE - LEU - Asp - CYS - CYS - Asn - TYR -

- ARG - ALA - Ala - ILE - SER - LEU - GLY - Pro - Arg - CYS - Ile - LYS - ALA - PHE - Thr - Glu - CYS - CYS - Val - Val -

40   50

- ILE - Ala - Lys - LEU - ARG - Gln - GLN - HIS - Ser - ARG - Asn - Lys - Pro -

- ILE - Thr - Glu - LEU - ARG - Arg - GLN - HIS - Ala - ALA - ARG - Ala - SER - HIS -

- Ala - Ser - Gln - LEU - ARG - ALA - Asn - Ile - SER - HIS - Lys - Asp - Met - Gln - LEU - Gly -
                                   |
                                 CHO

60   70

- LEU - GLY - LEU - ALA - ARG • COOH   77

- LEU - GLY - LEU - ALA - ARG • COOH   74

- LEU - Gly - ARG • COOH   74

70

Fig. 4.11 Complete primary structures for porcine C3a(I), human C3a(II), and human C5a(III). Identities among the sequences are shown in pink. There are 54 common residues between human and porcine C3a. Human C3a and C5a share 29 homologous sites, if four gaps and one indentation are introduced to optimize the alignment. The six half-cystine positions in human C3a appear to have been conserved in both porcine C3a and human C5a. Conservation of half-cystine indicates their probable role in dictating a preferred conformational arrangement of these molecules. A single carbohydrate attachment site exists at position 64 in human C5a. (Modified with permission from Hugli, 1978.)

The C5b fragment (generated by either of the activation sequences) binds C6 and C7 in a 1:1:1 ratio and the resultant trimolecular complex then binds one molecule of C8. The C8 component initiates damage to the bimolecular lipid layer of adjacent cell membranes. In the final step of the complement cascade up to six C9 molecules may bind to C8 in the C5−8 complex, although the C8 is usually unsaturated with respect to C9. It has been shown that the carboxy-terminal portion of C9 produces lysis by inserting itself into the lipid layer, thus producing the characteristic lesions, which may be visualized by electron microscopy. The cDNA-derived sequence of the 573 amino acids comprising the C9 component have revealed a relatively hydrophilic amino-terminal portion and a relatively hydrophobic carboxy-terminal portion as would be expected from the above properties.

Control proteins of the complement system

Indiscriminate activation of the complement system could be extremely damaging to the host and a number of regulatory proteins have evolved to keep the system in homeostatic balance. Five such proteins are known at present. Others almost certainly exist. Some of the characteristics of these proteins are summarized in Table 4.4.

C1 esterase inhibitor (syn: C1-INH, α_2-neuraminoglycoprotein)

This is believed to be one of the most highly glycosylated proteins in serum, the carbohydrate content being approximately 35%. The protein is an α-globulin comprising a single peptide chain with a mol. wt. of 100 000. C1-INH is a potent inhibitor of the enzyme activities of a number of serum proteins, e.g. C1r, plasmin, kallikrein, activated Hageman factor. However, as far as is known, it is the only inhibitor of C1s̄. The inhibitor removes both C1s and C1r from the activated C1 complex so that the collagenous regions of the residual C1q are free to interact with cell surface receptors. Deficiency of C1-INH is the most common genetically determined deficiency of the complement system (see Chapter 14) and may present as either a quantitative deficiency with low levels of the protein or in a rarer form

ein (synonym)	Serum concentration (μg/ml)	Mol. wt.	Mol. wt. of peptide chains after reduction (unactivated)	Substrate
sterase inhibitor INH)	200	110 000	110 000	C1s
or I (C3b inactivator)	20	88 000	50 000 + 38 000	C3b, C4b, C5b
or H (β₁H globulin, NA accelerator)	650	150 000	—	C3bBb
m carboxypeptidase	35	310 000	8 × 36 000	C3a, C5a
inding protein	250	550 000	7 × 70 000	C4b

in which levels are normal but the protein is functionally abnormal.

Factor I (C3bINA, C3b inactivator)

Factor I is a β-globulin of mol. wt. 88 000 which is able to inactivate C3b and probably also C4b and C5b. In the inactivation of C3b it is probable that the control protein functions as a protease and cleaves the α chain at two sites. Structurally, Factor I consists of two chains having mol. wts. of 50 000 and 38 000.

Factor H (β_1H globulin)

H is a single-chain β-globulin of mol. wt. 150 000. It is a glycoprotein and contains 15−20% carbohydrate. It appears to function as a regulator protein by enhancing Factor I activity. It does not appear to have enzymatic activity but displaces Bb from the complex C3bBb and thereby permits access of Factor I.

Serum carboxypeptidase B

The two major anaphylatoxins liberated in the complement cascade, C3a and C5a, induce smooth muscle contraction and increase vascular permeability, even at low serum concentrations. Such potent biological activity requires efficient control, and a serum enzyme, initially described as a kininase or anaphylatoxin inactivator, has been shown to function as an inactivator of both C3a and C5a. It is a serum carboxypeptidase able to remove the essential C-terminal arginyl residue from both of the anaphylatoxins, thereby inactivating their spasmogenic activity. The serum carboxypeptidase B appears to be similar to pancreatic carboxypeptidase B, has a mol. wt. of 310 000 and probably consists of eight similar peptide subunits of mol. wt. 36 000.

C4b-binding protein

A C4b-binding protein has been described. As its name implies it interacts with the C4b fragment of C4 and can displace C2 from the C4b2a enzyme complex. It also acts as a co-factor in the degradation of C4b by Factor I. It appears to be a multichain protein with approximately seven to eight disulphide-linked polypeptide chains, each of mol. wt. 70 000.

Biosynthesis of complement components

The complement components appear to be synthesized in several different sites. Lachmann and Peters (1982) summarized some of these as follows: Factor B, C3, C6, C8, C9 and C1 esterase inhibitor appear to be of hepatic origin; C1q, C1r and C1s are derived from intestinal epithelium; and many others, including C2, C4, C3, C5, Factor D, Factor B, Factor I and Factor H, are synthesized by macrophages. More recent data suggests that most of the components are probably synthesized in the liver and that macrophage synthesis may provide a source for continuing complement activation at sites of inflammation.

In 1977 Hall and Colten showed that guinea pig C4 (a three-chain molecule) was derived from a single-chain precursor called pro-C4. Subsequently pro-C4 was identified in human serum and single-chain precursors of the two-chain proteins C3 and C5 were described in several species.

It is possible that the pro-C3, pro-C4 and pro-C5 molecules are converted into C3, C4 and C5 respectively either during translation or post-translationally before secretion. The conversion process is probably similar to that described for the conversion of pro-insulin to insulin and involves cleavage of one or more peptide bonds (see Figure 4.12). Although no large peptide appears to be excised at the time of translation there is probably loss of an N-terminal pre-peptide of some 15–30 residues from each of these three complement components. This, if it is confirmed, would add these proteins to a growing list of secretory proteins for which

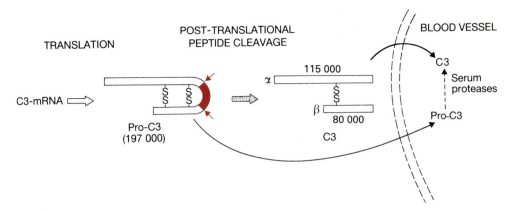

Fig. 4.12 Biosynthesis of C3 showing the translation of C3-mRNA into a single-chain pro-C3 molecule of mol. wt. 197 000. Most pro-C3 is modified by proteolytic enzymes which excise a small peptide (shown in red) before the molecule is secreted. Some pro-C3 escapes this process and trace levels may be detected in fresh serum. C4 and C5 are believed to undergo similar processing. (Modified with permission from Minta *et al.*, 1979.)

pre-proteins have been described. In every case the signal (pre-)peptide appears to be necessary for efficient transfer of the protein across the rough endoplasmic reticulum of the cell and, its function complete, it is removed by the action of a microsomal peptidase.

Recent studies of a cDNA-derived sequence for human C3 have revealed a 22 residue signal peptide followed by 645 residues corresponding to the β chain, 4 intervening basic residues and finally 992 residues corresponding to the α chain. The 4 basic residues are presumably those lost during conversion of pro-C3 to the serum form.

Structural relationships between complement components

Several lines of evidence indicate that there are 'subfamilies' of complement components. Three such groups of related proteins will be described, but further examples are emerging as more structural detail becomes known.

C1r and C1s

As previously noted, C1r and C1s each consists of a single polypeptide chain having a mol. wt. of approximately 83 000, but in solution C1r occurs as a dimer whereas C1s remains monomeric. The proteins have similar amino acid compositions and homologous sequences (see Figure 4.5), strongly suggesting a common genetic precursor. Upon activation both proteins are cleaved into a large and small peptide chain, still linked by a disulphide bridge (see Figure 4.6). In both cases the exposed active site appears to be on the smaller peptide chain.

C2 and Factor B

Both of these proteins form Mg^{2+}-dependent reversible protein−protein complexes (C2 with C4b and B with C3b). Both are cleaved by a serine esterase to give unequal fragments, the larger of which forms complexes having esterase activity (i.e. C4b2a and C3bBb). Both of these enzyme complexes cleave the α chain of C3 at an arginyl−X peptide bond and are able to bind additional C3b and so modify their substrate specificity from C3 to C5. Furthermore, Factor B, like C2, is coded for by a gene in the MHC and it seems very probable that both proteins arose by a process of gene duplication.

C3, C4 and C5

Although each of these is a multichain protein it is now clear that each is synthesized as a single peptide chain (see p. 86) which then undergoes proteolytic cleavage just before secretion. Complete (C3, C4) or partial (C5) amino acid sequences of these proteins have recently been derived by cDNA cloning and all share strong sequence homology both with each other and also with α_2-macroglobulin (see Reid, 1985).

These developments are the satisfying culmination of work which began when the two anaphylatoxins, C3a and C5a, derived from the amino-terminal ends of the α chains of C3 and C5 respectively, were directly sequenced and compared with each other (see Figure 4.11). Such comparisons showed that only 29 of the 74 residues in human C5a are homologous to corresponding residues in human C3a. However, this similarity was sufficient, according to Hugli (1978), to permit one to conclude that C3a and C5a must share a common genetic ancestry (the probability that the observed similarity was due to a random alignment was less than 0.1%).

Similarities between C3 and C4 were next revealed by

the studies of Sim and Sim (1981). They were able to show that haemolytically active forms of C3 and C4 undergo an unusual form of autolytic cleavage under certain denaturing conditions. In the case of C3 the α chain is split into fragments of 74 000 and 46 000 apparent mol. wt., whereas C4 gives rise to fragments of 53 000 and 41 000 apparent mol. wt. These reactions do not occur with inactivated C3 or C4. The characteristic features of this denaturation-cleavage are strikingly similar to those which have been described for the active form of α_2-macroglobulin. It is known that the active forms of C3, C4 and α_2-macroglobulin are able transiently to form covalent bonds after activation by limited proteolysis. In the case of C3 it has been proposed that covalent binding occurs by transfer of an electrophilic carbonyl group from a thiol ester bond in

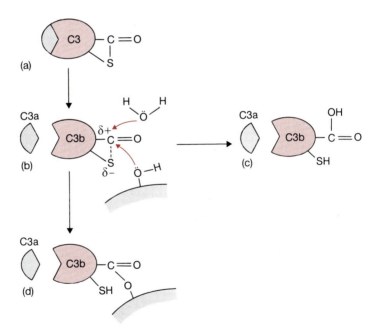

Fig. 4.13 Mechanism of covalent bond formation by nascent C3b fragments released from activated C3 molecules. Native C3 possesses a moderately reactive esterified carbonyl group (a) and cleavage of C3 to C3a and C3b causes polarization or possible fission of a bond to this carbonyl group, making the latter more electrophilic. The reactive carbonyl group may then be attacked by mild nucleophiles (such as OH groups) on nearby binding surfaces resulting in the formation of an ester bond (b)–(d) or, alternatively, may be discharged by reaction with water (b)–(c). (Adapted from Sim et al., 1981.)

C3 to form an ester or amide bond with a suitable binding surface (see Figure 4.13). It appears likely that the reactive groupings involved in the denaturation-cleavage reaction are the same as those involved in the formation of covalent bonds. Interestingly, C5, although structurally similar to C3 (see above), does not appear to undergo denaturation-induced cleavage or share with C3 the ability to form covalent bonds.

Sim and Sim (1981) have proposed a pattern of alignment for the large fragments of C3, C4, C5 and α_2-macroglobulin (see Figure 4.14). This is based on the relative positions of the protease-sensitive sites of all four proteins, and in the case of C3, C4 and α_2-macroglobulin the denaturation-sensitive cleavage sites and known regions involved in covalent binding. This alignment predicts an order for the known chains of C3, C4 and C5 in the pro forms (i.e. prior to secretion).

Although the functional roles of the complement proteins and α_2-macroglobulin are apparently unrelated, it is possible to discern areas of similarity. The principal role of α_2-macroglobulin is believed to be the trapping

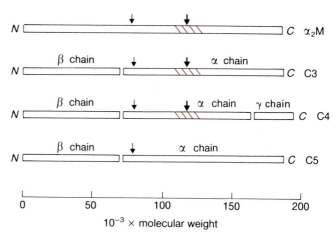

Fig. 4.14 Structural similarities of α_2-macroglobulin, C3, C4 and C5. The polypeptide chain segments corresponding to the α and β chains of C3 and C5 and of the α, β and γ chains of C4 are indicated. The light arrows indicate the sites of activation of these proteins by limited proteolysis and the heavy arrows indicate the suggested sites of denaturation-induced cleavage. The red shading indicates the regions of the polypeptide chains which may be involved in covalent bond formation. N, N-terminus; C, C-terminus. (Modified with permission from Sim & Sim, 1981.)

of cellular and exogenous endopeptidases by a covalent binding mechanism. Thus, the major roles of all the proteins under discussion may be the clearance of exogenous material from circulation.

Although C3 and C5 are not thought to be linked to the MHC (unlike C4), there are sound structural reasons for regarding these three proteins as representing a group of related molecules. The extent of the structural similarities will, however, require much further investigation.

Complement receptors

In recent years it has become clear that there are cell-membrane proteins which interact with various complement components and frequently play a major modulatory role in the biological aspects of complement activation. Advances in our knowledge of their structure, genetics and associated deficiencies have been reported with increasing frequency in the last three years.

Three major types of receptor are currently recognized and each will be described briefly.

The CR1 receptor

Sometimes known as the immune adherence receptor, this is found on human erythrocytes, phagocytic cells, B lymphocytes and kidney podocytes. The CR1 receptor was the first to be isolated and structurally characterized as a membrane glycoprotein having a mol. wt. of approximately 200 000. Subsequently it has become clear that CR1 receptors are a rather heterogeneous group of polymorphic molecules with no clear evidence of any functional differences between the variants. The major role of CR1 molecules appears to be to act as surface bound co-factors for Factor I in the breakdown of C3bi to C3c and C3dg. In a more general sense, however, CR1 shares many functional roles with both Factor H and the C4-binding protein. Each of these proteins appears to be capable of binding to C3b and C4b, to act as co-factors for Factor I in the inactivation of C3b and C4b and to dissociate both C3- and C5-convertase enzymes.

The CR2 receptor

This is the major complement receptor of B lymphocytes and B-lymphoblastoid cells and has specificity for C3d. Its precise functional significance remains unclear, but it is known that it binds Epstein–Barr (EB) virus. Structurally, CR2 appears to be a membrane glycoprotein with a mol. wt. of 140 000.

The CR3 receptor

This receptor is also known as the macrophage 1 (Mac-1) glycoprotein (syn: OKM1) and is a cell-surface molecule with specificity for C3bi. Structurally, Mac-1 is known to be related to the lymphocyte function-associated 1 (LFA-1) molecule. The proteins have α subunits of apparent mol. wt. 170 000 and 180 000 respectively and these are non-covalently associated with β subunits having an apparent mol. wt. of 95 000. The β chains appear to be identical whereas the α chains, though related, differ in amino acid sequence. N-terminal sequencing has revealed 33% homology between Mac-1 and LFA-1 and a recently reported computer search (Springer et al., 1985) also showed that both proteins share significant homologies with the α interferon family. It is suggested that the α subunits of Mac-1, LFA-1 and the interferon family may have arisen from a common ancestral gene and constitute another superfamily resembling the immunoglobulin superfamily.

The specificity of the CR3 receptor is for C3bi and the receptor binding site appears to be localized to the α subunit.

Other complement receptors

Other receptors for complement proteins are known to exist but much less is known about them. A CR4 receptor, with specificity similar to CR2, has been described on both neutrophils and mononuclear cells and reports of a group of distinct C3b binding membrane proteins are beginning to appear in the literature.

References

Campbell R.D., Bentley D.R. & Morley B.J. (1984) The Factor B and C2 genes. *Phil. Trans. Roy. Soc. Lond.* B **306**, 367.

Hall R.E. & Colten H.R. (1977) Cell-free synthesis of the fourth component of guinea pig complement (C4). Identification of a precursor of serum C4 (pro-C4). *Proc. natn. Acad. Sci. USA* **74**, 1707.

Hugli T.E. (1978) Classical aspects of the serum anaphylatoxins. In *Contemporary Topics in Molecular Immunology*, Vol. 7, (eds. Reisfeld R.A. & Inman F.P.), p. 181. Plenum Press, New York.

Knobel H.R., Villiger W. & Isliker H. (1975) Chemical analysis and electron microscopy studies of human C1q prepared by different methods. *Eur. J. Immunol.* **5**, 78.

Lachmann P.J. & Peters D.K. (1982) Complement. In *Clinical Aspects of Immunology*, 4th edn., (eds. Lachmann P.J. & Peters D.K.), p. 18. Blackwell Scientific Publications, Oxford.

Minta J.O., Ngan B.-Y. & Pang A.S.D. (1979) Purification and characterisation of a single chain precursor C3-protein (Pro-C3) from normal human plasma. *J. Immunol.* **123**, 2415.

Reid K.B.M. & Porter R.R. (1981) The proteolytic activation systems of complement. *Ann. Rev. Biochem.* **50**, 433.

Sim R.B. & Sim E. (1981) Autolytic fragmentation of complement components C3 and C4 under denaturing conditions, a property shared with α_2-macroglobulin. *Biochem. J.* **193**, 129.

Sim R.B., Twose T.M., Paterson D.S. & Sim E. (1981) The covalent-binding reaction of complement component C3. *Biochem. J.* **193**, 115.

Springer T.A., Teplow D.B. & Dreyer W.J. (1985) Sequence homology of the LFA-1 and Mac-1 leukocyte adhesion glycoproteins and unexpected relation to leukocyte interferon. *Nature* **314**, 540.

Turner M.W. (1983) Complement. In *Paediatric Immunology*, (eds. Soothill J.F., Hayward A.R. & Wood C.B.S.), p. 21. Blackwell Scientific Publications, Oxford.

Further reading

Fearon D.T. & Austen K.F. (1977) Immunochemistry of the classical and alternative pathways of complement. In *Immunochemistry: An Advanced Textbook*, (eds. Glynn L.E. & Steward M.W.), p. 365. John Wiley & Sons, Chichester.

Hugli T.E. (1978) Chemical aspects of the serum anaphylatoxins. In *Contemporary Topics in Molecular Immunology*, Vol. 7, (eds. Reisfeld R.A. & Inman F.P.), p. 181. Plenum Press, New York.

Lachmann P.J. & Peters D.K. (1982) Complement. In *Clinical Aspects of Immunology*, 4th edn., (eds. Lachmann P.J. & Peters D.K.), p. 18. Blackwell Scientific Publications, Oxford.

Pangburn M.K. & Müller-Eberhard H.J. (1984) The alternative pathway of complement. *Springer Semin. Immunopath.* **7**, 163.

Porter R.R. & Reid K.B.M. (1978) The biochemistry of complement. *Nature* **275**, 699.

Reid K.B.M. (1985) Application of molecular cloning to studies on the complement system. *Immunology* **55**, 185.

Reid K.B.M. & Porter R.R. (1981) The proteolytic activation systems of complement. *Ann. Rev. Biochem.* **50**, 433.

Ross G.D. & Atkinson J.P. (1985) Complement receptor structure and function. *Immunol. Today* **6**, 115.

Stroud R.M., Volanakis J.E., Nagasawa S. & Lint T.F. (1979) Biochemistry and biological reactions of complement proteins. In

Immunochemistry of Proteins, Vol. 3, (ed. Atassi M.Z.), p. 167. Plenum Press, New York.

Whaley K. & Ferguson A. (1981) Molecular aspects of complement activation. *Mol. Aspects Med.* **4**, 209.

Part 2
Genes of the
Immune System

Chapter 5
Classical genetics: immunoglobulins

Introduction

Polymorphic variation in serum proteins is extremely common. In the past two decades high resolution electrophoretic techniques used to study individual protein systems of both man and other animals have almost invariably shown some degree of polymorphism. Haptoglobin (in 1955) and transferrin (in 1957) were the first serum proteins in which such variation was demonstrated but the list has now been extended to include the Gc components, ceruloplasmin, α_1-antitrypsin, orosmucoid, α_2-macroglobulin, β-lipoprotein, complement components (see Chapter 11) and, of course, the immunoglobulins. Many polymorphisms are revealed by serological detection of antigenic differences. In the immunoglobulin family both electrophoretic and antigenic differences have been detected and not only do we have population gene frequencies for most of the alleles but, in many cases, we also have some knowledge of the structural features correlating with the presence of particular allotypes.

Allotypic variation is, however, not the only form of antigenic variation seen in the immunoglobulins. It is one of three major sources of heterogeneity. These are *isotypy, allotypy* and *idiotypy* and each will now be considered, although greatest attention will be devoted to allotypy. Human variations are considered first and data for the mouse and rabbit are discussed separately.

Isotypy in human immunoglobulins

The expression of immunoglobulin constant region genes represents a particularly well developed example of the state known as *isotypy*. This is in essence the *simultaneous* expression within the *same* individual of *multiple copies of related structural genes*. The associated gene products are called *isotypes*. The five heavy chain classes γ, α, μ, δ and ϵ are thus regarded as isotypes and the various IgG and IgA subclasses are similarly classified as isotypic variants.

In the human IgG subclasses a range of subclass-specific isotypic antigens are present in all sera from healthy individuals and are particularly well characterized by sensitive haemagglutination-inhibition techniques. Yet other isotypes are shared by several subclasses (Table 5.1).

Using haemagglutination-inhibition assay systems and various proteolytic fragments of the immunoglobulin subclasses it is possible to localize many of the isotypic antigens to one or other of the constant homology regions. Some of this data is summarized in Table 5.2.

In the case of the light chain constant regions isotypy is less extensive. In man there is only one known isotype of κ chain and relatively simple amino acid substitutions account for the various λ chain isotypes — Oz, Kern and Mcg. The known amino acid substitutions associated with the λ chain isotypes are listed in Table 5.3 and the location of the Oz and Kern antigens in the three-dimensional structure of the C_λ domain are indicated in

Table 5.1 Human IgG subclass isotypes

Subclass-specific isotypes	Shared isotypes
IgG1$_\kappa$Fab	IgG1-2-3 Fc
IgG1 F(ab')$_2$	IgG2-3 Fc
IgG1 Fc	IgG2-4 Fc
IgG2$_\kappa$Fab	IgG1-3 Fc
IgG2 Fab	
IgG2 Fc	
IgG3 Fab	
IgG3 F(ab')$_2$	
IgG3 Fc	
IgG4$_\kappa$Fab	
IgG4 Fc	

Reproduced with permission from Natvig & Kunkel, 1973.

Table 5.2 Structural location of selected isotypes of human IgG

Immunoglobulin subclass	Constant region domain		
	C_H1	C_H2	C_H3
IgG1	γ_1Fab, γ_1F(ab')$_2$	γ_1Fc	γ_{1-2-3}Fc
IgG2	γ_2Fab	γ_2Fc, γ_{2-3}Fc, γ_{2-4}Fc	γ_{1-2-3}Fc
IgG3	γ_3Fab, γ_3F(ab')$_2$	γ_3Fc, γ_{2-3}Fc	γ_{1-2-3}Fc
IgG4	γ_4Fab	γ_4Fc$_I$, γ_{2-4}Fc	γ_4Fc$_{II}$

Reproduced with permission from Natvig & Kunkel, 1973.

Table 5.3 Isotypic variants of human λ chains

	C$_\lambda$ sequence number				
	112	114	152	163	190
Kern(−), Oz(−), Mcg(−)	Ala	Ser	Ser	Thr	Arg
Kern(+), Oz(−), Mcg(−)	Ala	Ser	Gly	Thr	Arg
Kern(−), Oz(+), Mcg(−)	Ala	Ser	Ser	Thr	Lys
Kern(−), Oz(−), Mcg(+)	Asn	Thr	Ser	Lys	Arg

Figure 5.1. It is of interest that two of the five sequence positions correspond to residues in the κ chain which are associated with Km allotypic variation (see next section).

Allotypy in human immunoglobulins

Introduction

Allotypic antigens of human immunoglobulins were

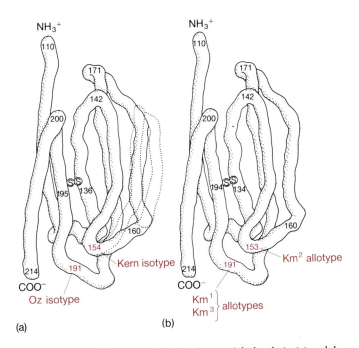

Fig. 5.1 Model of C-terminal half of a human λ light chain (a) and, by extrapolation, a similar model for human κ light chain (b) showing the basic 'immunoglobulin fold' in both. The approximate location of residues which determine Oz and Kern isotypes and the Km allotypes are indicated. The dotted line indicates the position of the extra loop in the V$_L$ region. (Modified with permission from Poljak, 1975.)

first described by Grubb (1956) and Grubb and Laurell (1956). The test system used, based on haemagglutination-inhibition, remains in wide use today (see Figure 5.2). Rh+ red cells are first coated with so-called incomplete (i.e. non-agglutinating) anti-Rh antibodies. The coated red cells are agglutinated by an appropriate antibody preparation which may be derived from a patient with rheumatoid arthritis (a so-called *Ragg* reagent) or from a healthy donor sensitized by transfusion or pregnancy (a *Snagg* ('serum normal') reagent). The serum of any individual may then be introduced into this system as an inhibitor and its titre determined by setting up serial dilutions. (Grubb and Laurell originally noted that the sera of patients with rheumatoid arthritis reacted with cells coated with certain anti-Rh antibodies but not with cells coated with other anti-Rh preparations.) Many minor variations of this assay system have been described as the complexity and range of specificities became clear. For example, some classes and

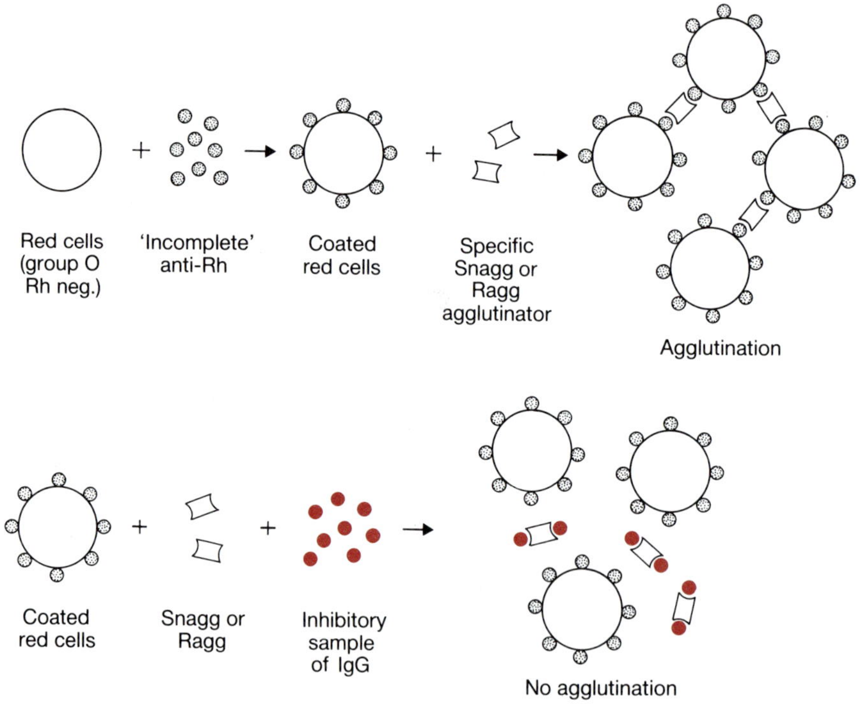

Red cells (group O Rh neg.) + 'Incomplete' anti-Rh → Coated red cells + Specific Snagg or Ragg agglutinator → Agglutination

Coated red cells + Snagg or Ragg + Inhibitory sample of IgG → No agglutination

Fig. 5.2 Schematic illustration showing principle of the haemagglutination-inhibition assay system used for much Gm typing. If the IgG in the test sample reacts with the specific agglutinator no haemagglutination occurs and the sample is typed as positive (see text for further details).

subclasses of immunoglobulin rarely occur in incomplete anti-Rh antibody preparations and for antigens confined to such isotypes it is necessary to coat the red cells with isolated monoclonal (myeloma) proteins. This is usually achieved with a coupling agent such as bis-diazotized benzidine or chromic chloride.

It may be necessary for several antibody preparations to be tested in parallel if only Ragg reagents are available, since these frequently contain a spectrum of antiglobulin specificities. Snagg reagents are generally preferred because they are usually monospecific and show less tendency to elicit prozone phenomena. It is possible to raise appropriate typing reagents by immunizing rabbits and primates with human myeloma proteins and immunoglobulin fragments. Such antisera are then absorbed with sera or immunoglobulin lacking the particular allotypic antigen under study. Reagents specific for G1m(z) and G2m(n) have been exclusively obtained by such techniques.

Nomenclature

In the test system originally described by Grubb, inhibitory sera were designated Gm(a+) and non-inhibitory sera Gm(a−) (Gm indicating genetic marker of γ-globulins). Subsequently, using similar techniques, a range of genetic antigens has been described for human immunoglobulins. As shown in Table 5.4 three major systems are currently recognized: Gm antigens are characteristic of γ heavy chains, Am antigens of α heavy chains and Km antigens of κ light chains.

A large number of Gm antigens are now recognized and these are listed in Table 5.5, together with the IgG subclass in which they occur. Originally an alphameric terminology was used and then a numeric terminology

Table 5.4 Allotypic systems associated with human immunoglobulin chains

	Heavy		Light
	α	γ	κ
Gm		+	
Am	+		
Km*			+

*Previously designated Inv.

Table 5.5 Allotypes of human immunoglobulins

Class	Subclass	Chain	Previous nomenclature		Present nomenclature
			Alphameric	Numeric	
IgG	IgG1	γ_1	Gm(a)	Gm(1)	G1m(a)
			Gm(x)	Gm(2)	G1m(x)
			Gm(f)	Gm(3)(4)	G1m(f)
			Gm(z)	Gm(17)	G1m(z)
	IgG2	γ_2	Gm(n)	Gm(23)	G2m(n)
	IgG3	γ_3	Gm(g)	Gm(21)	G3m(g)
			Gm(b^0)	Gm(11)	G3m(b^0)
			Gm(b^1)	Gm(5)(12)	G3m(b^1)
			Gm(b^3)	Gm(13)(25)	G3m(b^3)
			Gm(b^4)	Gm(14)	G3m(b^4)
			Gm(b^5)	Gm(10)	G3m(b^5)
			Gm(c^3)	Gm(6)	G3m(c^3)
			Gm(c^5)	Gm(24)	G3m(c^5)
			Gm(s)	Gm(15)	G3m(s)
			Gm(t)	Gm(16)	G3m(t)
			Gm(Pa)		G3m(u)
			Gm(Ray)		G3m(v)
IgA	IgA2	α_2	Am(1)	Am_2	A2m(1)
					A2m(2)

Table 5.6 Allotypes of human κ light chains

Previous nomenclature		Present nomenclature
Alphameric	Numeric	
InV, Inv(1)	Inv(1)	Km(1)
(a)	(2)	(2)
(b)	(3)	(3)

was proposed by the WHO (World Health Organization). Both still coexist in the world literature, but many investigators now prefer the more rational variation of alphameric, which also indicates subclass (see Table 5.5). This also has the advantage that it is readily adapted for antithetic antigens (see later).

The terminology (past and present) of the Km antigens found on κ light chains is given in Table 5.6.

General characteristics of Gm antigen expression

It was realized at an early stage that the Gm factors obey Mendelian laws of inheritance and represent the products of autosomal co-dominant genes. One of the consequences of this is that individuals may be either homozygous or heterozygous for a particular antigen. In practice this means that in the case of the G1m(z) antigen the homozygous individual expresses this antigenic specificity on all of his IgG1 molecules, whereas the antigen occurs on a fraction (theoretically 50%) of the IgG1 molecules of the heterozygote. The remaining IgG1 molecules will express the allelic antigen G1m(f). G1m(z) and G1m(f) differ in amino acid sequence at a single position (see Figure 5.3), and so a single molecule can express only one of the two alleles.

Although all the Gm antigens are directly localized to the constant regions of γ heavy chains some of them, notably G1m(f) and G1m(z), which are in the C_H1 region, require the presence of light chains for full antigenic expression. Most other antigens are Fc region located and are expressed on isolated fragments and chains.

In any given molecule certain Gm antigens may be simultaneously expressed. The G3m(b) mosaic of the IgG3 subclass is one example of such multiple expression where little is known concerning the structural

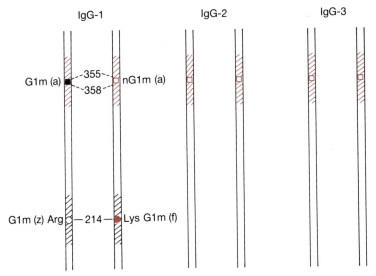

Fig. 5.3 Probable origin of isoallotypes ('non-markers'). Mutations giving rise to new antigens may be expressed in stretches of peptide chain which are shared by several subclasses (general γ chain regions shown in red cross-hatching) or in stretches which are characteristic of a particular subclass (IgG1 subclass region shown in black cross-hatching). In the figure the single point mutation giving rise to the alleles G1m(z) and G1m(f) occurred in a region of peptide characteristic of γ₁ chains only, whereas the mutation underlying G1m(a) expression was in a portion of peptide chain common to γ₁, γ₂ and γ₃ chains. The nG1m(a) (or 'non-a') antigen is thus an allotype in the IgG1 subclass but an isotype in the IgG2 and IgG3 subclasses.

Table 5.7 Most common Gm−Am gene complexes in different population groups

	IgG1	IgG2	IgG3	IgG4	IgA2
Caucasian	G1m(f), nG1m(a)	G2m(n)	G3m(b^1)(b^0)(b^3)(b^4), nG3m(g)	4b	A2m(1)
	G1m(f), nG1m(a)	G2m(n−)	G3m(b^1)(b^0)(b^3)(b^4), nG3m(g)	4a	A2m(1)
	G1m(a), G1m(z)	G2m(n−)	G3m(g), nG3m(b^1)	4a	A2m(1)
	G1m(a), G1m(x), G1m(z)	G2m(n−)	G3m(g), nG3m(b^1)	4a	A2m(1)
Mongoloid	G1m(a), G1m(f)	G2m(n)	G3m(b^1)(b^0)(b^3)(b^4)	4a	A2m(1)
	G1m(a), G1m(z)	G2m(n−)	G3m(g)	4a	A2m(1)
	G1m(a), G1m(z)	G2m(n−)	G3m(b^0)(b^3)		
Negroid	G1m(a), G1m(z)	G2m(n−)	G3m(b^1)(b^0)(b^3)(b^4)		
	G1m(a), G1m(z)	G2m(n−)	G3m(b^1)(c^3)(b^0)		
	G1m(a), G1m(z)	G2m(n−)	G3m(b^1)(c^3)(b^0)(b^4)		
	G1m(a), G1m(z)	G2m(n−)	G3m(b^0)(b^3)		
Japanese	G1m(a), G1m(z)		G3m(g)		
	G1m(a), G1m(x), G1m(z)		G3m(g)		
	G1m(x), G1m(z)		G3m(g)		
	G1m(a), G1m(z)		G3m(b^3)		
Melanesian	G1m(a), G1m(f)		G3m(b^1)(b^3)(b^4)		
	G1m(a), G1m(z)		G3m(b^1)(b^3)(b^4)		
	G1m(a), G1m(z)		G3m(g)		
	G1m(a), G1m(x), G1m(z)		G3m(g)		

Table 5.8 Frequencies of various Gm haplotypes in different population groups

	Caucasian	Negroid	Japanese	Australian aborigine
Gm(za; n−; g)	0.187	—	0.407	0.087
Gm(zax; n−; g)	0.098	—	0.164	—
Gm(za; n−; b^{0}s t b^3 b^5)	0.001	—	0.277	—
Gm(fa; n+; b*)	—	—	0.152	—
Gm(f; n+; b*)	0.450	—	—	—
Gm(f; n−; b*)	0.249	—	—	—
Gm(za; n+; b*)	—	—	—	0.913
Gm(za; n−; b*)	0.004	0.618	—	—
Others	0.011	0.382	—	—

b* = b^0 b^1 b^2 b^3 b^4 b^5.

basis for each specificity. Similarly, IgG1 molecules may also express G1m(a), G1m(x) and G1m(z) antigens. In this case it is established that the (a) and (z) antigens are widely separated, distinct epitopes on the molecule but are genetically linked (see Figure 5.3).

The three subclasses of IgG that carry recognized Gm antigens (IgG1, IgG2 and IgG3) occur in all healthy individuals and it might be supposed that a totally random assortment of Gm antigens would be detected in any given population. However, this is not so. It appears that a small number of gene combinations are characteristic of different racial groups and that these complexes remain stable over several generations (see Tables 5.7 and 5.8). The retention of such gene complexes suggests that the individual genes are very closely linked with very low rates of recombination. It also follows that the identification of the rare individual with an unusual gene combination is of great significance. Estimates of the chromosomal cross-over frequency in the Gm region may be calculated and the probable order of the constant region genes may be directly derived from such information. These aspects are discussed further below.

Isoallotypes

One of the greatest areas of confusion in the field of human immunoglobulin genetics concerns the antigens which were given the unfortunate nomenclature of 'non-markers'. These behave as an allotype within one subclass but are also to be found on *all* the molecules of one or two additional subclasses. Hence they are now more logically known as isoallotypes. Their existence is a direct result of the very high level of structural homology between the human IgG subclasses. If a mutation occurs in the nucleotide sequence coding for a subclass-specific region a classic allotypic variant would result, such as is represented by G1m(z) and G1m(f) of IgG1. However, most mutations arise in nucleotide regions coding for γ chain peptide sequences common to two or more subclasses. For each of the allotypes G1m(a), G3m(g), G3m(b^0) and G3m(b^1) there exist corresponding structural antigens which, when present within the same subclass behave as an allotype, or, when present in another subclass behave as an isotype. Naturally, it is assumed that these mutations have occurred since the duplications of the ancestral γ chain gene into the present day genes for γ_1, γ_2, γ_3 and γ_4 chains. This concept is illustrated in Figure 5.3 and some of the known isoallotypes are listed in Table 5.9, together with data on their subclass distribution.

Table 5.9 Isoallotypes of human immunoglobulin G

Previous nomenclature	Present nomenclature	IgG subclasses			
		IgG1	IgG2	IgG3	IgG4
Non-a	nG1m(a)	Allo-	Iso-	Iso-	—
Non-g	nG3m(g)	—	Iso-	Allo-	—
Non-b^0	nG3m(b^0)	Iso-	Iso-	Allo-	—
Non-b^1	nG3m(b^1)	Iso-	Iso-	Allo-	—
4a	nG4m(a)	Iso-	—	Iso-	Allo-
4b	nG4m(b)	—	Iso-	—	Allo-

Structural location of Gm and Km allotypes

Some progress has been achieved in the structural localization of both allotypes and isoallotypes, and some of this data is summarized in Table 5.10. Peptide mapping and primary sequence analysis are, of course, essential to pinpoint the correlative determinants of allotypic antigens but other approaches are also necessary when multiple allotypic antigens occur on the same chain. For example, G1m(a) and G1m(z) are both found on the same γ chain of most Caucasians but papain cleavage of IgG shows that G1m(a) is an Fc antigen whereas G1m(z) is found in the Fab fragment.

Table 5.10 Probable structural location of some human immunoglobulin allotypes and isoallotypes

Allotype or isoallotype	Chain	Homology region	Sequence	Amino acid
G1m(a)	γ_1	$C_\gamma 3$	355–358	Arg-Asp-Glu-Leu
nG1m(a)	$\gamma_1, \gamma_2, \gamma_3$	$C_\gamma 3$	355–358	Arg-Glu-Glu-Met
G1m(x)	γ_1	$C_\gamma 3$	431	Gly
G1m(f)	γ_1	$C_\gamma 1$	214	Arg
G1m(z)	γ_1	$C_\gamma 1$	214	Lys
G3m(g)	γ_3	$C_\gamma 2$	296	Tyr
nG3m(g)	γ_2, γ_3	$C_\gamma 2$	296	Phe
G3m(b°)	γ_3	$C_\gamma 3$	436	Phe
nG3m(b°)	$\gamma_1, \gamma_2, \gamma_3$	$C_\gamma 3$	436	Tyr
nG4m(a)	$\gamma_1, \gamma_2, \gamma_4$	$C_\gamma 2$	309	Val-Leu-His
nG4m(b)	γ_2, γ_4	$C_\gamma 2$	309	Val-His
A2m(1)	α_2	$C_\alpha 1$	212....221	Pro....Pro
A2m(2)	α_2	$C_\alpha 1$	212....221	Ser....Arg
Km1	κ	C_κ	153....191	Val....Leu
Km1, 2	κ	C_κ	153....191	Ala....Leu
Km3	κ	C_κ	153....191	Ala....Val

Most Gm antigens are, in fact, Fc antigens and the dissection of the G3m(b) mosaic cannot be achieved by simple papain digestion. However, mild pepsin digestion under defined conditions for each subclass releases the $C_\gamma 3$ domain and permits further structural localization within the Fc region. The methods and results have been described in detail (Turner *et al.*, 1970; Natvig & Turner, 1971) and Figure 5.4 summarizes the data obtained. Antigens that were not detected in the $C_\gamma 3$ region are assumed to be present in the $C_\gamma 2$ region but, alternatively, they may require for their expression the presence of both the $C_\gamma 2$ and the $C_\gamma 3$ domains.

Further proteolytic cleavage of the isolated $C_\gamma 3$ domain has furnished data on the determinants of antigenicity for several allotypes and isoallotypes. Trypsin removes some 7 amino acid residues from the N-terminus of the $C_\gamma 3$ fragment called pFc′ without affecting the expression of G1m(a), nG1m(a), G1m(x) or G3m(b°). However, when papain and chymotrypsin are used, a C-terminal peptide of about 14 amino acid residues is lost in addition to the N-terminal peptide. In this case the resultant fragments are no longer able to inhibit appropriate detection systems for these four antigens. This may be because the antigen itself is located near the C-terminus of the γ chain (e.g. G3m(b°) antigenicity appears to be directly correlated with residue number

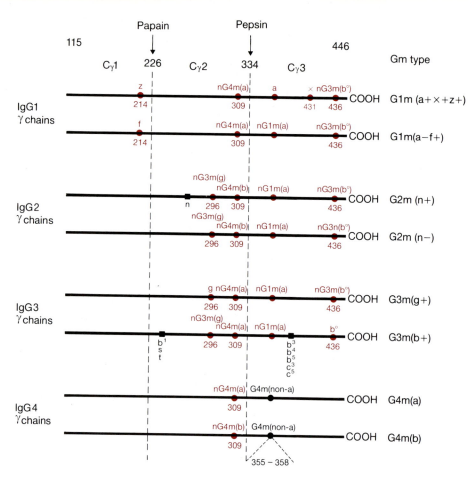

Fig. 5.4 Schematic representation of the three constant homology regions of the four human IgG subclasses. The structural locations of allotypes and isoallotypes which are well established are indicated by red spots and sequence numbers. The allotypes G3m(b³, b⁴, b⁵, c³,c⁵) are associated with the peptic fragment pFc' of these two subclasses (equivalent to the C$_\gamma$3 region) but their precise structural location is not known. The antigens G3m(b¹, s, t) were not detected in the pFc' fragment of appropriate IgG3 proteins and are provisionally located in the C$_\gamma$2 homology region. The G4m(non-a) antigen is an IgG4 isotype occupying the same structural region as G1m(a) and nG1m(a). (Modified with permission from Natvig & Turner, 1971.)

436 — see Table 5.8) or, alternatively, the C-terminal peptide may play a role as a non-correlative determinant of antigenicity. This concept was elaborated in detail by Todd (1972) and is illustrated schematically in Figure 5.5. In the Gm system one sees this phenomenon clearly involved in the expression of the G1m(a) and nG1m(a) antigens. These specificities are known to be associated with particular amino acids in the tetrapeptide between residues 355 and 358 (see Table 5.10). Although this region of the γ chain is well separated from the C-

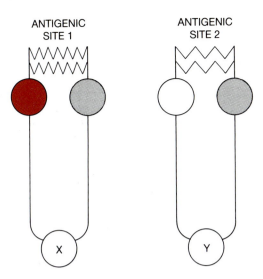

Fig. 5.5 Schematic diagram of two hypothetical antigenic sites such as are recognized by anti-allotype reagents. The sites are located on two similar proteins and certain amino acids contributing to the structural expression of both sites are indicated in grey and would be designated *non-correlative determinants*. Other residues contributing to the specificity of the site, shown in red and white respectively, are called *correlative determinants*. Located at a distance from the site are residues (X) and (Y) which coincidentally correlate with specificities 1 and 2 but since they play no role in the expression of antigenicity they are designated *correlative non-determinants*. (Modified with permission from Todd, 1972.)

terminal peptide in terms of the linear sequence, in the three-dimensional structure the C-terminal peptide is closely juxtaposed to the tetrapeptide. Residues of the C-terminal peptide may not themselves constitute contact residues of the antigenic site, but they may exert a critical modulating influence on the conformation of the site. Such non-correlative determinants are probably very common and at our present level of knowledge they go largely unrecognized. Their existence elsewhere is suggested by the dependence of Km antigenicity on the presence of a heavy chain and the dependence of G1m(f) antigenicity on the presence of a light chain. The other peptide chain is only essential for full antigenicity and this may again be regarded as a modulatory effect.

The Km system of antigens is now understood in considerable detail following the investigations of Milstein *et al.* (1974). Two Km alleles (Km[1,2] and Km[3]) are common and one (Km[1]) is very rare in Caucasians (frequency 0.001). Studies on an unusual Bence Jones protein, which was antigenically typed as Km1+, 2−,

3−, have provided the data necessary to understand the antigenicity of the whole Km system. Figure 5.6 represents schematically the interaction of the Km antigens and antisera specific for them.

Amino acid residues at positions 153 and 191 of the chain are the most critical determinants for the expression of Km antigens and, extrapolating from crystallographic models of the λ chain (see Figure 5.1), it is likely that in the κ chain these residues occur on the surface of adjacent loops, are accessible to reagents and are separated by less than 1 nm (10Å). It is suggested that antisera for the detection of Km(1) recognize a leucine residue at position 191 but do not encompass an alanine residue at position 153. In contrast, antisera to Km(2) recognize the leucine residue at position 191 but also encompass the alanine residue at position 153. If the alanine residue is replaced by valine then the anti-Km(2) reagent is no longer able to recognize the complete antigenic determinant, possibly because of steric hindrance by the larger valine residue. The third Km allotype is expressed when a valine residue occurs at position 191 and appears to be independent of residue 153, although this is not certain. We may, therefore, regard leucine at position 191 and valine at position 191 as correlative determinants for Km(1) and Km(3) respectively. Alanine at position 153 is (for Km(1,2)) a correlative determinant for Km(2), whereas leucine at position 191 is, in this case, a non-correlative determinant.

Evidence for the importance of non-correlative determinants in allotypy has also been obtained using chemical probes. For example, ethyl acetimidate hydrochloride has been used for the amidination of the ε-amino groups of lysine residues. Human G-myeloma proteins were tested for the expression of various Gm and Km antigens before and after such treatment and it was shown that G1m(a), nG1m(a), G1m(z) and Km(1) antigen reactivities were lost after this treatment.

Similarly, it is possible to selectively substitute the amino groups of lysine residues by citraconylation and show that G1m(a) antigen activity is lost after such treatment. Lysine residues are either directly involved or adjacent to the amino acids which correlate with these antigens. There is a lysine residue at position 360, close to the tetrapeptide correlated with G1m(a) and nG1m(a) antigen activity. Similarly, there are lysine

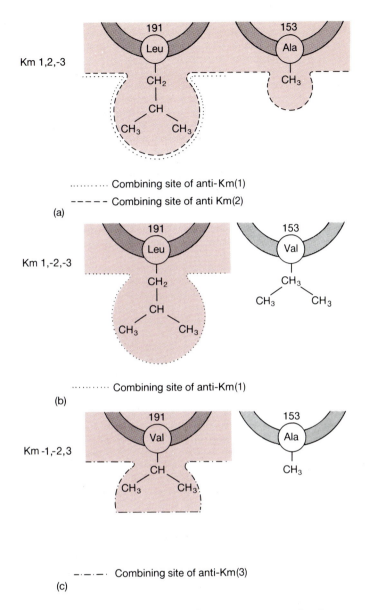

Km 1,2,-3

191
Leu
CH_2
CH
CH_3 CH_3

153
Ala
CH_3

·········· Combining site of anti-Km(1)
- - - - - Combining site of anti Km(2)

(a)

Km 1,-2,-3

191
Leu
CH_2
CH
CH_3 CH_3

153
Val
CH_3
CH_3 CH_3

·········· Combining site of anti-Km(1)

(b)

Km -1,-2,3

191
Val
CH
CH_3 CH_3

153
Ala
CH_3

—·—·— Combining site of anti-Km(3)

(c)

Fig. 5.6 Km antigens and amino acid sequence. Antisera that detect the Km(1) antigen probably interact with a leucine residue at position 191 but do not encompass residue 153. In contrast, antisera to Km(2) recognize the leucine residue at position 191, but also encompass an alanine residue at position 153 (a). If the alanine residue is replaced by valine the anti-Km(2) reagent is no longer able to recognize the complete antigenic determinant, possibly because of steric hindrance by the larger valine residue (b). The third Km allotype, Km(3), is expressed when a valine residue occurs at position 191 and appears to be independent of residue 153, although this is uncertain (c).

residues at positions 190 and 188 of the κ chain and these may be involved in the antigenicity of Km(1). The G1m(z) antigen, however, is correlated with a lysine residue at position 214 and in this case the reagent may be assumed to exert its effect directly on the correlative determinant.

Structural location of Am allotypes

Early studies suggested that the Am allotypic antigen of the IgA2 subclass was structurally located near the Fc−Fab junction. Most IgA2 molecules lack the usual heavy−light chain disulphide bridges but exist as non-covalently stabilized dimers of α_2 and L_2. These molecules type as A2m(1) positive. A smaller proportion of the IgA2 molecules have disulphide bonds between the α and light chains and type as A2m(2) (Figure 5.7). Two amino acid exchanges appear to correlate with the expression of the A2m allotypes (see Table 5.10). Two proline residues in the A2m(1) proteins are replaced by a serine and an arginine. It seems probable that these differences correlate with the heavy−light S−S bridge patterns described above.

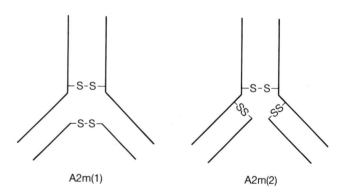

A2m(1)   A2m(2)

Fig. 5.7 The two allotypic forms of IgA2 molecule are associated with different inter-chain disulphide bond patterns.

Occurrence of Gm antigens in higher primates

Although the Gm antigens are of relatively recent origin there is a good deal of circumstantial evidence to suggest that some of the allotypic variants had arisen before the divergence of the ancestral hominids from other higher primates about 50 million years ago. IgG3 antigens (particularly the G3m(b) mosaic) have been detected in most cercopithecoids (Old World monkeys) and in all apes. However, these antigens, when present, do not appear to behave as allotypes. In the IgG of the Old World monkey a peptide has been identified which is structurally intermediate between the G1m(a) peptide and the isoallotype nG1m(a) (Figure 5.8).

	355				359
G1m(a) peptide	Asp–	Glu–	Leu–	Thr–	Lys
Old World monkey peptide	Glu–	Glu–	Leu–	Thr–	Lys
nG1m(a) peptide	Glu–	Glu–	Met–	Thr–	Lys

Fig. 5.8 Sequence comparison of peptides from the C_H3 domain of human IgG molecules bearing the G1m(a) allotype or the nG1m(a) isoallotype and the corresponding peptide from the IgG of an Old World monkey.

Rare gene complexes, cross-overs and recombination

As previously discussed, the Gm gene complexes of the various ethnic groups are remarkably stable and certain antigen combinations are almost completely unknown. Nevertheless, studies by Natvig *et al.* (1967) of over 5000 sera revealed two families in which a recent cross-over event had occurred, giving rise to two unusual gene complexes.

In one of these families the gene combination G1m^{fy} G3m^g G2m^{n-} was detected. The G3m(b) mosaic of antigens, which are almost always to be found in association with G1m^{fy} in Caucasians (see Table 5.7), were not detected. This unusual gene complex (haplotype) was absent from the parents of the propositus but appears to have arisen as a fresh cross-over in the mother (see Figure 5.9) in the region of the IgG3 and the IgG1 genes. Moreover, this new complex was transmitted to one child showing that the gene complex was stable and inheritable.

In a second family the propositus was again detected because he had the rare gene complex G1m^f G3m^g G2m^n. The G3m(b) mosaic usually associated with G1m^f was not detected. In this family the two children inherited the gene complex G1m^{zax} G3m^g G2m^{n-} from their mother and one of the children inherited the expected G1m^f G3m^b G2m^{n-} complex from her father. The unusual Gm typing result in her brother could only

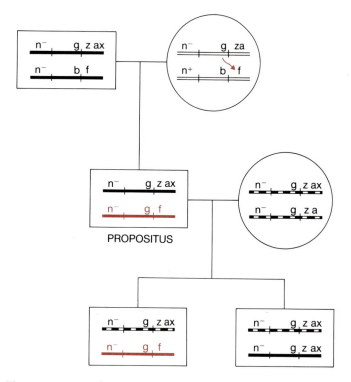

Fig. 5.9 An unusual Gm gene complex G2m^{n-} G3m^g G1m^f was detected in the propositus and also in his son (as indicated in red). It is believed that a chromosomal cross-over occurred in the mother giving rise to this rare complex (curved arrow). (Modified with permission from Natvig et al., 1967.)

be explained by a recombination in the father involving the rare G1m^{za} G3m^g G2m^n gene complex with the IgG2-IgG3 genes of this complex recombining with the IgG1 G1m^f gene of the other chromosome (see curved arrow in Figure 5.10).

Unequal homologous cross-over in the Gm gene complexes

Occasionally the unusual Gm complex of an individual is found to arise from unequal homologous cross-over of chromosomal material at meiosis. Three different events will be described.

Deletions

Whole genes may be deleted so that the haplotypes lack all common alleles of that particular subclass. Figure 5.11 shows a family in which the IgG3 gene was

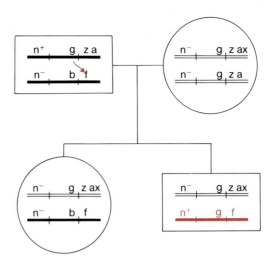

Fig. 5.10 The unusual gene complex G2m^{n+} G3m^g G1m^f observed in the propositus could only be explained by a recombination in the father involving the rare Gm^{za} Gm^g Gm^n gene complex, with Gm^g Gm^n recombining with Gm^f as shown by the curved red arrow. (Adapted with permission from Natvig *et al.*, 1967.)

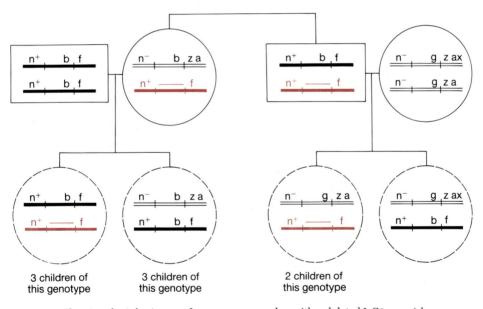

3 children of this genotype   3 children of this genotype   2 children of this genotype

Fig. 5.11 Showing the inheritance of a rare gene complex with a deleted IgG3 gene (shown in red). This chromosome was present in both a brother and a sister and was inherited by two children in the next generation. (Modified with permission from Natvig & Kunkel, 1973.)

involved and a complex (G1m^f) (——) (G2m^{n+}) was inherited in the usual Mendelian manner. All individuals with this complex had half-normal levels of IgG3 in their sera. Similar structural deletions of IgG1 genes have been reported.

Duplications

Families have been described in which there appeared to be two IgG1 genes on one chromosome. This was indicated by the stable inheritance of the rare complex $G1m^{fza}$. The homoalleles G1m(f) and G1m(z) are associated with residue 218 of the γ chain and would not normally both be encoded on the same chromosome. Physicochemical studies of the serum IgG indicated a complete duplication of the IgG1 gene with two genetically different, intact IgG1 polypeptide products. The possible mechanism underlying these observations is shown schematically in Figure 5.12.

Hybridizations

Unequal homologous cross-overs may also give rise to molecular hybrids. A well-documented example of this was the so-called Lepore-type hybrid. A serum was found to lack all the expected Gm antigens and extensive investigations showed that the individual was probably homozygous for a rare gene complex that was inherited in a Mendelian fashion through the family. Analysis of the patient's IgG showed that IgG1 C_H1 domain antigens and IgG3 C_H2 and C_H3 antigens were absent, whereas IgG1 C_H2 and C_H3 antigens and IgG3 C_H1 antigens were present. Moreover, antisera specific for the IgG3 Fab region also precipitated IgG1 Fc antigens, suggesting that both were carried on the same molecule. All the evidence suggested that the molecule was an IgG3 – IgG1 hybrid being of IgG3 class in the C_H1 region and IgG1 class in the Fc region. The molecule did type positively for the isoallotype nG1m(a) — as would be expected. No defects in the light chain were detected. The concentration of the hybrid protein was far higher than the usual IgG3 level, but below the normal mean for IgG1, suggesting that the rate of synthesis was influenced both by the exon coding for IgG3 Fab and the exon coding for IgG1 Fc. As shown in Figure 5.13 an unequal homologous cross-over involving mispairing of heavy chain genes would readily explain the deletion of the Gm allotypes. The figure depicts diagrammatically the two major haplotypes found in

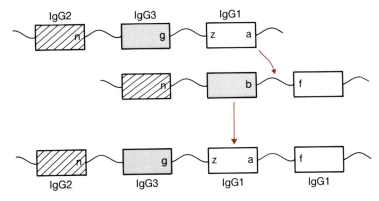

Fig. 5.12 Schematic representation of immunoglobulin gene duplication in man. The IgG constant region genes are mispaired (top) and an inter-genic cross-over occurs at meiosis — curved red arrow. Two IgG1 genes occur in the resultant chromosome (bottom) and are associated with the expression of different allotypes. (Reproduced with permission from Natvig & Kunkel, 1973.) Note that the IgG2 gene is now thought to be on the 3' side of IgG1.

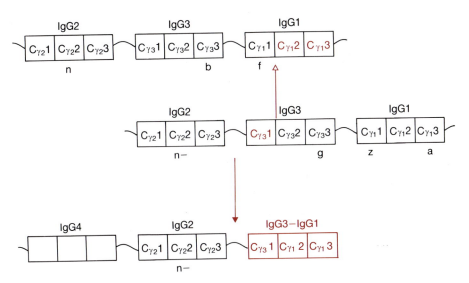

Fig. 5.13 Diagram of the consequence of mispairing of immunoglobulin genes at meiosis followed by intra-genic cross-over leading to an IgG3–IgG1 hybrid gene. Note that the hybrid gene product would express none of the allotypic specificities of IgG1 or IgG3 normally found in serum. (Reproduced with permission from Kunkel *et al.*, 1969.) Note that IgG2 and IgG4 genes are now thought to be on the 3' side of IgG1.

Caucasians with the IgG3 and IgG1 genes placed in the order determined by observed recombination frequencies. A mispairing of the IgG1 and IgG3 genes at meiosis followed by an intra-genic cross-over would result in a gamete with the genotype indicated at the bottom of Figure 5.13. The figure shows how a single cross-over could result in a hybrid gene lacking all known Gm allotypes. This particular case was probably homozygous for the hybrid gene since other family members appeared to have a similar 'silent' allele representing the heterozygous state.

Another hybrid protein recorded in the literature was a myeloma protein detected in the serum of a Negroid patient. Laboratory investigations showed that this protein was a hybrid of IgG4 and IgG2 molecules with a probable cross-over point between the C_H2 and C_H3 domains, i.e. the C_H1 and C_H2 regions were derived from an IgG4 molecule and the C_H3 region from an IgG2 molecule.

Both of these hybrid molecules are reminiscent of the $\delta\beta$ chain hybrids established for Lepore-type haemoglobins. In the first patient there was evidence that the hybrid molecules retained functional haemagglutinating antibody activity but it seems more likely that effector functions such as C1q and macrophage binding would be modified by such a cross-over. Homozygous individuals with such defects might have normal levels of immunoglobulin and functional antibodies (in the antigen binding sense) and yet immunity deficiency might be manifest if effector functions were compromised. Since IgM is presumably normal in such individuals and, moreover, extremely effective in complement activation, deficiencies of the IgG system might be concealed.

Order of the IgG heavy chain genes

Studies of the Lepore-type of IgG hybrids and other rearrangements of the immunoglobulin genes permitted a provisional ordering of the genes on the chromosome. The IgG3 origin of the N-terminal end of the molecule and the IgG1 origin of the C-terminal end suggested that the IgG3 gene lay on the N-terminal side of the IgG1 gene. Moreover, since the IgG2 and IgG4 subclasses were not deleted, it was highly likely that the IgG3 and IgG1 genes were adjacent. Similarly, the data obtained for the IgG2−4 hybrid suggested that the IgG4 gene was N-terminal and adjacent to the IgG2 gene. A sequence of IgG4−IgG2−IgG3−IgG1 was suggested two decades ago (see Figure 5.14). Although recent studies using cDNA probes have confirmed the close linkage of IgG4 with IgG2, and of IgG3 with IgG1, it is now believed that the overall sequence is IgG3−IgG1−IgG2−IgG4 (see Chapter 7, pp. 183−4 and Figure 7.30).

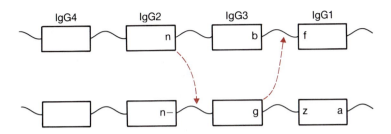

Fig. 5.14 An early proposal for the order of human IgG genes. The two major allelic gene complexes of Caucasians are shown. The positions of observed cross-overs which indicated the probable order are shown by the red arrows. (Reproduced with permission from Natvig *et al.*, 1967). Compare with Figure 7.30.

Influence of Gm antigens on subclass concentrations

A relationship between Gm gene expression and quantitative levels of individual IgG subclasses has been clearly demonstrated. Individuals who are homozygous for the G3m(b) allotype have IgG3 levels which are twice those of individuals who are homozygous for the G3m(g) allotype. Heterozygous individuals expressing both (b) and (g) allotypes have intermediate levels. Similarly, individuals who are homozygous for the G2m(n) allotype have higher concentrations of IgG2 than do individuals who are homozygous for G2m(n−).

Idiotypy in human immunoglobulins

The hypervariable regions of immunoglobulin V domains include contact residues in antigen binding reactions and so by definition are accessible portions of the whole antibody molecule. It is not surprising, therefore, that these same regions can themselves be antigenic when the immunoglobulin is used as an immunogen either in a foreign species or within the same species. Initially, in the 1950s, it was shown by Kunkel *et al.* that human myeloma proteins possessed such individual antigenic specificities. Later, in the 1960s, it became clear that this was also a characteristic of non-pathological antibodies such as the human anti-blood group antibodies. In general, idiotypic antibodies are more easily demonstrated following intraspecies immunization (see also Chapter 6, p. 142). After immunization of a foreign species extensive absorption will be required to remove antibodies to constant region epitopes.

The surface feature of the immunoglobulin molecule which makes contact with an anti-idiotype antibody is called the *idiotope* and some workers have found it convenient to subdivide anti-idiotype antibodies into various categories based on the nature of the idiotope involved. This may, however, be difficult since the structural features of most idiotopes are still largely unknown. In some cases CDR regions alone may be involved but more often framework regions would also make a contribution. In some cases the idiotope might be confined to the V_H or V_L region whereas in others it would span both. Much of the detailed work on idiotype specificity and idiotope structure has been performed with hapten-binding mouse myeloma proteins and will be discussed in a later section. Much of the work on idiotypes in man has made use of the so-called cold agglutinins of the IgM class which are frequently detected in the sera of patients with acquired haemolytic anaemia. These antibodies have specificities for the blood group antigens I and i and agglutinate red cells at low temperature. The presence of a common antigenic specificity in a group of molecules with similar function has been called 'cross-idiotypy' by Kunkel. Studies using antisera against intact cold agglutinins suggest that both H and L chains are required for the full expression of these antigens, although there is evidence that the H chain is the major determinant.

Amino acid sequence analysis of the amino-terminal regions of ten such proteins (reviewed by Capra & Kehoe, 1975) showed that seven of the light chains belonged to the $V_\kappa III$ subgroup which normally accounts for less than 40% of human κ chains. The hypervariable regions of the proteins were not sequenced in this study. Subsequently, Feizi et al. (1977) described a heavy chain variable region marker ($V_H Mar$) associated with the cross-reactive cold agglutinin idiotypes but also found in normal pooled IgG and some unselected myeloma proteins.

Cross-idiotypic specificity also exists among human

IgM proteins with antiglobulin activity. Such proteins were divided by Kunkel into two groups (called after the prototype proteins Wa and Po) on the basis of cross-idiotypic specificity — the proteins in one group cross-reacted with one another but not with proteins of the other group. Sequence studies on the Wa group of proteins in the 1970s showed that the light chains of all such proteins belonged to the $V_\kappa III$ subgroup. Limited studies of isolated heavy chains suggested that the proteins were probably of the $V_H II$ subgroup with blocked N-terminal residues.

More extensive sequence data is available on two proteins belonging to the Po group of IgM anti-γ-globulins. Light chain sequencing showed that, unlike the Wa group proteins, these belonged to different V_κ subgroups ($V_\kappa I$ and $V_\kappa II$ respectively). However, the heavy chain V regions were strikingly similar (see Figure 5.15) within the hypervariable regions. Only three residues differed between the two proteins and two of these were in the hypervariable region of H chains which has no complementarity determining function. These observations strongly implicate the hypervariable regions in cross-reactive idiotypic expression.

A particular aspect of idiotypic antigens which has aroused much interest is the so-called network theory proposed by Niels Jerne. This envisages a network of interactions which begin with the production of anti-idiotypic antibodies after primary antigen exposure and the generation of 'new' antibodies by the host. The anti-idiotype is, in essence, an autoantibody and it is suggested that it has a role in the suppression of immune responses. Animal studies (Chapter 6, p. 142) suggest that the network may be extended through several stages and that anti-anti-anti-idiotypes may be generated in some responses.

Allotypy in rabbit immunoglobulins

Immunoglobulin allotypes were described in the rabbit by Oudin in 1956. His methods were those of classical serology. Rabbit allotypes are designated by a lower case letter for each locus, or allotypic group, together with a number specifying the allele. Allotypic markers are known for most of the rabbit immunoglobulin genes (Table 5.11).

The allelic differences are known at the amino acid

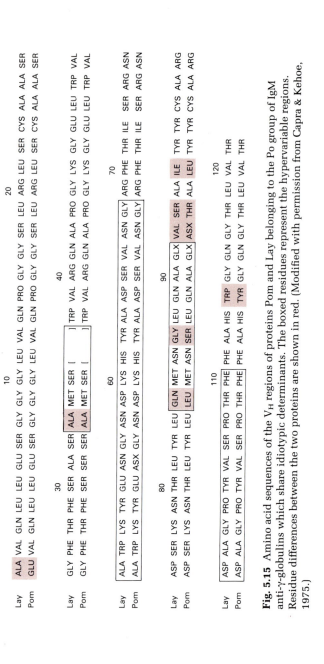

Fig. 5.15 Amino acid sequences of the V_H regions of proteins Pom and Lay belonging to the Po group of IgM anti-γ-globulins which share idiotypic determinants. The boxed residues represent the hypervariable regions. Residue differences between the two proteins are shown in red. (Modified with permission from Capra & Kehoe, 1975.)

Table 5.11 Major allotypes of rabbit immunoglobulins

Molecular location	Allotypic group	Allotypes
H chain		
V region	a	1, 2, 3
	x	32
	y	33
C region-IgG	d	11, 12
	e	14, 15
C region-IgA	f	68, 70, 71, 72, 73
C region-IgA	g	74, 75, 76, 77
C region-IgM	n	81, 82
κL chain		
C region	b	4, 5, 6, 9
λL chain		
C region	c	7, 21
Secretory component	t	61, 62

Table 5.12 Amino acid sequence differences correlating with rabbit C_H allotypes

Allotypic group	Allotypes	Correlates
d	11, 12	d11-Methionine (225)
		d12-Threonine (225)
e	14, 15	e14-Threonine (309)
		e15-Alanine (309)

sequence level for the d and e loci (Table 5.12). These alleles are single amino acid changes explicable by point mutations.

Two unusual features of rabbit allotypy are the V region allotypes and the complex group b allotypes of κ chains; interlinked with these phenomena is the finding of 'latent' allotypes discussed below.

V_H allotypes

Of mouse, man and rabbit, the latter is the only species in which allotypic markers have been found on the variable regions of immunoglobulin heavy chains. In the face of the sequence diversity of V_H regions the allelic determinants are not readily defined. Suffice it to say, they are complex, i.e. multiple amino acid differences.

The group a allotypes played an important part in the elucidation of the genetics of antibodies. In 1963 Todd reported the finding that rabbit IgM and IgG share the group a V region allotypes. He suggested that this sharing was indicative of V_H genes being used by both C_μ and C_γ genes, a prediction validated by our present molecular knowledge of V−C joining and C_H gene switching (see Chapters 7 and 8). V_H allotypes have also been demonstrated on rabbit IgA and IgE.

Inheritance of group a allotypes according to Mendelian principles was taken as an argument for a single V_H gene in the rabbit and this was a major plank in the case for somatic mutations as the origin of antibody diversity. The finding of distinct V_H loci (e.g. groups x and y) showed that rabbits inherit at least a small number of V_H genes (Figure 5.16). The finding of 'latent' allotypes has blurred the genetic arguments still further. Following the inheritance of group a allotypes together with C_H markers, linkage between V_H and C_H was demonstrated. Also, in rabbit family studies a cross-over frequency of 0.3% has been observed for V_H and C_H genes. This estimate by Mage, confirmed by Kindt, was the first demonstration of the genetic distance between V and C genes in the germ-line DNA.

Complex group *b* allotypes

At the amino acid sequence level the four b allotypes of κ chains differ at multiple positions. This was an unexpected finding for constant region genes that might be expected to show simple point mutation differences. One explanation is that the 'alleles' are not true alleles but are isotypes under control of an allelic expression mechanism. The finding of 'latent' allotypes is consistent with the latter hypothesis.

Latent allotypes

Conventional genetics indicates that a maximum of two alleles at each locus can be expressed in a single animal. In a rabbit defined by serology as being a^1a^3/b^4b^5, Strosberg (1977) elicited with *Micrococcus lysodeikticus* antibodies with allotypes a2 and b6. Subsequently a survey of the rabbit colony in Kindt's laboratory showed, by use of a sensitive radioimmune assay to study non-immune sera, the presence of group a allotypes not detected by serological typing or predicted by the parental phenotypes (Kindt & Yarmush, 1981).

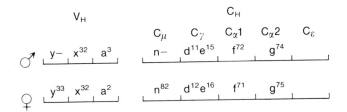

Fig. 5.16 The complete heavy chain typing of a single rabbit. The allogroup inherited from the paternal side is at the top and that from the maternal side is at the bottom.

The serologically determined allotype, followed via inheritance studies, is termed the 'nominal' allotype and other allotypes, found either in low levels in non-immune sera or occasionally at higher levels in specific antibody populations, are termed 'latent' allotypes. The expression of latent allotypes in non-immune sera of individual rabbits is transitory and sporadic. The concentration of latent allotype in serum is about 100-fold lower than that of nominal allotype except in the few instances when a clone of B cells expressing latent allotype is selected and expanded in a particular antibody response.

That rabbits carry genes encoding each of the so-called alleles of the group a locus seems certain. The mechanism whereby one allele per haplotype behaves in a Mendelian fashion in inheritance studies remains to be elucidated.

Allotypy in murine immunoglobulins

In mice, allotypic markers have only been detected serologically for the C_H regions. However, the mouse has numerous advantages for immunogenetic study. The size and short gestation period led to the study of transplantation in mice (see Chapter 9). The discovery and exploitation, by Potter, of transplantable plasmacytomas (myelomas) that are readily inducible in BALB/C and NZB strains allowed extensive study of individual murine immunoglobulins. Idiotypic determinants (see Chapter 6, p. 143) on myeloma proteins were shown to be expressed as allotype markers of individual V_H genes. Thus, linkage and recombinant studies of V_H and C_H genes have been carried out in mice to confirm the rabbit data (though recombination frequencies for individual V_H genes with C_H vary from 0.2 to 12%).

Table 5.13 Distribution of alleles of the *Igh-C* loci in the *Igh-C* haplotypes

Haplotype	Prototype strain	Locus and chain					
		Igh-1 γ2a	*Igh-2* α	*Igh-3* γ2b	*Igh-4* γ1	*Igh-5* δ	*Igh-6* μ
a	BALB/c	a	a	a	a	a	a
b	C57BL	b	b	b	b	b	b
c	DBA/2	c	c	a	a	a	—
d	AKR	d	d	d	a	a	—
e	A	e	d	e	a	e	e
f	CE	f	f	f	a	a	—
g	RIII	g	c	g	a	a	—
h	SEA	h	a	a	a	a	—
j	CBA/H	j	a	a	a	a	a
k	KH-1*	k	c	a	a	—	—
l	KH-2*	l	c	a	a	—	—
m	Ky*	m	b	b	b	—	—
n	NZB	e	d	e	a	a	e

*Wild mice.

Table 5.14 Antigenic specificities controlled by *twelve allelic genes* at the *Igh-1* locus

Haplotype	Prototype strain	Antigenic specificities	Allele
a	BALB/c	1, 6, 7, 8, 26, 28, 29, 30	a
b	C57BL	2, 27, 29	b
c	DBA/2	3, 8, 29	c
d	AKR	4, 6, 7, 8, 26, 29	d
e	A/J. NZB	4, 6, 7, 8, 26, 28, 29, 30	e
f	CE	5, 7, 8, 26, 30	f
g	RIII	3, 8, 26	g
h	P/J	1, 6, 7, 8, 28, 30	h
j	CBA/J	1, 6, 7, 8, 28, 29, 30	j
k	KH-1	3, 5, 7, 8	k
l	KH-2	3, 5, 8	l
m	Ky	1, 2, 6, 7, 8	m

The allotypes of murine C_H genes are designated by a name for the locus, e.g. Igh-1, Igh-2, etc., with a letter denoting the allele or haplotype (Table 5.13).

Each haplotype is defined by a pattern of individual antigenic specificities encoded at each Ig locus. The extent of polymorphism varies. The Igh-1 (γ2a) locus has the highest degree of polymorphism with twelve haplotypes defined by an assortment of more than forty antigenic specificities (Table 5.14). In most cases an antigenic specificity is shared by numerous haplotypes; such antigens are said to be 'public'. Those specificities characteristic of one or a very few haplotypes are said to be 'private'. The 'public' and 'private' terminology stems from that used in the murine H-2 genetic nomenclature (see Chapter 9, p. 128).

The *Igh* loci are tightly linked, as shown by the fact that no cross-over between Igh genes was detected in the study of over 5000 progeny from three to four factor crosses.

References

Capra J.D. & Kehoe J.M. (1975) Hypervariable regions, idiotypy and the antibody combining site. In *Advances in Immunology*, Vol. 20, (eds. Dixon F.J. & Kunkel H.G.), p. 1. Academic Press, New York.

Feizi T., Lecomte J. & Childs R. (1977) An immunoglobulin heavy chain variable region (V_H) marker associated with cross-reactive idiotypes in man. *Clin. exp. Immunol.* **30**, 233.

Grubb R. (1956) Agglutination of erythrocytes coated with 'incomplete' anti-Rh by certain rheumatoid arthritic sera and some other sera. The existence of human serum groups. *Acta path. microbiol. scand.* **39**, 195.

Grubb R. & Laurell A.B. (1956) Hereditary serological human serum groups. *Acta path. microbiol. scand.* **39**, 390.

Kindt T.J. & Yarmush M. (1981) Expression of latent immunoglobulin allotypes and alien histocompatibility antigens: relevance to models of eukaryotic gene regulation. *CRC Crit. Rev. Immunol.* **2(4)**, 297.

Kunkel H.G., Natvig J.B. & Joslin F.G. (1969) A 'Lepore' type of hybrid gamma-globulin. *Proc. natn. Acad. Sci. USA* **62**, 144.

Milstein C.P., Steinberg A.G., McLaughlin C.L. & Solomon A. (1974) Amino acid sequence change associated with genetic marker Inv(2) of human immunoglobulin. *Nature* **248**, 160.

Natvig J.B. & Kunkel H.G. (1973) Human immunoglobulins: classes, subclasses, genetic variants and idiotypes. *Adv. Immunol.* **16**, 1.

Natvig J.B., Kunkel H.G. & Litwin S.D. (1967) Genetic markers of the heavy chain subgroups of human γG globulin. *Cold Spring Harb. Symp. quant. Biol.* **XXXII**, 173.

Natvig J.B. & Turner M.W. (1971) Localization of Gm markers to different molecular regions of the Fc fragment. *Clin. exp. Immunol.* **8**, 685.

Oudin J. (1956) L' "allotypie" de certaines antigenes proteidiques du serum. *CR Acad. Sci.* **242**, 2606.

Poljak R.J. (1975) Three-dimensional structure, function and genetic control of immunoglobulins. *Nature* **256**, 373.

Strosberg A.D. (1977) Multiple expression of rabbit allotypes: the tip of the iceberg? *Immunogenetics* **4**, 499.

Todd, C.W. (1963) Allotypy in rabbit 19S protein. *Biochem. Biophys. Res. Comm.* **11**, 170.

Todd C.W. (1972) Genetic control of H chain biosynthesis in the rabbit. *Fedn. Proc.* **31**, 188.

Turner M.W., Natvig J.B. & Bennich H. (1970) Isolation and antigenic characterization of pFc' fragments from six genetically different human G myeloma proteins. *FEBS Lett.*, **6**, 3.

Further reading

Capra J.D. & Kehoe J.M. (1975) Hypervariable regions, idiotypy and the antibody combining site. In *Advances in Immunology*, Vol. 20, (eds. Dixon F.J. & Kunkel H.G.), p. 1. Academic Press, New York.

Eichmann K. (1978) Expression and function of idiotypes on lymphocytes. In *Advances in Immunology*, Vol. 26, (eds. Dixon F.J. & Kunkel H.G.), p. 195. Academic Press, New York.

Grubb R. (1970) The genetic markers of human immunoglobulins. In *Molecular Biology, Biochemistry and Biophysics*, Vol. 9, (eds. Kleinzeller A. Springer G.F. & Wittman H.G.). Springer-Verlag, Berlin.

Janeway C., Sercarz F.E. & Wigzell H. (1981) *Immunoglobulin Idiotypes*. ICN-UCLA Symposia on Molecular and Cellular Biology, Vol. XX. Academic Press, New York.

Kehoe J.M. & Seide-Kehoe R. (1979) Antigenic features of immunoglobulins. In *Immunochemistry of Proteins*, Vol. 3, (ed. Atassi M.Z.), p. 87. Plenum Press, New York.

Lieberman R. (1978) Genetics of IgCH (allotype) locus in the mouse. *Springer Semin. Immunopath.* **1**, 7.

Loghem E.V. (1986) Allotypic markers. In *Basic and Clinical Aspects of IgG subclasses*, (ed. Shakib F.). Monographs in Allergy No. 19. Karger, Basel.

Natvig J.B. & Kunkell H.G. (1968) Genetic markers of human immunoglobulins. The Gm and Inv systems. *Ser. Haematol.* **I(1)**, 66.

Natvig J.B. & Kunkel H.G. (1973) Human immunoglobulins: classes, subclasses, genetic variants and idiotypes. *Adv. Immunol.* **16**, 1.

Potter M. (1972) Immunoglobulin-producing tumours and myeloma proteins of mice. *Physiol. Rev.* **52**, 631.

Schanfield M.S. & Loghem E.V. (1986) Human immunoglobulin allotypes. In *Handbook of Experimental Immunology*, 4th edn., Vol. 3, (eds. Weir D.M., Herzenberg L.A., Blackwell C. & Herzenberg L.A.), p. 45.1. Blackwell Scientific Publications, Oxford.

Chapter 6
Antibody diversity

Introduction

Antibody diversity provides the basis for recognition, by the humoral immune system, of self and non-self. Antibodies recognize and distinguish between an almost limitless number of molecular structures. The structural explanation of antibody specificity lies in the amino acid sequences of the particular V_L and V_H regions making up any given antibody combining site. The structure of the antibody combining region is introduced in Chapter 2. The cellular expression of antibody diversity is described in Chapter 13. Chapters 7 and 8 deal with the genetic basis of antibody diversity. This chapter describes the diversity of the antibody phenotype.

Repertoires

The antibody phenotype is impressive in its complexity. The diversity of the antibody phenotype can be demonstrated either in terms of the overall variety of different specificities that can be elicited (the total antibody repertoire) or in terms of the extent of heterogeneity of antibodies elicited by a single antigen (a specific antibody repertoire). It is important to understand that the term 'specific antibody' generally refers to a population of distinct molecules each having in common specificity for the eliciting antigen. The diverse antibody molecules comprising a specific repertoire can be sorted and distinguished by many methods (described below) and the size of specific repertoires can be estimated. The extent of the total repertoire of antibodies can be gauged by measuring what proportion any specific repertoire is of the total repertoire. The initial source of wonder at the antibody phenotype came from the ability to respond to the unexpected.

Antibody response to the unexpected

All vertebrates are capable of producing antibodies. An

antibody response was initially recognized and under-stood as a defensive reaction to a foreign organism. It seemed reasonable to expect that an animal would in-herit the ability to defend itself against specific 'pre-dictable' pathogens. That the antibody-producing system can go much further than this and can respond specifically with antibodies against totally new and un-expected chemical groupings caused a revolution in immunological thought. Landsteiner challenged the immune systems of rabbits with a wide variety of foreign antigens. As unexpected antigens he presented synthetic chemicals attached as side chains (haptens) to carrier proteins. In each case specific antibodies against the novel chemical grouping were elicited.

The use of substituted phenyl ring structures as hap-tens revealed the exquisite specificity of the antibody response. Antibodies can discriminate small and subtle structural changes in hapten structure; both the chemi-cal nature and the stereochemistry of the hapten in-fluences binding to the antibody combining site. Cross-reactions, where they occur, are partial; e.g. antibody raised against a meta-sulphonated benzene ring will show some binding with an ortho-sulphonate hapten and a little binding with a meta-arsonate hapten. Partial cross-reactions are now understood in terms of over-lapping specific antibody repertoires, i.e. some anti-bodies in the population raised against hapten 1 would also bind hapten 2 and would be included in the popu-lation of antibodies elicited by presentation of hapten 2 (Figure 6.1).

Size of the antibody combining site

Hapten binding studies gave the first indications of the shape of the antibody combining site and they also gave us knowledge of the size of that site. Kabat pioneered these measurements by using antibodies raised against dextran, a linear polymer of glucose residues. Oligo-mers of glucose, containing one to seven units, provi-ded a series of haptens whose quantitative binding to the anti-dextran antibodies was measured by the extent of inhibition of the dextran−anti-dextran precipitation reaction. The efficacy of inhibition increases with the increasing length of the oligomer up to a hexasac-charide; the heptasaccharide showed inhibition equal to the hexasaccharide (Figure 6.2). The conclusion drawn

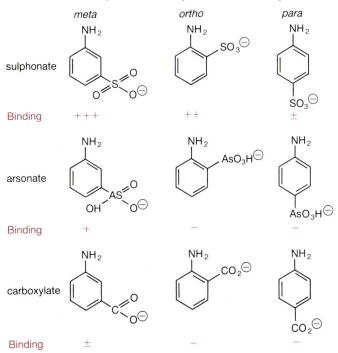

Fig. 6.1 This experiment of K. Landsteiner revealed the capacity of the immune system to respond to totally new and unexpected chemicals when presented as haptens. It is also a demonstration of the exquisite specificity of antibodies for the eliciting hapten.

is that the combining site of anti-dextran antibodies is filled by an isomaltohexose unit having extended dimensions of approximately 3.4 × 1.2 × 0.7 nm.

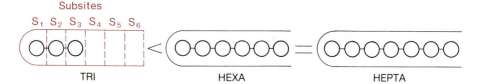

Subsites
S₁ S₂ S₃ S₄ S₅ S₆

TRI   HEXA   HEPTA

Fig. 6.2 The binding of haptens to antibodies allows an estimate to be made of the size of antigenic determinants and hence of the dimensions of the antibody combining site. In the dextran anti-dextran system shown here, the hexasaccharide is a better ligand for antibody than smaller oligosaccharides and equivalent to the heptasaccharide. The interpretation illustrated shows the hexasaccharide just filling the antibody combining site while the additional sugar residue of the heptasaccharide lies outside the site. The hexasaccharide thus achieves maximum binding energy and defines a size estimate for the antibody combining site.

Similar types of hapten binding studies have shown that antibodies to a variety of synthetic and natural antigens (Table 6.1) have comparable maximal binding capacity.

Three-dimensional structure of the antibody combining site

The heterogeneity of antibodies has prevented their crystallization. Antigen binding myeloma proteins were, therefore, studied. X-ray crystallographic analysis of two such proteins, one human and one murine in

Table 6.1 Comparison of the sizes of antigenic determinants seen by antibodies produced in different species

Antigen	Species	Determinant
Dextran	Human	Isomaltohexose
Dextran	Rabbit	≥ Isomaltohexose
Poly-γ-glutamic acid (killed *B anthracis*)	Rabbit	Hexaglutamic acid
Polyalanyl-bovine serum albumin	Rabbit	Pentaalanine
Polylysyl-rabbit serum albumin	Rabbit	Penta-or hexa-lysine
Polylysyl-phosphoryl-bovine serum albumin	Rabbit	Pentalysine
α-Dinitrophenyl-(lysine)₁₁	Guinea pig	α-Dinitrophenyl-heptalysine
α-Dinitrophenyl-polylysine	Guinea pig	α-Dinitrophenyl-trilysine
(D-Ala)ₙ − Gly − RNase	Rabbit	Tetrapeptide
Denatured DNA	Human*	Pentanucleotide

*Sera from patients with systemic lupus erythematosus.

origin, has given us a direct determination of the size of the antibody combining site.

In the human myeloma protein NEW the putative combining site can be defined as a shallow cleft lying between the V_H and V_L regions. This cleft has the approximate dimensions 1.6 nm long, by 0.7 nm wide, by 0.5 nm deep. Fourier difference analysis reveals that the hapten β-hydroxy vitamin K_1, which binds to protein NEW with $K_A = 1.7 \times 10^5$ M^{-1}, is bound to this cleft. The hapten makes contact with some 10 or 12 amino acid residues distributed along both the V_H and V_L regions. The hypervariable loops play a major part in defining both the size and the shape of the hapten-binding cleft (Figure 6.3a).

The murine myeloma protein MOPC603 also has a cleft bordered by V_H and V_L and measuring approximately 2.0 nm long, by 1.5 nm wide, by 1.2 nm deep. MOPC603 binds the hapten phosphorylcholine but this molecule does not fill the cleft and cannot be considered to be a complete determinant (Figure 6.3b).

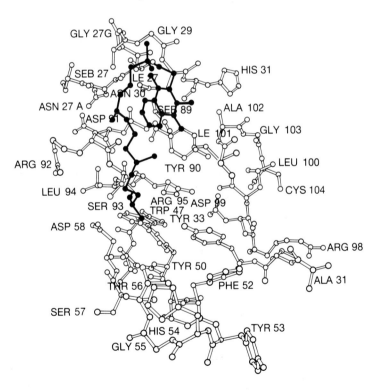

Fig. 6.3a Schematic representation of vitamin K_1OH bound to the combining region of the human myeloma protein NEW. (From Amzel & Poljak, 1979.)

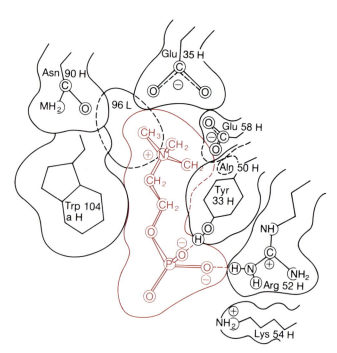

Fig. 6.3b Schematic representation of the specific interactions between phosphorylcholine and the protein side groups in M603. The binding cavity is located in the cleft between the L and H chains. Choline binds in the interior while the phosphate group is towards the exterior of the cavity. (From Amzel & Poljak, 1979.)

Specific antibody heterogeneity

The heterogeneity of an antibody specific for a selected antigen proved a source of trouble to immunochemists studying the structure of antibodies. The existence of such heterogeneity illustrates that the immune system has an extensive antibody diversity repertoire and can produce many different antibody combining sites, each having specificity for a single presented antigen. That specifically purified antibodies are heterogeneous populations of combining sites was first clearly shown by affinity measurements.

The binding of a hapten to an antibody can be measured quantitatively and expressed as an affinity constant, K_A:

$$K_A = 1/K$$

where K = molar concentration of free hapten

given 50% saturation of the antibody combining site at equilibrium.

A Scatchard plot of hapten–antibody binding data should, for a homogeneous antibody, give a straight line whose slope will be determined by K_A for the hapten. In practice the Scatchard plot for a hapten-specific antibody is usually curved (Figure 6.4). The curve is indicative of the heterogeneity of antibody affinities present in the hapten-specific population. Each curve could be transformed into a straight line (a Sips plot) by using an empirical index, α. The value of α is referred to as the heterogeneity index but it bears no direct relationship to the extent of heterogeneity of an antibody.

Experimental control of antibody heterogeneity

Limiting the extent of heterogeneity of specific antibody populations is an important step towards defining the

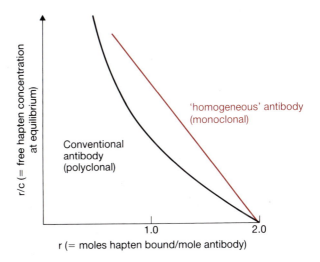

Fig. 6.4 A Scatchard plot of antibody-hapten binding. The binding of low molecular weight hapten to antibody is determined by equilibrium dialysis at a range of initial hapten concentrations. The plot of bound (r) divided by free (c) hapten concentrations versus r is governed by the equation: $r/c = nK - rK$ where K is the affinity of the antibody and is defined by the slope of the line and n is the valency of the antibody.

For conventional polyclonal antibody the line usually deviates from straight. The average affinity for the population of antibodies (when n = 2) is r/c at $r = 1$. An index of heterogeneity (a) for the antibody population can be calculated by plotting $\log(r/n - r)$ against log C. This Sips plot obeys the equation: $\log(r/n - r) = a \log c - a \log K$. A monoclonal antibody yields a straight line in the Scatchard plot, as shown, and a heterogeneity index $a = 1$ in a Sips plot.

phenotypic diversity of antibodies. The most important methods are as follows: antigen (epitope) selection, physicochemical separation, biological cloning of antibody-producing cells, fusion of antibody-secreting cells with suitable myeloma cells to obtain hybridoma cell lines.

Antigen (epitope) selection

The number of different antibody combining sites elicited by an antigen can be limited by choosing antibodies combining with an epitope of sufficient size to fill the available space in the $V_H - V_L$ cleft. The small aromatic haptens such as dinitrophenyl (DNP) are immunogenic and easy to use in affinity measurements. The effectiveness of DNP is due to the fact that it constitutes the immunodominant portion of a complete antigenic determinant. An immunodominant hapten contributes disproportionately to the binding energy when compared with the proportion of the space occupied by that hapten within the antibody combining site (Figure 6.5). The measured diversity of anti-hapten antibodies is maximized by the use of only the small immunodominant hapten in immunochemical procedures such as affinity measurements. Large haptens, or defined epitopes on natural macromolecules such as proteins or polysaccharides, are not as easy to study as small haptens. Where comparisons have been made the specific antibody repertoire against a complete epitope is less diverse than the repertoire selected against a small hapten (see below).

Physicochemical separation

The extent of heterogeneity of specific antibody is generally beyond the resolving power of mobility electrophoresis or ion-exchange chromatography, though partial fractionation can be achieved. Some fractionation on the basis of affinity differences can be achieved by stepwise or gradient elution of antibodies from solid-phase antigen.

Isoelectric focusing (i.e.f.) was first used to analyse antibodies by Awdeh et al. (1968). This equilibrium separation technique provides a resolving power comparable with the diversity of specific antibodies. Even using analytical i.e.f. it is usually necessary to limit the heterogeneity of the antibody population by biological methods prior to focusing. However, in comparative

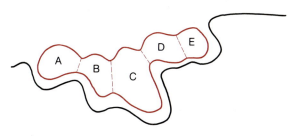

Fig. 6.5 A hypothetical antibody combining site filled by a complete epitope. Parts of the epitope are defined as A, B, C, D and E. Group C is the immunodominant portion of the epitope. Thus, C might be a hapten attached to a carrier protein.

analysis of different subpopulations, of the order of 10^4 antibody molecules can be differentiated. Each monoclonal antibody exhibits a microheterogeneous spectrum of isoelectric bands (as do most proteins, except when newly synthesized). This spectrum, termed a spectrotype, can be used as a phenotypic marker for diversity measurements, for breeding studies or for somatic cell genetics (Figure 6.6).

Biological cloning of antibody-producing cells

The cellular basis of antibody diversity is clonal. B lymphocytes are committed to production of a single antibody specificity and antibody receptors are displayed on the cell surface. Antigen-driven clonal expansion generates memory cells (differing from their progenitor 'virgin' B cells by being relatively long lived) and antibody-secreting cells. Memory cells (but not antibody-secreting cells) can be transferred from an immune mouse to a syngeneic, lethally irradiated mouse and, in this secondary host, antigen will effect clonal expansion, generating from each memory cell both more memory cells and antibody-secreting cells. Transferring memory cells at limiting dilution yields single clones of antibody-secreting cells in a proportion of the secondary hosts; the clonal complexity can be defined by analysis of the secreted antibody by i.e.f. Selected clones can be further expanded by serial passage of memory cells to syngeneic mice (Figure 6.7). Alternatively, the spleen of an irradiated secondary recipient of memory cells can be removed 24 h after cell transfer and cut into small pieces, each of which is cultured with antigen to elicit antibody secretion. A proportion

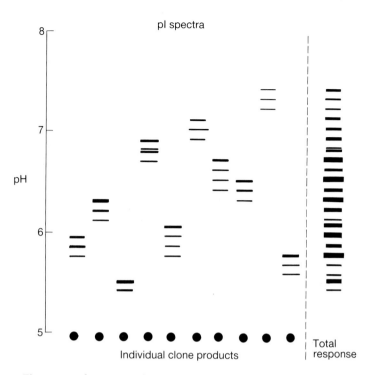

Fig. 6.6 A schematic analysis of a heterogeneous antibody response in terms of the isoelectric spectra of individual clonal products and of the total antibody. The 'total response' is drawn to resemble the spectrum of a typical heterogeneous anti-DNP antibody. On the left are drawn ten individual spectrotypes each corresponding to the product of a single antibody-forming cell clone (the multiple bands in each spectrotype are referred to as microheterogeneity and result from post-translational modification). Superimposition of the ten spectra yield the total response spectrum. It is not possible to reverse this process in practice and predict the ten individual spectrotypes from the total spectrum.

of organ fragments secrete monoclonal antibody in culture (Figure 6.8). These techniques of cloning *in vivo* and *in vitro* have been used to analyse antibody diversity.

Hybridoma production

Fusion of an antibody-secreting cell with a plasmacy-toma cell can generate a continuously growing clone of hybrid cells secreting the original antibody. Köhler and Milstein first reported the production of hybridoma cell lines by such a fusion procedure in 1975. The method is outlined in Figure 6.9. Cloning the antibody-secreting cells of a specific repertoire by generating hybridoma

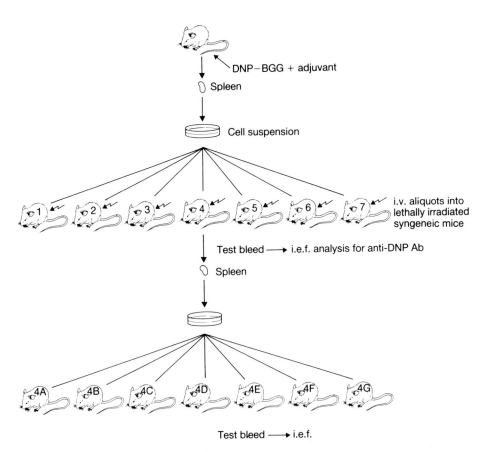

Fig. 6.7 The cloning and propagation of antibody forming cell clones *in vivo* as devised by Askonas *et al.* (1970). Spleen cells from a donor mouse, previously immunized against the antigen (e.g. DNP-BGG), are injected intravenously at limiting dilution into a series of lethally irradiated, syngeneic mice together with soluble antigen. Analysis of serum, taken from individuals a week after cell transfer, is made by isoelectric focusing with specific detection of anti-DNP antibody by overlay with radiolabelled hapten followed by radioautography. In serum from each mouse either zero, one or two monoclonal antibody spectrotypes are found (see Fig. 6.6). A selected mouse (4) expressing a single antibody forming cell clone is rechallenged with antigen, then the spleen is removed and the cells used to repopulate a series of lethally irradiated, syngeneic mice. Antigen elicits expression of identical monoclonal antibody in each of these recipients 4A–4G. Serial transfer of spleen cells from any of these mice to further recipients, together with antigen, propagates the antibody forming cell clone. Such clones cannot be propagated indefinitely; a single B cell clone has been estimated to have a finite life-span estimated at about 80 rounds of cell division.

cell lines allows each antibody to be produced in un-limited amounts and allows the genes encoding the heavy and light chains to be cloned and sequenced using recombinant DNA techniques (see Chapter 7). These combined techniques permit the most extensive analysis of antibody diversity. The availability of un-

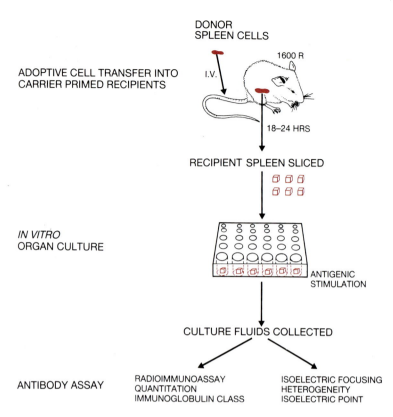

DONOR
SPLEEN CELLS

ADOPTIVE CELL TRANSFER INTO
CARRIER PRIMED RECIPIENTS

1600 R

I.V.

18–24 HRS

RECIPIENT SPLEEN SLICED

IN VITRO
ORGAN CULTURE

ANTIGENIC
STIMULATION

CULTURE FLUIDS COLLECTED

ANTIBODY ASSAY

RADIOIMMUNOASSAY
QUANTITATION
IMMUNOGLOBULIN CLASS

ISOELECTRIC FOCUSING
HETEROGENEITY
ISOELECTRIC POINT

Fig. 6.8 The fragment culture method for obtaining and analysing antibody-forming cell clones. In this method devised by Klinman (1969) unprimed spleen cells are transferred at limiting dilution into mice that have been primed with carrier protein and then lethally irradiated. The spleen of this host is collected the next day and cut into small ($\sim$1 mm^3) pieces each of which is cultured with antigen, e.g. a hapten presented on a carrier protein. The aim is to distribute precursor B cells such that no splenic fragment contains more than one (thus many fragments contain none) while ensuring a sufficient level of helper cells to allow the precursor B cell to respond to antigenic stimulation in organ culture.

limited amounts of monoclonal antibodies has revolutionized the use of antibodies in a wide range of analyses including the use of antisera in classical immunogenetics (see Chapter 12).

Phenotypic markers defining individual antibody specificity

Four different phenotypic markers have been extensively used to identify individual antibodies in breeding studies and in somatic cell genetics. These markers are: fine specificity, i.e.f. spectrotype, idiotype, amino acid sequence (V_H and V_L).

1 The fine specificity of antibody combining sites is

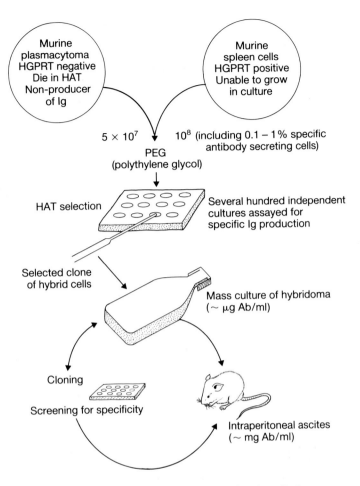

Fig. 6.9 A scheme for the production of monoclonal antibodies as devised by Köhler & Milstein (1975). Hybridoma cell lines are generated by fusion of murine plasmacytoma cells with specific antibody secreting cells (from immune mice) followed by selection and cloning. The original immunizing antigen need not be pure provided only that a specific assay for antibodies against the antigen is available. Hybridoma clones may be propagated, either *in vivo* or *in vitro*, indefinitely. HGPRT, hypoxanthine (guanine) phosphoribosyl transferase; HAT, culture fluid containing hypoxanthine, aminopterin and thymine.

measured by determining the affinity constants for a series of structurally related haptens. A subset of an antibody population raised against one hapten (e.g. NP) will cross-react with a structurally related hapten (e.g. NIP). This subset can be further defined by testing against a third related hapten (e.g. NNP) (Figure 6.10). Measuring the affinities for each of these haptens can be used to define a particular antibody combining site, or a

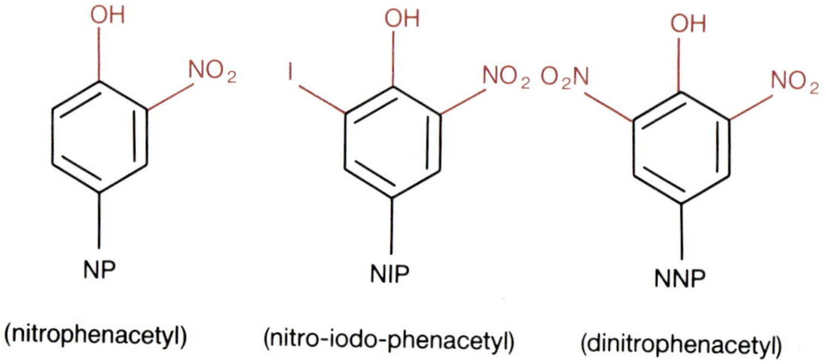

Fig. 6.10 Chemical structures of three related haptenic groups used in analysis of antibody diversity.

small subset of similar sites. Cross-reactivity of an antibody with structurally unrelated (i.e. not obviously related) antigens can also be observed and used to define a combining site. Screening with random antigens is a fortuitous approach and has been of most value in determining the specificity of myeloma immunoglobulins, each one a monoclonal antibody of undefined specificity. Random screening revealed that the combining site of a myeloma protein may bind a series of antigens, some of which are not obviously structurally related. The concept emerged of an antibody combining site having multiple shared specificities. Each antibody may thus be polyfunctional. The multispecificity of each combining site is compatible with the exquisite specificity of a population of antibodies present in a conventional antiserum because specificity for the eliciting antigen is the common feature of the population. A monoclonal antibody might show unpredictable cross-reactivities but these can be defined experimentally and required specficity maintained by the control of reaction conditions such as pH and ionic composition of the medium.

2   The spectrotype of an antibody is revealed by i.e.f. in thin-layer polyacrylamide gels, followed by overlay with antigen to detect the spectrum of antibody bands (see Figure 6.2).

3   The word *idiotypy* was coined by Oudin (1966) to describe an antigenic marker peculiar to an antibody raised by a particular antigen in an individual outbred rabbit and detected by immunization (with the first antibody) of an allotypically matched rabbit. A similar phenomenon of *individual antigenic specificity* of human pathological immunoglobulins was demonstrated by Kunkel et al. (1963) who extensively cross-absorbed antisera raised in a heterologous species to

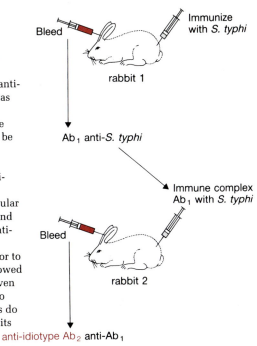

Fig. 6.11 The protocol of the experiment originally performed by Oudin (1966) in which he demonstrated the formation of anti-idiotype antibodies. A similar protocol was used by Oudin to raise anti-allotype antibodies except that in the anti-idiotype protocol rabbits 1 and 2 were matched to be identical for known immunoglobulin allotypes. The experimental difference between anti-idiotype antiserum and anti-allotype antiserum was that the former reacted serologically only with the particular anti-S. typhi antibody raised in rabbit 1 and used to immunize rabbit 2; by contrast anti-allotype antiserum reacted with immunoglobulins taken from rabbit 1 prior to immunization. Note that later studies showed that in some cases very low levels of a given idiotype can be detected in serum prior to immunization; these 'inherited' idiotypes do not invalidate the concept of idiotypy or its utility.

derive antibody specific for the individual immunizing immunoglobulin (see also pp. 119–21).

The term *idiotype* is now in general use to describe the antigenic properties peculiar to an individual immunoglobulin or antibody molecule. Each antibody molecule has several individual antigenic determinants, each of which is referred to as an *idiotope*. A set of idiotopes comprises an idiotype. An idiotype may be expressed on either a V_H or a V_L region and in some instances only on particular $V_H V_L$ pairs.

Oudin distinguished between an idiotype and a V-region allotype by two main criteria: (1) the idiotype was not detectable in pre-immune serum of the rabbit donating the first antibody; and (2) the idiotype was not an inherited characteristic.

Subsequent studies have shown that neither of these criteria need apply to idiotypy. For example, the T15 idiotype (described in more detail below) is characteristic of anti-phosphorylcholine antibodies elicited in any BALB/C mouse, and a low level of the T15 idiotype can be detected in the serum of BALB/C mice prior to experimental immunization (environmental immunogens are the likely cause). In breeding studies the T15 idiotype is inherited as a private V_H allotype. There remains no clear theoretical distinction between an allotype and an idiotype; experimentally we find a continuum of more or less public or private antigenic determinants. Nevertheless, the concept of an idiotype as a definition of an individual antibody combining site has been, and continues to be, very important in studies of antibody diversity. A physiological function for idiotype—anti-idiotype regulatory networks suggested by Jerne (1974) has stimulated considerable interest in cellular immunology. Following the studies of Nisonoff et al. (1975), idiotypes can be classified according to the degree to which the idiotype—anti-idiotype interaction can be inhibited by the specific hapten. Hapten-inhibitable idiotypes are explained by the proximity of certain idiotopes to the antibody combining site.

4 The amino acid sequences of the V_H and V_L regions of an antibody provide the ultimate phenotypic property defining the combining site of that antibody. When myeloma proteins were the only source of monoclonal immunoglobulins V_H and V_L sequences were accumulated and patterns of variability were defined (see Chapter 2).

Within each family of V regions the definition of *groups (and subgroups)* of sequences, by the criterion of greater sequence homology between members of a group (or subgroup) than with members of any other groups, showed that variability is discontinuous.

In comparisons of randomly selected V regions *hypervariable regions* were recognized as short linear sequences of amino acids where sequence differences are most frequently seen. Hypervariable sequences are, therefore, the sequences that are most peculiar to individual molecules.

Correlation of antibody specificity, idiotypy and hypervariable regions

The molecular basis for antibody specificity and, therefore, for antibody diversity reveals the interrelationship between the various phenotypic markers described above.

1 The information necessary for the construction of an antibody combining site is contained entirely within the sequences of the V_H and V_L regions that combine to form that site.

2 Diverse sites are generated by *combinatorial association* of V_H and V_L regions.

3 The sequences and lengths of hypervariable regions

of both V_H and V_L are major factors in determining the size and specificity of the antibody combining site. *Complementarity-determining residues* are frequently, but not exclusively, in hypervariable regions, although the terms are used interchangeably by some authors. In a given combining site not all hypervariable regions nor all residues in any particular hypervariable region contribute complementarity-determining side chains.

4 Idiotopes are frequently, but not exclusively, constituted by hypervariable region sequences.

Quantitative estimates of the extent of diversity of the antibody phenotype

Each of the phenotypic markers described above has been used either separately or in various combinations to quantify the diversity of the antibody phenotype. The general approach has been to estimate: (1) the size of a specific antibody repertoire (or in some cases the size of a myeloma protein repertoire); and (2) what fraction of the total antibody repertoire each specific repertoire comprises.

Myeloma protein repertoires

An extensive number of myeloma protein V region amino acid sequences have been determined and catalogued. The use of such information in estimating repertoire size is best illustrated by the analysis of the sequences of V regions of light chains obtained from BALB/C and NZB myeloma tumours.

Murine immunoglobulin has about 95% κ chains and 5% λ chains; this ratio is reflected in the products of induced plasmacytomas. Various criteria have been used to group V_κ sequences. Taking sequence identity in the first framework region (1−23) to define members of a group results in about thirty groups with no tendency to plateau as the data increase. Divergence within other framework regions has been used to define subgroups. Two identical V_κ sequences have been found in the first fifty proteins examined; statistically we can estimate the V_κ myeloma repertoire to contain 700−10 000 sequences (90% confidence limits). It is not possible to say what fraction of the total V_κ repertoire is represented by the V_κ myeloma phenotype.

By the criteria applied to V_κ sequences, murine $V_{\lambda I}$

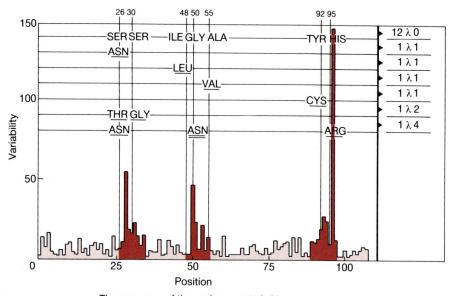

The summary of the replacements in V_λ.

Fig. 6.12 The amino acid replacements found in the first 18 $V_{\lambda I}$ sequences to be analysed. There are 12 identical sequences, designated $\lambda 0$ and postulated to correspond to the germ-line encoded $V_{\lambda I}$ gene sequence. The summary is superimposed on a variability plot, constructed according to the Wu & Kabat (1970), for murine V_k sequences.

sequences all fit within one group with no subgroups. Of the first eighteen $V_{\lambda I}$ regions sequenced, twelve are identical while the remainder differ by one or two amino acid substitutions at positions corresponding to the hypervariable regions of κ chains (Figure 6.12). The prediction was made by Weigert and Cohn that a single germ-line gene encodes $V_{\lambda I}$; sequences varying from the most common sequence were postulated to be derived by somatic mutation. By analogy, each group of V_κ sequences could be the products of a set of germ-line genes with at least one gene per subgroup. The accuracy of these predictions from phenotype has been shown by molecular genetic studies (see Chapter 7).

Specific antibody repertoires

To estimate the size of a specific antibody repertoire, experiments have been devised that aim to measure the frequency with which identical antibody molecules are elicited in independent events. Prior to the advent of hybridoma technology, monoclonal antibodies could not be extensively compared by amino acid sequence analysis.

Antibody spectrotypes have been compared as an approximation to sequence comparison. Combining this analytical comparison with limiting cell dilution in a spleen cell transfer system to separate antibody-forming cells from NIP-BGG immunized CBA/H mice, Kreth and Williamson (1973) observed 337 spectrotypes from four donor mice: between donors only five pairs of spectra were indistinguishable. Analysis of these data yields a statistical estimate for the minimum number of different $V_H V_L$ pairs specific for NIP of 5000 (2700 − 16 000 with 90% confidence limits). This method has been used to examine several other antibody specificities (Table 6.2).

Estimation of the proportion of the total antibody repertoire occupied by any specific repertoire has been made in one of two ways: (1) the fraction of specific B lymphocytes in the total population can be estimated from the limiting cell-dilution assay; and (2) the proportion of 'natural' specific antibody present in a pool of normal serum immunoglobulin can be measured. A consensus minimum size for the total repertoire is approximately 10^7 antibodies.

Hybridoma repertoires

The hybridoma technique allows the comparison, at a complete amino acid sequence level, of a set of mono-clonal antibodies generated against the same antigen. Gearhart *et al.* (1981) prepared hybridomas from BALB/C mice immunized with phosphorylcholine-haemocyanin

Table 6.2 Specific antibody repertoires

	Antibody specificity		
Eliciting antigen	Testing antigen	Mouse Strain	Repertoire size (minimum)
NIP BGG	NIP	CBA/H	5000
DNP BGG	TNP	CBA/H and C_3H/He	500
β-galactosidase	β-galactosidase	BALB/C	6000
β-galactosidase	Mutant enzyme activated by antibody	BALB/C	1200
Group A streptococci	Group A carbohydrate	BALB/C	40

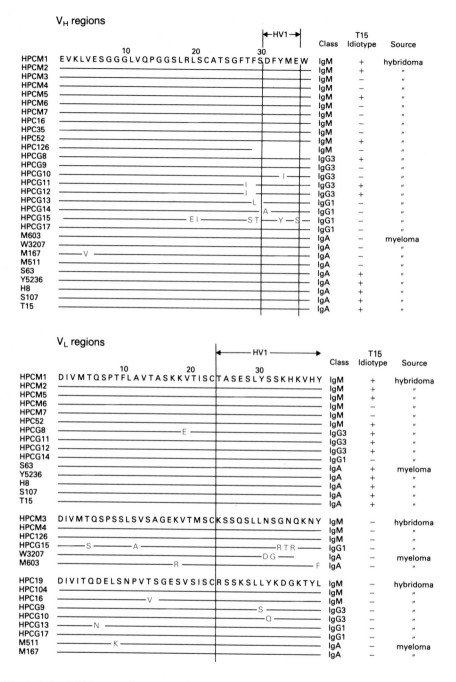

Fig. 6.13 Partial N-terminal amino acid sequences of the V_H and V_L regions of anti-phosphorylcholine monoclonal antibodies and phosphorylcholine binding myeloma proteins, all from BALB/C mice. Identity to the prototype HPCM1 sequence is indicated by a line; sequence differences are shown in red. (From Gearhart et al., 1981.)

and compared sixteen monoclonal antibodies specific for the phosphorylcholine hapten. The amino-terminal sequences of both V_L and V_H regions for each of these antibodies and for nine phosphorylcholine binding myeloma proteins are shown in Figure 6.13. The complete sequences of many of the V_H regions are shown in Figure 6.14.

The most striking feature overall is the limited extent of diversity coupled with the distribution of sequence differences. All eleven of the V_H regions in IgM antibodies fit the criteria for a single group (1−30 identity) and five have been shown to be identical up to the third hypervariable region (HV3). The amino-terminal V_L sequences fall into three groups and with one exception the V_L sequences of IgM antibodies are identical with one of the group prototype sequences. The V_L and V_H sequences available for IgG and IgA antibodies or myeloma proteins show limited variation from group prototype sequences, and in V_H regions this limited diversity occurs in different positions scattered across HV1, HV2 and the first three framework regions. The HV3 region (corresponding to the D segment, see Chapter 7) shows extensive variation in sequence. The fourth framework region (corresponding to the J_H segment, see Chapter 7) is identical between all sixteen V_H regions completely sequenced.

This analysis of amino acid sequences of a specific antibody repertoire is the ultimate level of phenotypic study. The implications drawn from these data are: (1) a small number of germline genes encode the V_H and V_L regions expressed in IgM anti-phosphorylcholine antibodies; (2) part of the phenotypic diversity of the repertoire could be due to the use of different D region segments; (3) part of the phenotypic diversity of the repertoire is probably due to somatic mutations; and (4) the phenotypic diversity (other than D region diversity) is almost exclusively in the V_H and V_L regions of IgG and IgA proteins. These conclusions from the phenotype are strengthened by molecular genetic analysis of the anti-phosphorylcholine system.

Similar analyses of the hybridoma and myeloma repertoires of immunoglobulins binding arsonate, dextran and nitrophenyl have yielded data comparable to that from the phosphorylcholine repertoire.

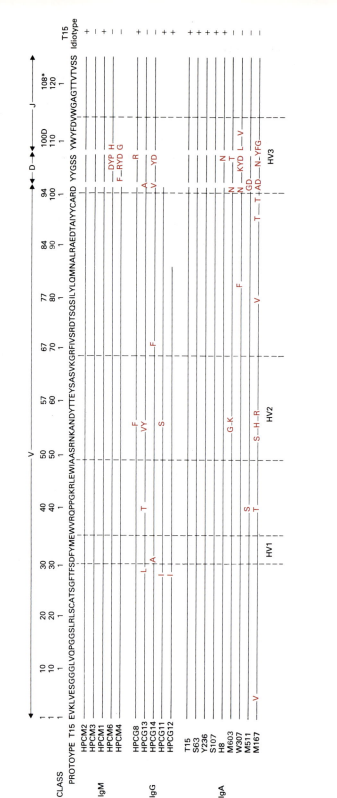

Fig. 6.14 Amino acid sequences of the V_H regions of anti-phosphorylcholine binding myeloma proteins. The sequences encoded by V, D and J genes are indicated at the top. The location of hypervariable regions (HV1, HV2 and HV3) for mouse heavy chains determined according to Wu & Kabat (1970) (see Chapter 2) are indicated by dotted lines. Sequence identity to the prototype T15 myeloma protein are shown by a line; differences are shown in red. The presence or absence of the T15 'inherited idiotype is recorded at the right of each sequence. There is an excellent correlation between the sequence of HV3 (comprising the D region and the N-terminal six residues of the J region) and the expression of the T15 idiotype.

References

L.M. Amzel & R.J. Poljak (1979) Structure of Immunoglobulins. *Ann. Rev. Biochem.* **48**, 961.

Askonas B.A., Williamson A.R. & Wright B.E.G. (1970) Selection of a single antibody-forming cell clone and its propagation in syngeneic mice. *Proc. natn. Acad Sci. USA* **67**, 1398.

Awdeh Z.L., Williamson A.R. & Askonas B.A. (1968) Isoelectric focusing in polyacrylamide gel and its application to immunoglobulins. *Nature* **219**, 66.

Gearhart P.J., Johnson N.W., Douglas R. & Hood L. (1981) IgG antibodies to phosphorylcholine exhibit more diversity than their IgM counterparts. *Nature* **291**, 29.

Jerne N.K. (1974) Towards a network theory of the immune system. *Ann. Immunol. Inst. Pasteur (Paris)* **125C**, 373.

Klinman N.R. (1969) Antibody with homogeneous antigen binding produced by splenic foci in organ culture. *Immunochemistry* **6**, 757.

Köhler G. & Milstein C. (1975) Continuous cultures of fused cells secreting antibody of predefined specificity. *Nature* **256**, 495.

Kunkel H.G., Mannik M. & Williams R.S. (1963) Individual antigen is specificity of antibodies. *Science* **140**, 1218.

Kreth H.W. & Williamson A.R. (1973) The extent of diversity of anti-hapten antibodies in inbred mice: anti-NIP (3-nitro, 4-hydroxy, 5-iodophenacetyl) antibodies in CBA/H mice. *Europ. J. Immunol.* **3**, 141.

Nisonoff A., Hopper J.E. & Spring S.B. (1975) Idiotype specificities of immunoglobulins. In *The Antibody Molecule*, (eds. Dixon F.J. Jr & Kunkel H.), p. 444. Academic Press, New York.

Oudin J. (1966) The genetic control of immunoglobulin synthesis. *Proc. Roy. Soc. Lond. B* **166**, 207.

Wu T.T. & Kabat E. (1970) An analysis of the sequences of the variable regions of Bence Jones proteins and myeloma light chains and their implications for antibody complementarity. *J. exp. Med.* **132**, 211.

Further reading

Burnet F.M. (1959) *The Clonal Selection Theory of Immunity.* Vanderbilt Press, New York.

Ehrlich P. (1900) On immunity with special reference to cell life. *Proc. Roy. Soc. Lond. B* **66**, 424.

Landsteiner K. (1946) *The Specificity of Serological Reactions.* Harvard University Press, Cambridge, Mass.

Loh D.Y., Bothwell A.L.M., White-Scharf M.E., Imanishi-Kari T. & Baltimore D. (1983) Molecular basis of a mouse strain-specific anti-hapten response. *Cell* **33**, 85.

Manser T., Huang S.-Y. & Gefter M.L. (1984) Influence of clonal selection on the expression of immunoglobulin variable region genes. *Science* **226**, 1283.

McMichael A.J. & Fabre J.W. (Eds.) (1982) *Monoclonal Antibodies in Clinical Medicine.* Academic Press, London.

Milstein C. (1980) Monoclonal antibodies. *Sci. Am.* **243**, 56.

Potter M. (1977) Antigen-binding myeloma proteins of mice. *Adv. Immunol.* **25**, 141.

Chapter 7
Molecular genetics of immunoglobulins

Introduction

Recombinant DNA techniques have rapidly accelerated the rate of accumulation of molecular genetic information. For the immunoglobulins this has meant that data is now available that could not have been obtained in any other way. The cloning of expressed gene sequences in the form of complementary DNA molecules and their comparison at the nucleotide sequence level with cloned germ-line genes has answered many of the questions raised by classical genetics and study of the antibody phenotype. The complexity of the immunoglobulin gene families and their somatic rearrangements is greater than could have been foreseen.

This chapter begins with a brief outline of the fundamentals of recombinant DNA techniques, followed by their application to the murine λ_I gene system, the simplest Ig gene system. The molecular genetics of the Igk and Igh loci in the mouse are then described. The map of Ig loci in the human genome is shown. Finally a molecular explanation of V allotypes and latent allotype expression in the rabbit is given.

Recombinant DNA technology: genetic engineering

This technology comprises methods for converting messenger RNA (mRNA) sequences into complementary DNA (cDNA) sequences, precisely cutting genomic DNA into conveniently sized lengths, inserting either double-stranded cDNA or genomic DNA fragments into autonomously replicating genetic elements of any microorganism or cultured cell, and thereby cloning and amplifying the inserted sequence. Techniques for controlled expression of cloned genes in the new host organism, for analysis of gene structure and arrangement, and for primary sequencing, are crucial to the usefulness of genetic engineering.

Complementary DNA (cDNA)

Retroviruses contain an enzyme, reverse transcriptase, capable of using an RNA molecule as a template, together with a suitable deoxyoligonucleotide primer to synthesize a complementary DNA molecule (Figure 7.1). Starting with mRNA one can use an oligo-dT primer from the poly (A) tail. The resultant RNA–DNA duplex can be separated by alkaline hydrolysis of the RNA, leaving a single-stranded cDNA molecule that has, at its 3' end, a short hairpin structure, capable of priming DNA polymerase I catalysed synthesis of a second DNA strand to yield double-stranded cDNA in a hairpin form. The single-stranded nuclease S_1 cuts the hairpin loop to yield double-stranded cDNA.

Restriction endonucleases

These enzymes are site-specific deoxyribonucleases derived from bacteria. Each strain of bacteria encodes a host modification and restriction system that functions to prevent the replication in the bacterium of inappropriate foreign DNA. Restriction enzymes cleave unmodified double-stranded DNA at specific, short palindromic

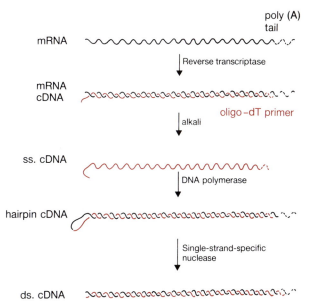

Fig. 7.1 The steps in the synthesis of single-stranded complementary DNA (ss cDNA) using reverse transcriptase to copy mRNA and then the construction of double-stranded cDNA.

153 MOLECULAR GENETICS OF IMMUNOGLOBULINS

sequences, i.e. having two-fold rotational symmetry. Some restriction enzymes cut both DNA strands at the same site:

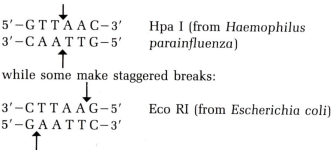

5′−G T T A A C−3′   Hpa I (from *Haemophilus*
3′−C A A T T G−5′   *parainfluenza*)

while some make staggered breaks:

3′−C T T A A G−5′   Eco RI (from *Escherichia coli*)
5′−G A A T T C−3′

leaving short, single-stranded sequences at the end of the two new duplex DNA fragments. Identical complementary ends are generated when any DNA molecule is cut by such a restriction enzyme. The restriction fragments from different DNA molecules can, therefore, be cross-hybridized by the complementary or cohesive

Recombining DNA molecules

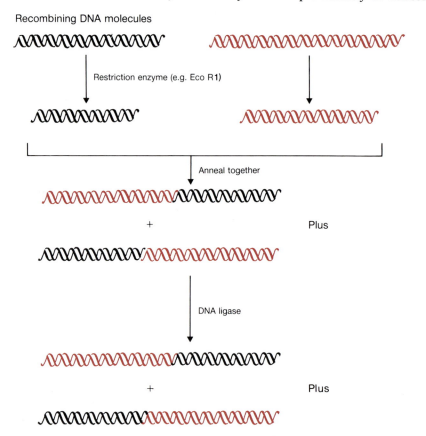

Fig. 7.2 Recombination of DNA molecules can be effected by cutting with a restriction endonuclease that generates cohesive single-stranded ends, annealing and covalently joining by the action of DNA ligase.

ends and joined covalently by the action of DNA ligase. These reactions are used in the cloning of DNA fragments by insertion into plasmid or phage DNA molecules (Figure 7.2).

Host modification of DNA to prevent the unwanted action of restriction enzymes involves methylation of bases in the restriction site.

The recognition sites for a selection of restriction enzymes can be conveniently tabulated according to the site sequence (Table 7.1).

Cloning vehicles or vectors

The vector DNA molecules into which other DNA is inserted for cloning are autonomously replicating DNA genomes such as plasmids or phages. Plasmids are circular, double-stranded DNA molecules that replicate in bacteria independently from the bacterial chromosome

Table 7.1 The recognition sites of some commonly used restriction endonucleases

Sequence (5'→3')	Tetra-nucleotides	Hexanucleotides A —— T	T —— A	C —— G	G ———— C	
AATT	Eco RI*				Eco RI	
ATAT						
ACGT						
AGCT	Alu I	Hind III		(Bam I)		
TATA						
TTAA					Hpa I	} Hinc II and Hind II
TCGA	Taq I			Xho I	Sal I, Ava I	}
TGCA				Pst I		
CATG						
CTAG			Xba I			
CCGG	Hpa II			Sma I	Ava I	
CGCG	(Bce R)			(Sac II, Sst II)		
GATC	(Mbo I)	Bgl II	Bcl I		Bam HI	
GTAC					Kpn I	
GCGC	Hha I	Hae II			Hae II	
GGCC	Hae III	(Hae I)	(Hae I, Bal I)			

A large number of restriction endonucleases recognize palindromic tetranucleotide sequences. The sixteen palindromic tetranucleotide sites which are possible are listed vertically and the sixty-four palindromic hexanucleotide sites are represented by four columns, each headed by the additional outer pair of bases. (Only bases in one of the two strands are shown.) Restriction endonucleases for twenty-seven of the total eighty sites are shown in the appropriate positions in the table. (After an idea from L.W. Coggins.)

and may, therefore, be present in high copy numbers (naturally up to approximately twenty per cell but experimentally hundreds or thousands of copies per cell).

Essential properties of a plasmid cloning vector are:

1 The presence of a unique restriction site allowing the conversion of the circular DNA to a linear DNA by a single cut.

2 The unique restriction site must be in a part of the plasmid that is not essential for plasmid function.

In addition, it is desirable that the plasmid vector carry a marker, such as drug resistance, suitable for selection and that the plasmid can achieve a high copy number. Wild-type plasmids have been modified by restriction enzymology to yield highly suitable plasmid cloning vectors (e.g. pBR322 shown in Figure 7.3). There are practical limits to the size of DNA that can be cloned in plasmids — ~ 10 kb (kilobases) — and for larger DNA fragments bacteriophages are more suitable vectors. Phage lambda, with a high replication number per lytic cycle, is a suitable starting point for construction of vectors (Figure 7.4). Multiple restriction sites are reduced to unique sites by mutation and restriction enzyme cleavage to remove non-essential phage genes. The requirement for the size of the phage DNA to be within certain limits (minimum and maximum length to allow packaging into a phage particle) can be used to select for phage containing inserted DNA sequences.

For the cloning of very large DNA fragments (~ 40 kb) cosmid vectors have been constructed.

Tailing

Since double-stranded cDNA molecules lack cohesive ends for insertion into vectors, a tailing technique can be used to generate suitable ends. As shown in Figure 7.5, single-stranded tails of homopolymeric nucleotides are added to the 3′ ends of the insert DNA and the linearized vector DNA using the enzyme terminal deoxynucleoside transferase. Using complementary tails, i.e. poly (dA)-poly(T) or poly(dC)-poly(dG), the insert and vector can be annealed together and covalently ligated.

Screening

Vectors containing any insert DNA can be selected by

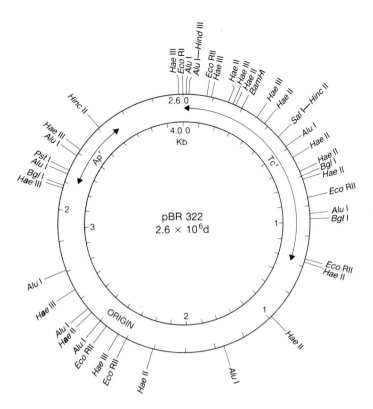

Fig. 7.3 Physical map of the plasmid pBR322. This is a plasmid designed for use in cloning DNA molecules in *E. coli*. The site of cleavage of several restriction enzymes are indicated. Unique sites (e.g. Eco R1, Hind III and Pst 1) are suitable for the introduction of foreign DNA. pBR322 contains two genes encodiong resistance to antibiotics, ampicillin (Ap^r) and tetracycline (Tc^r). (Bolivar *et al.*, 1977.)

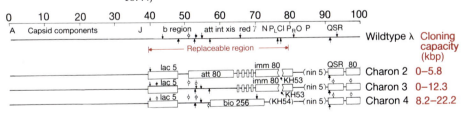

Fig. 7.4 Physical map of λ and three Charon λ phages. The Charon λ bacteriophages have been developed as vectors for cloning. The phage λ map, estimated to be 49 400 ± 600 bp (S.E.M.), is drawn to scale with the replaceable region indicated relative to essential genes for lytic growth. Beneath it are the Charon phage vectors with portions from λ (indicated by lines) aligned with the λ map. Boxes indicate substitutions, but the lengths of DNA substituted are not shown to scale. Downward and upward arrows are Eco RI and Hind III sites, respectively, The ø symbol shows known Sst I sites. lac5 and bio256 are substitutions from Lac and Bio regions of *E. coli*. The boxes labelled att80, imm80, QSR80 are portions of phage ø 80. The region shown by four small boxes at around 70 in some phages is from ø 80; it is partially homologous to λ. The parenthesized KH53, KH54, and nin5 are deletions. Since λ capsids can only accommodate DNA molecules within the size range 38−53 kb pairs (kbp) each phage vector has a maximum, and some have a minimum, cloning capacity. (From Blattner *et al.*, 1977.)

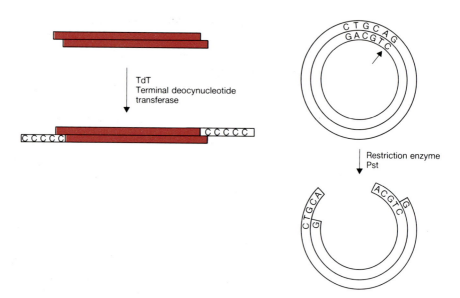

Fig. 7.5 Insertion of a DNA molecule into a plasmid can be facilitated by tailing the foreign DNA and the linearized plasmid with runs of complimentary homopolynucleotides. The single-stranded poly C and poly G additions are added using the enzyme terminal deoxynucleotide transferase.

simple devices such as cloning into an antibiotic resistance gene on the vector. Screening for the nature of the insert DNA is most conveniently done by using a radioactive nucleic acid hybridization probe. A print of bacterial colonies or phage plaques is taken from an agar plate onto a nitrocellulose membrane. After fixation of the DNA to the membrane, hybridization to a suitable probe is detected by autoradiography.

The probe might be labelled mRNA, a synthetic oligonucleotide (the sequence being deduced from the known protein sequence) or a cross-hybridizing cloned DNA sequence.

In the absence of a suitable probe the membrane-bound cloned DNA can be identified by using it to select specifically mRNA from a heterogeneous mRNA population. The mRNA hybridizing to the cloned DNA can be eluted and identified by translation, either in a suitable cell-free protein synthesizing system (e.g. a messenger dependent rabbit reticulocyte lysate), or by injection into an oocyte of the frog *Xenopus laevis*; the product is identified immunologically.

Cloned fragments of mammalian genomic DNA can be directly screened for their coding capacity by using them to transfect suitable cells in culture and detecting the product serologically. Mouse L cell mutants have been used successfully in this procedure. The fluorescence activated cell sorter (FACS) has been used to screen and select for expression of cell surface products encoded in cloned DNA.

Blotting: restriction endonuclease analysis

Restriction maps of DNA genomes or fragments thereof can be constructed by cutting the DNA with many different restriction enzymes (separately and in combination) and analysing the fragments by gel electrophoretic separation according to size. Specific fragments may be identified by hybridization, after transfer from the gel to either nitrocellulose or chemically reactive paper. This blotting technique was devised by Southern and the resultant prints are referred to as Southern blots (see Figure 7.9). RNA molecules separated by gel electrophoresis can be blotted onto chemically reactive paper for subsequent hybridization to DNA probes: this useful technique is termed 'northern' blotting.

Electron microscopic analysis of cloned DNA

The presence and position of a coding sequence within a cloned DNA molecule can be visualized by the R-loop method. An RNA probe is annealed with the DNA molecule under conditions favouring formation of RNA−DNA hybrids relative to DNA−DNA hybrids. The hybrids are prepared by protein binding and metal shadowing for electron microscopic visualization. The stretches of DNA complementary to the RNA probe are seen as loops along the DNA strand (see Figure 7.10). Heteroduplex analysis is used to compare two different cloned DNA molecules to determine the extent of shared sequences. This is illustrated in Figure 7.11.

DNA sequencing

Complete sequencing of cloned DNA molecules can be rapidly accomplished by one of two methods.
1   The chemical method devised by Maxam and Gilbert (Figure 7.6). DNA fragments obtained by restriction enzyme cleavage are radiochemically (^{32}P) labelled at one end. Random, partial, base-specific chemical cleavages are then used to generate four sets of DNA fragments, each labelled at one end and terminated in a

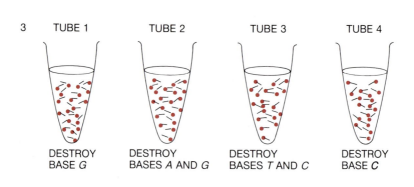

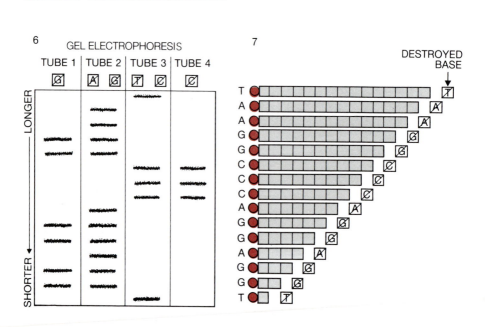

specific base at the other. Size separation of these fragments by gel electrophoresis yields a band corresponding to each occurrence of a specific base in the DNA sequence. Four such base-specific banding patterns are read in conjunction to determine a complete sequence of about 200–300 nucleotides per experiment. The rate-limiting step for longer sequences is the generation of restriction fragments.

2 The enzymic method devised by Sanger (Figure 7.7). This method also generates base-specific banding patterns but uses a biosynthetic approach. The DNA molecule is copied using DNA polymerase and radioactive nucleotides. In different reactions one of the four dideoxynucleotides is present to generate a random, base-specific series of termination points. This method also allows a sequence of 200–300 nucleotides to be read in one experiment. However, the use of this method has been adapted for sequencing large DNA molecules by 'shot-gun' cloning of partial restriction fragments in the single-stranded DNA vector phage M13. Randomly selected cloned fragments are sequenced by copying directly from the insert in M13. Many different sequences are rapidly generated and aligned by overlap to give the sequence of the original large DNA molecule.

Genomic DNA libraries

The entire genomic DNA of an organism (e.g. a mouse) is cut into suitable lengths for insertion into the chosen vector, and a sufficient population of insert-containing vector molecules is grown to ensure a complete representation of the mouse genome fragments; such a collection is termed a mouse genomic library (Figure 7.8). Random partial restriction enzyme cleavage is the most useful way of generating a library because this allows the isolation of overlapping cloned fragments.

Fig. 7.6 Sequencing of DNA by the chemical degradative method devised by Maxam & Gilbert (1977). A[^{32}P]-phosphate group is added as a radioactive label to the 5′-end of each DNA strand (1) and the strands are then separated (2). Each strand is sequenced by treating a separate portion of the labelled preparation with four distinct chemical agents to destroy randomly different bases, G, A and G, T and C or C as shown (3). In each tube a set of fragments is generated, each terminating at the position of a given base. Under controlled conditions a set of fragments (5) is generated corresponding to the positions of each base in the sequence (4). Separation of the four sets of fragments by electrophoresis on a polyacrylamide gel (6) allows them to be ordered according to size and the base sequence of the DNA to be read directly (7). Sequencing both strands of the DNA molecule affords a check on accuracy since exactly complementary sequences should be obtained.

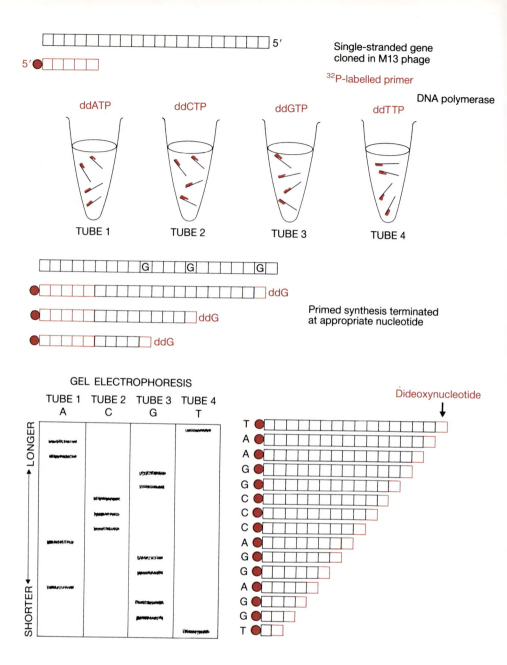

Fig. 7.7 Sequencing DNA by the dideoxynucleotide method of Sanger *et al.* (1977). The DNA molecule to be sequenced is cloned into the single-stranded DNA phage M13. A [P^{32}]-labelled oligonucleotide primer, complementary to the M13 sequence adjacent to the clone insertion site, is used to initiate synthesis of a complementary DNA strand. In each of four separate reactions a different dideoxynucleotide triphosphate is added at a concentration that affords random chain termination. A set of molecules is synthesized differing in length according to the positions of a given base. The four sets are separated by electrophoresis on a gel and the base sequence of the DNA read in a manner similar to that used in the chemical degradative method (Fig. 7.6).

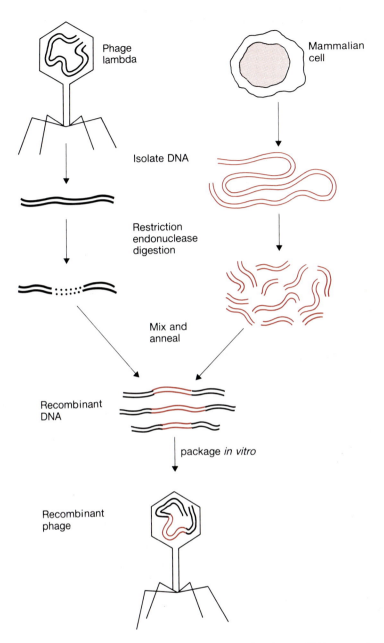

Fig. 7.8. A genome DNA library is generated by cloning DNA molecules, cut with restriction endonuclease, into bacteriophage lambda. The specially constructed Charon phages (see Fig. 7.4) allow insertion of different sizes of DNA molecules into the phage DNA. Addition of phage proteins to the recombined DNA molecules allows self assembly of lambda phage that can be grown in E. *coli* hosts. Each individual phage plaque affords a single cloned DNA molecule and the complete population is a library of molecules comprising the complete mammalian genomic DNA.

From a single identified cloned DNA fragment overlapping sequences extending to either side can be selected. Repetition of such selection is a method for chromosomal 'walking' by which families of genes can be ordered at the molecular level.

The murine *Igl* locus

The amino acid sequences of mouse plasmacytoma λ_I light chains showed a simple pattern of diversity consistent with the hypothesis that the λ_I V region is encoded by a single germ-line gene. The germ-line gene sequence was postulated to encode the amino acid sequence common to twelve of the eighteen λ_I V regions sequenced.

Genomic libraries generated from either mouse embryo DNA or DNA from a λ_I-producing plasmacytoma (HOPC2020) were screened with λ_I specific probes. From the embryonic DNA library were isolated separate cloned DNA fragments containing $V_{\lambda I}$ and $C_{\lambda I}$ sequences; from the plasmacytoma library came a single cloned fragment containing both the V and C sequences of λ_I. These cloned fragments correspond to those visualized on Southern blots (Figure 7.9). Analysis by R-looping shows the location of V and C coding regions

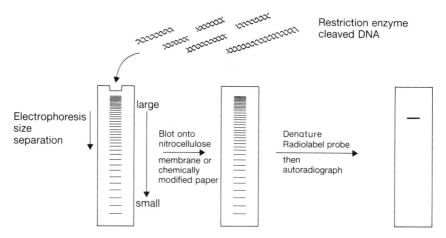

Fig. 7.9 Southern blotting. Detection of individual DNA molecules by the use of complementary probes. DNA restriction fragments are separated electrophoretically (in an agarose or polyacrylamide gel) according to size, then blotted onto nitrocellulose membrane (or chemically reactive paper), and fixed there in the denatured state. The dried blot is exposed to a radiolabelled DNA probe that hybridizes to any complementary fragment which can then be visualized by autoradiography. A single blot can be washed and successively re-utilized with a series of different probes. The method was devised by Southern (1975).

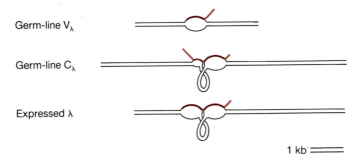

Germe-line V_λ

Germe-line C_λ

Expressed λ

1 kb

Fig. 7.10 Schematic representation of R loops formed by hybridizing
λ mRNA (——) with cloned V_λ, C_λ and joined $V_\lambda - C_\lambda$ DNA fragments.
The R loops were visualized, after metal shadowing, by electron
microscopy. (From Brack *et al.*, 1978.)

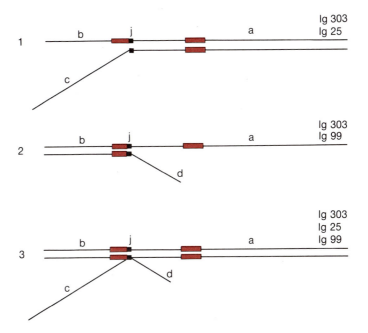

Fig. 7.11 Schematic representation of heteroduplex molecules
formed by hybridizing two or three cloned DNA molecules together.
The hybrids are visualized, after metal shadowing, by electron
microscopy.
Ig303 = expressed V_λ joined to C_λ
Ig25 = germ-line C_λ
Ig99 = germ-line V_λ
    The position of the J_λ region is indicated (■). The lengths of the
labelled pieces of DNA are as follows:
a ~ 5.5 kb
b ~ 2.0 kb
c ~ 2.9 kb
d ~ 1.5 kb
    The positions of V and C genes (■) are deduced from R loop
experiments (Fig. 7.10). (From Brack *et al.*, 1978.)

165 MOLECULAR GENETICS OF IMMUNOGLOBULINS

with each of the cloned DNA fragments (Figure 7.10). The first visual evidence for the J (joining) segment and for the intervening sequence (intron) separating J and C segments came from R-loop mapping and heteroduplex analysis (Figure 7.11). Complete DNA sequencing through the coding regions and introns revealed the V, J and C segments and a segment (P) encoding most of the hydrophobic signal peptide (Figure 7.12).

The P segment is separated from the V segment by an intron, 93 bp (base pairs) long. The V segment encodes only 95 residues of the 110 amino acids designated as the V region of the λ_I polypeptide. The remaining V-region codons are in the J segment placed 1250 bp 5′ to the $C_{\lambda I}$ segment. In embryonic DNA the $V_{\lambda I}$ and J-$C_{\lambda I}$ segments are separated at an unknown distance (on chromosome 16). In HOPC2020 DNA a somatic recombination event has joined $V_{\lambda I}$ contiguously to $J_{\lambda I}$ to allow expression of a λ_I chain. In the expressed genome, introns I_1 and I_2 remain and are transcribed (Figure 7.13). A contiguous colinear mRNA encoding PVJC is generated by RNA splicing (see Chapter 10). The sequence of the germ-line single $V_{\lambda I}$ gene translates to give the most commonly found $V_{\lambda I}$ polypeptide. The sequence of the HOPC2020 $V_{\lambda I}$ shows one base pair change from the germ-line sequence, and that change accounts for the known amino acid sequence of HOPC2020 $V_{\lambda I}$. No other coding region differences are seen but there is one base pair difference in I_2 between germ-line and HOPC2020.

Further light on the signals controlling V−J joining came from similar analysis of the more complex *Igk* and *Igh* loci.

The murine *Igk* locus

The extensive sequence diversity of murine κ chain V regions contrasts with the limited variability of $V_{\lambda I}$ regions. Multiple germ-line V_κ genes were postulated to encode V_κ regions of different length and diverse sequence. Arrangement of V_κ sequences into some thirty groups with further classification into subgroups allowed the interpretation of one V_κ gene per subgroup. Molecular genetic analysis could be expected to reveal multiple V_κ genes so that common features involved in V−J joining could be deduced.

−10
MetAlaTrp

Germl-line V$_\lambda$
GCCCAGCCCAGCCCATACTAAGAGTTATATTATGTCTGTCTCACAGCGCTGCTGCTGACCAATATTGAAAAGAATAGACCTGGTTTGTGAATTATGGCCTGG
CGGGTCGGGTCGGGTATGATTCTCAATATAATACAGACAGAGTGTCGGACGACGACTGGTTATAACTTTTCTTATCTGGACCAAACACTTAATACCGGACC

Expressed V$_\lambda$
GCCCAGCCCAGCCCATACTAAGAGTTATATTATGTCTGTCTCACAGCCTGCTGCTGACCAATATTGAAAAGAATAGACCTGGTTTGTGAATTATGGCCTGG
CGGGTCGGGTCGGGTATGATTCTCAATATAATACAGACAGAGTGTCGGACGACGACTGGTTATAACTTTTCTTATCTGGACCAAACACTTAATACCGGACC

−4
IleSerLeuIleLeuSerLeuLeuAlaLeuSerSerGly
ATTTCACTTATACTCTCTCTCCTGGCTCTCAGCTCAGTCAGCAGCCTTTCTACACTGCAGTGGGTATGCAACAATGCGCATCTTGTCTCTGATTTGCT
TAAAGTGAATATGAGAGAGAGGACCGAGAGTCGAGTCCAGTCGTCGGAAAGATGTGACGTCACCCATACGTTGTTACGCGTAGAACAGAGACTAAACGA

ATTTCACTTATACTCTCTCTCCTGGCTCTCAGCTCAGTCAGCAGCCTTTCTACACTGCAGTGGGTATGCAACAATGCGCATCTTGTCTCTGATTTGCT
TAAAGTGAATATGAGAGAGAGGACCGAGAGTCGAGTCCAGTCGTCGGAAAGATGTGACGTCACCCATACGTTGTTACGCGTAGAACAGAGACTAAACGA

−4
GlyAlaIleSerGlnAlaValValThrGlnGluSerAlaLeuThrThrSerProGlyGluThrValThr
ACTGATGACTGGATTTCTCATCTGTTTGCAGGGGCCATTTCCCAGGCTGTTGTGACTCAGGAATCTGCACTCACCACATCACCTGGTGAAACAGTCACA
TGACTACTGACCTAAAGAGTAGACAAACGTCCCCGGTAAAGGGTCCGACAACACTGAGTCCTTAGACGTGAGTGGTGTAGTGGACCACTTTGTCAGTGT

ACTGATGACTGGATTTCTCATCTGTTTGCAGGGGCCATTTCCCAGGCTGTTGTGACTCAGGAATCTGCACTCACCACATCACCTGGTGAAACAGTCACA
TGACTACTGACCTAAAGAGTAGACAAACGTCCCCGGTAAAGGGTCCGACAACACTGAGTCCTTAGACGTGAGTGGTGTAGTGGACCACTTTGTCAGTGT

20   30   40   50
LeuThrCysArgSerSerThrGlyAlaValThrThrSerAsnTyrAlaAsnTrpValGlnGluLysProAspHisLeuPheThrGlyLeuIleGlyGly
CTCACTTGTCGCTCAAGTACTGGGGCTGTTACAACTAGTAACTATGCCAACTGGGTCCAAGAAAAACCAGATCATTTATTCACTGGTCTAATAGGTGGT
GAGTGAACAGCGAGTTCATGACCCCGACAATGTTGATCATTGATACGGTTGACCCAGGTTCTTTTTGGTCTAGTAAATAAGTGACCAGATTATCCACCA
HV 1

Thr   Gly
CTCACTTGTCGCTCAACTACTGGGGCTGTTACAACTGGTAACTATGCCAACTGGGTCCAAGAAAAACCAGATCATTTATTCACTGGTCTAATAGGTGGT
GAGTGAACAGCGAGTTGATGACCCCGACAATGTTGACCATTGATACGGTTGACCCAGGTTCTTTTTGGTCTAGTAAATAAGTGACCAGATTATCCACCA
HV 1

60   70   80
ThrAsnAsnArgAlaProGlyValProAlaArgPheSerGlySerLeuIleGlyAspLysAlaAlaLeuThrIleThrGlyAlaGlnThrGluAspGlu
ACCAACAACCGAGCTCCAGGTGTTCCTGCCAGATTCTCAGGCTCCCTGATTGGAGACAAGGCTGCCCTCACCATCACAGGGGCACAGACTGAGGATGAG
TGGTTGTTGGCTCGAGGTCCACAAGGACGGTCTAAGAGTCCGAGGGACTAACCTCTGTTCCGACGGGAGTGGTAGTGTCCCCGTGTCTGACTCCTACTC
HV 2

Germ-line C$_\lambda$
GATCCTGGGAAGAAGGATCTTTCAGTGATGTCACCACCTTCCAAGAATTACCAGGAGCTGCATACATCACAGATGCAACTTGAGAATAAAATG
CTAGGACCCTTCTTCCTAGAAAGTCACTACAGTGGTGGAAGGTTCTTAATGGTCCTCGACGTATGTAGTGTCTACGTTGAACTCTTATTTTAC

ACCAACAACCGAGCTCCAGGTGTTCCTGCCAGATTCTCAGGCTCCCTGATTGGAGACAAGGCTGCCCTCACCATCACAGGGGCACAGACTGAGGATGAG
TGGTTGTTGGCTCGAGGTCCACAAGGACGGTCTAAGAGTCCGAGGGACTAACCTCTGTTCCGACGGGACTGGTACTGTCCCCGTGTCTCACTCCTACTC
HV 2

90   97
AlaIleTyrPheCysAlaLeuTrpTyrSerAsnHis
GCAATATATTTCTGTGCTCTATGGTACAGCAACCATTTCCACAATGACATGTGTAGATGGGGAAGTAGATCAAGAACACTCTGGTACAGTCTCATAACT
CGTTATATAAAGACACGAGATACCCATGTCGTTGGTAAAGGTGTTACTGTACACATCTACCCCTTCATCTAGTTCTTGTGAGACCATGTCAGAGTATTGA
HV 3

CATGCAAGGTTTTTGCATGAGTCTATATCACAGTGCTGGGTGTTCGGTGGAGGAACCAAACTGACTGTCCTAGGTGAGTCACTCCTTCCTCCTTTGTTA
GTACGTTCCAAAAACGTACTCAGATATAGTGTCACGACCCACAAGCCACCTCCTTGGTTTGACTGACAGGATCCACTCAGTGAGGAAGGAGGAAACAAT

GCAATATATTTCTGTGCTCTATGGTACAGCAACCATTTCCACAATGACATGTGTAGATGGGGAAGTAGATCAAGAACACTCTGGTACAGTCTCATAACT
CGTTATATAAAGACACGAGATACCCATGTCGTTGGTAACCCACAAGCCACCTCCTTGGTTTGACTGACAGGATCCACTCAGTGAGGAAGGAGGAAACAAT
HV 3

96
TrpValPheGlyGlyGlyThrLysLeuThrValLeuGly
ACCACTTCTTAACAGGTGGCTACATCTCCCTAGTCTGTTCTCTTTTACTATAGAGAAATTTATAAAAGCTGTTGTCTCAATCAATAAAAAGTTTTATT
TGGTGAAGAATTGTCCACCGATGTAGAGGGATCAGACAAGAGAAAATGATATCTCTTTAAATATTTTCGACAACAGAGTTAGTTATTTTTCAAAATAA

TTGTTCTCTCCAAGACTTGAGGTG CTTTTTGTTGTATACATTTCCCTTTCTGTATTCTGCTTCATACCTATACTTCACACTAGGTAAAGAATTTCTTTCT
AACAAGAGAGGTTCTGAACTCCACGAAAAACAACATATGAAAGGGAAAGACATAAGACGAAGTATGGATATGAAGTGTGATCCATTTCTTAAAGAAAGA

TTGTTCTCTCCAAGACTTGAGGTG CTTTTTGTTGTATACATTTCCCTTTCTGTATTCTGCTTCATACCTATACTTCACACTAGGTAAAGAATTTCTTTCT
AACAAGAGAGGTTCTGAACTCCAAGAAAAACAACATATGAAAGGGAAAGACATAAGACCAAGTATGGATATGAAGTGTGATCCATTTCTTAAAGAAAGA

CAACAAATTGTATAATTATGCCTTGATGACAAGCTTTGTTTACCAACTTGGCACAACATAGAATCATTGAGAAGAGAACC
GTTGTTTAACATATTAATACGGAACTACTGTTCGAAACAAATGGTTGAACCGTGTTGTATCTTAGTAACTCTTCTCTTGG

110   120
GlyGlnProLysSerSerProSerValThrLeuPheProProSerSerGluGluLeuThr
TCTCTCTGA...//...TTTTGACCTTCTCTTACTTCATCCTGCGGCCAGCCCAAGTCTTCGCCATCAGTCACCCTGTTTCCACCTTCCTCTGAAGAGCTCACT
AGAGAGACT...//...AAAACTGGAAGAGAATGAAGTAGGACGCCGGTCGGGTTCAGAAGCGGTAGTCAGTGGGACAAAGGTGGAAGGAGACTTCTCGAGTGA

TCTCTCTGA...//...TTTTGACCTTCTCTTACTTCATCCTGCGGCCAGCCCAAGTCTTCGCCATCAGTCACCCTGTTTCCACCTTCCTCTGAAGAGCTCACT
AGAGAGACT...//...AAAACTGGAAGAGAATGAAGTAGGACGCCGGTCGGGTTCAGAAGCGGTAGTCAGTGGGACAAAGGTGGAAGGAGACTTCTCGAGTGA

130   140   150
GluAsnLysAlaThrLeuValCysThrIleThrAspPheTyrProGlyValValThrValAspTrpLysValAspGly
GAGAACAAGGCCACACTCGTGTGAACGATCACTGATTTCTACCCAGGTGTGGTCACAGTGGACTGGAAGGTAGATGGA
CTCTTGTTCCGGTGTGAGCACACTTGCTAGTGACTAAAGATGGGTCCACACCACTGTCACCTGACCTTCCATCTACCT

GAGAACAAGGCCACACTCGTGTGAACGATCACTGATTTCTACCCAGGTGTGGTCACAGTGGACTGGAAGGTAGATGGA
CTCTTGTTCCGGTGTGAGCACACTTGCTAGTGACTAAAGATGGGTCCACACCACTGTCACCTGACCTTCCATCTACCT

Fig. 7.12 Nucleotide sequences of the regions of cloned DNA molecules encoding germ-line V$_{\lambda I}$, germ-lines J−C$_{\lambda I}$ and expressed λ$_I$ gene. The deduced amino acid sequence is shown in the top line where appropriate (Bernard *et al.*, 1978).

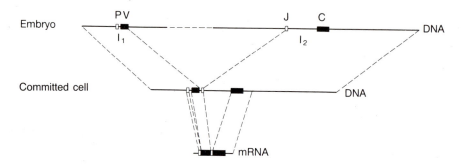

Fig. 7.13 Summary of the arrangement of exons P, V, J and C encoding the murine λ chain in embryonic (germ-line) DNA and in the committed cell. Expression of λ mRNA involves splicing of the transcribed RNA to remove intron sequences I_1 and I_2.

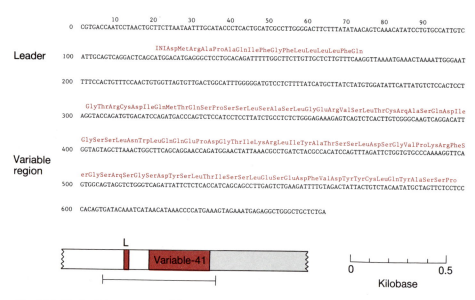

Fig. 7.14 Nucleotide sequence of the germ-line V_κ gene cloned from embryonic mouse DNA using κ chain cDNA from the murine plasmacytoma MOPC41. The signal or leader sequence (L) and the $V_\kappa - 41$ sequence are encoded in separate exons. Data from Seidman et al. (1979).

Starting with one plasmacytoma a specific VC_κ cloned cDNA sequence probe can be generated. Using such a probe to screen embryonic and plasmacytoma gene libraries a pattern of gene organization essentially similar to that found for the *Igl* locus is seen. The elements P, I_1, V, J, I_2 and C are all present in the *Igk* locus. Each germ-line V_κ encodes only up to residue 95 of the V_κ region (Figure 7.14). Not one but five J segments are present at approximately 3.7 kb 5′ to a single C_κ segment (Figure 7.15). The J segments are spaced about 300 bp apart. Comparison of the J segment coding sequences with known κ chains shows that only four of

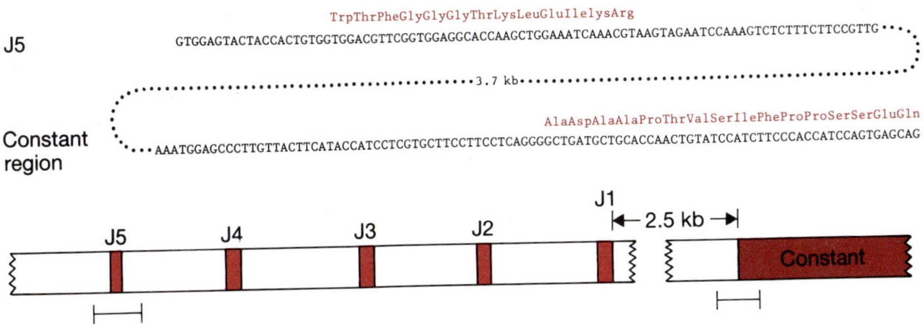

J5

TrpThrPheGlyGlyGlyThrLysLeuGluIleLysArg
GTGGAGTACTACCACTGTGGTGGACGTTCGGTGGAGGCACCAAGCTGGAAATCAAACGTAAGTAGAATCCAAAGTCTCTTTCTTCCGTTG···

···3.7 kb···

Constant
region

AlaAspAlaAlaProThrValSerIlePheProProSerSerGluGln
···AAATGGAGCCCTTGTTACTTCATACCATCCTCGTGCTTCCTTCCTCAGGGGCTGATGCTGCACCAACTGTATCCATCTTCCCACCATCCAGTGAGCAG

J1
← 2.5 kb →

J5   J4   J3   J2   Constant

Fig. 7.15 Arrangement of five J_κ genes relative to C_κ in germ-line DNA determined by nucleotide sequence analysis of a cloned fragment of embryonic mouse DNA. The sequence of J5 and the 5′-(amino-terminal) end of C_κ are shown above. Data from Seidman *et al.* (1979).

the five J segments are functional; $J_{\kappa 3}$ appears to be a non-functional pseudogene. Other examples of pseudo-genes have been found in the Ig families and in other multigene families.

The multiplicity of V_κ genes has been shown by restriction enzyme mapping with Southern blotting techniques and by cloning of a number of V_κ genes. The picture that emerges is of sets of V_κ gene segments with one to about ten genes per set (Figure 7.16); genes within a set are defined by cross-hybridization. A set of genes probably corresponds to a group of V_κ chains arranged by amino acid sequences. Several assumptions are involved in calculating the total number of V_κ genes per haploid gene, so the answers range from 100 to 1000; a consensus number would be of the order of 250 V_κ genes arranged in approximately fifty sets (see Chapter 8, p. 211). V_κ genes are spaced at a minimum distance of about 10 kb apart; this is similar to the spacing between genes in other multigene families (e.g. globin genes).

V−J joining: special recombination sequences

The sequences adjacent to J and V segments reveal patterns that probably play a critical role in the somatic recombination events joining V to J. Each of the four functional J_κ segments and the single J_λ segment has a self-complementary heptanucleotide palindrome at the 5′ end; a nonameric sequence GGTTTTTGT is located 12±1 or 23±1 bp 5′ to the heptameric palindrome. Similar, highly conserved nonamer and heptamer sequences with either 12 or 23 bp spacing are present at

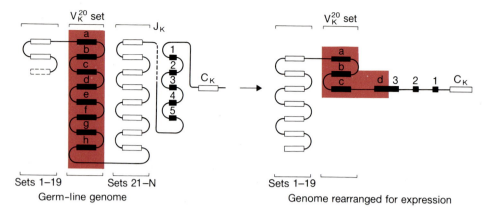

V_κ^{20} set J_κ

a
b
c
d
e
f
g
h

1
2
3
4
5

C_κ

Sets 1–19   Sets 21–N

Germ-line genome

V_κ^{20} set

a
b
c

d   3   2   1 C_κ

Sets 1–19

Genome rearranged for expression

Fig. 7.16 Summary of the arrangement and rearrangement for commitment and expression of murine or human κ genes. The number of V genes per set is variable. Each V gene has a leader sequence (not drawn separately) 5' to the V exon.

the 3' end of each V gene. A 12 bp spacer is about one turn of the DNA double helix and a 23 bp spacer is about two turns (Figure 7.17).

The heptamer — spacer — nonamer apparently constitutes a joining recognition site. The orientation of the heptamer and nonamer sequences next to a V gene are inverted relative to those associated with a J gene. For somatic recombination to occur, joining V and J, the two recognition sites involved should be one with a one-turn spacer and the other with a two-turn spacer. This one-turn — two-turn hypothesis has proved important in understanding the *Igh* locus rearrangement and in predicting the D gene segments (see p. 175). The special recombination process joining V to J must involve alignment of the recognition sites and a recombinase with two functionally distinct domains, one recognizing the palindromes with a one-turn spacer and the other recognizing the palindromes with a two-turn spacer. The same enzyme could mediate both κ and λ V to J joining and also the joining of T cell receptor gene segments (see Chapter 13).

The alignment process should involve *cis* elements according to our knowledge from the classical genetics of rabbit V and C region allotypes (see Chapter 5, p. 123). Joining of V and J elements on the same DNA strand could involve an inverted repeat stem structure; such a structure can be drawn (Figure 7.18) as cruciform DNA but would probably involve tetraplex (four-stranded) DNA.

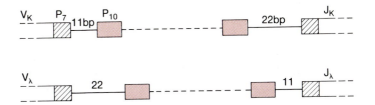

Fig. 7.17 Positions of self-complementary palindromes, P_7 and P_{10}, adjacent to each V_κ, V_λ, J_κ and J_λ exon. The spacing between P_7 and P_{10} is either $11-12$ bp (one turn of DNA double helix) or $22-23$ bp (two turns). In the κ system palindromes with a one-turn spacer are downstream from each V_κ and allow joining to palindromes with a two-turn spacing upstream from each J_κ. One- and two-turn spacings are reversed in the λ system. These findings revealed the one-turn/two-turn spacer rule that predicted that joining requires palindromes with a one-turn spacer to complement palindromes with a two-turn spacer. This rule has been found to apply in all immunoglobulin and T cell antigen receptor gene joining events.

When V−J joining is effected the DNA previously between the joined elements is no longer covalently linked to the chromosome. This intervening DNA (carrying unused V and J segments in many cases) could be reintegrated into the chromosome (but there is no obvious mechanism), or it could be lost. The loss of the intervening sequence has been indicated by negative results in Southern blotting experiments using germline DNA probes, corresponding to sequences downstream from V_λ and upstream from J_λ, to look at DNA from a λ chain producing plasmacytoma. Also in the κ chromosome, V genes downstream from the functional V_κ joined to J are missing, as judged by gene counting hybridizations. These experiments lend support to the deletion joining model. However, in some plasmacytoma lines producing κ chains, reciprocal recombination fragments have been identified. Such fragments are identified by the presence of the 3′ flanking sequences of a V gene linked head-to-tail to the 5′ flanking sequences of a J segment. It has been suggested that such fragments are formed when V−J joining occurs by unequal sister chromatid exchange. This mechanism is illustrated in Figure 7.19. It can be seen that the functional V−J chromosome and the reciprocal product of the recombination will segregate at mitosis. The presence of a reciprocal fragment in a plasmacytoma line can be explained by a second recombination joining another V to another J. Thus, the reciprocal fragment in the final cell would not correspond to functional

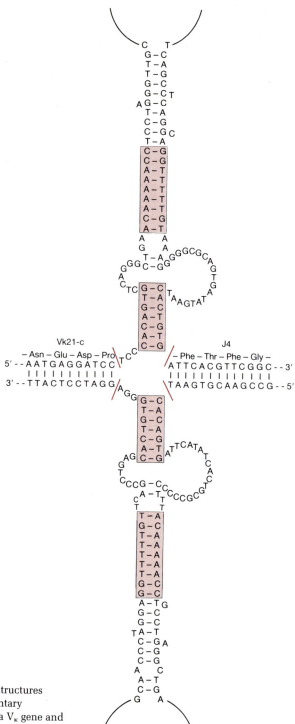

Fig. 7.18 Inverted repeat stem structures formed between self-complementary palindromes downstream from a V_κ gene and upstream from a J_κ gene. Data from Sakano *et al.* (1979).

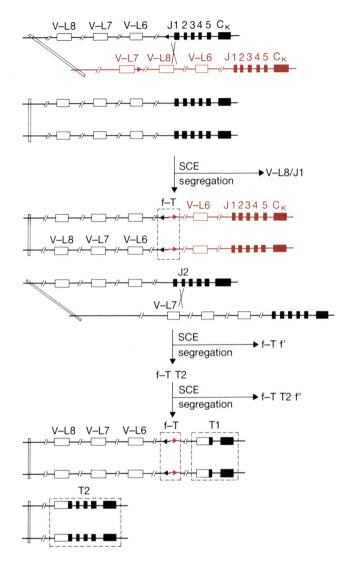

Fig. 7.19 Sister chromatid exchange has been proposed to be a rarely used manner of V−J joining. Usually each committed cell genome has one functionally rearranged V for light chain and one functionally rearranged V for heavy chain. The finding in a murine myeloma cell (T) of two rearranged V genes (T1 and T2) and of the flanking region recombination product (f−T) led to the hypothesis shown here. (From Hochtl *et al.*, 1982.)

V−J recombination yielding the product of that cell; the limited data available are consistent with the latter hypothesis.

We must conclude that there may be more than one mechanism operating to join V and J segments. Joining may be intra-chromosomal or (rarely) between sister chromatids; in either case the *cis* joining rule can be maintained.

Variability in the frame of somatic recombination

Whatever the mechanism of joining V to J, some variability in the joining frame can occur. Such variability explains the fact that position 96 in the amino acid sequence of BALB/C κ chains is the most hypervariable point in the V region (see Figure 6.12, p. 146). Indeed the sequences of many κ chains cannot be accounted for in terms of any known J_κ nucleotide sequence without introducing the idea that the frame of somatic recombination is variable. Cutting and splicing of DNA at several different positions can each form a contiguous V–J segment, but each join may specify a different codon for the amino acid residue at position 96 (Figure 7.20). Direct evidence for variation in the cross-over point in V–J joining comes from nucleotide sequence analysis of several plasmacytoma V–J DNA segments and comparison with the amino acid sequences of the products. In addition to the positional variation shown above, nucleotide deletion can also occur during V–J recombination. Deletion of one or two codons can occur, generating an in-phase but shortened J sequence; or deletion of one (or two) nucleotides can occur, giving rise to a cryptic gene with a nonsense sequence.

In conclusion, V–J recombination can occur at various cross-over points and may not always be in the proper phase.

The murine *Igh* locus

The *Igh* locus is by far the most complex of the three loci encoding immunoglobulin genes (Figure 7.21). In addition to a family of V_H genes and multiple J segments there are many C_H genes, each one encoding a class or subclass of heavy chain. Moreover, each C_H gene has several exons, one corresponding to each functional domain of the heavy chain protein. A single set of four J_H segments lies upstream from the C_H gene family, the proximal gene being C_μ. The generation of a functional V_H region coding segment involves at least two recombinant events (contrast the single recombination joining V_L to J_L) involving a third variable region

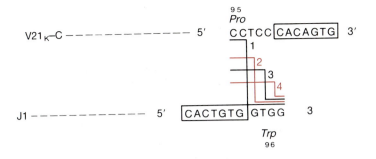

Fig. 7.20 Alternative recombinant events between one V_κ and one J_κ can generate different sequences at codon 96. Recombinations 1 and 2 give the sequence Pro(95) − Trp(96). Recombinations 3 and 4 give sequences Pro(95) − Arg(96) and Pro(95) − Pro(96) respectively. The heptameric palindromes are boxed. Data from Sakano et al. (1979).

element, the D (diversity) segment. A family of D segments, encoding sequences corresponding to the third hypervariable region of heavy chains, lies between the V_H gene family and the J_H cluster. The formation of a functional $V_H - D_H - J_H$ unit allows expression of a μ chain and/or a δ chain. Further DNA rearrangement precedes a switch of expression to any other classes of heavy chain; switch recombination involves S regions that appear to be high frequency general recombination sites (see Chapter 8, pp. 197−201).

The arrangement of V_H genes in sets corresponding to V_H region groups classified by amino acid sequence parallels the V_κ gene family, and there are similar numbers of V_H genes and V_κ genes in the mouse genome (see Chapter 8, p. 212). The novel genetic segment D increases the diversity of V_H region sequences.

The D_H segment

A D_H genetic element was hypothesized before being found. In the plasmacytoma MOPC141 the coding sequence for the third hypervariable region (14 amino acids requiring 42 bp of DNA sequence) is found neither in the germ-line V_H^{141} gene nor in the J_H segment that is used in MOPC141 H chain. It was proposed that the third hypervariable region of heavy chains is encoded in a segment of DNA distinct from V_H and J_H, and the new segment was termed D_H. Of the three hypervariable regions of heavy chains the third shows the most diversity in both sequence and length.

Several D_H segments have now been identified in

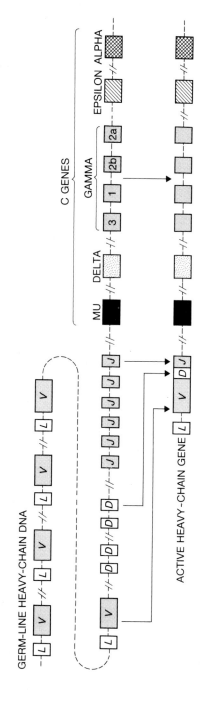

Fig. 7.21 A functional heavy chain gene is formed from four sets of genes, V, D, J and C. Somatic recombination events join V to D and D to J. Transcription and processing generate μ and δ mRNA.

germ-line DNA. The sequence of the D_H segment Q52 shows a 10 bp coding region flanked on both sides by the characteristic heptamer–spacer–nonamer sequence involved in the special recombination of V_L–J_L joining (Figure 7.22). The spacer separating the palindromes is 11 bp long on *both* sides of the D_H coding region. The symmetry of spacer length around D_H allows an explanation of the previous finding that the recognition sequences downstream from V_H and upstream from J_H each have 22 bp spacers (Figure 7.23). Both V_H–D_H and D_H–J_H recombinations conform to the one-turn–two-turn spacer rule. Two joining events could construct a heavy chain V region from V_H–D_H–J_H elements. However, a number of questions remain.

12   D_{Q52} 12

GCAAGGTTTTGACTAAGCGGAGCACCACAGTG CAACTGGGAC CACGGTGACGCGTGGCTCAACAAAAACCCTCTGTTTGG

Fig. 7.22 Nucleotide sequence of a murine D segment and flanking regions with recognition elements used in joining V_H to D and D to J_H. (From Sakano *et al.*, 1981.)

Must one D_H be used or can more than one D_H be joined with one V_H and one J_H? The extremes of length of the third hypervariable region occasion these questions. The joining of two D_H elements to each other should be prevented by the one-turn–two-turn spacer rule. However, some D_H coding regions contain a sequence homologous to the nonameric palindrome, thus allowing the possibility of a shorter D_H element being joined, in accordance with the recognition rules, to a second D_H. The occurrence of such D_H–D_H joining has not been proved. An alternative model has been proposed to account for third hypervariable regions longer

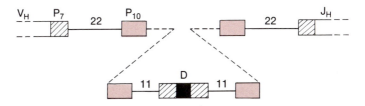

Fig. 7.23 The one-turn/two-turn spacer rule is obeyed in V_H–D and D–J_H joining.

than known D_H elements. This model invokes a step-wise joining process during which free D_H and J_H ends exist and to these ends (3') terminal deoxynucleotidyl transferase could add random deoxyribonucleotides (Figure 7.24). This model explains the existence of sequences, termed N regions, that are found either side of some D regions in functional VDJ genes. An attraction of this model lies in the useful role found for terminal transferase, an enzyme characteristic of thymus and bone marrow.

The use of different cross-over points in D_H-J_H joining can also add to diversity.

We do not know how many D_H genes are in the germ-line or indeed where they are. Like V genes, D_H genes are probably arranged in homologous sets. D_H genes probably lie between V_H and J_H (see Figure 7.21) and certainly one D_H, Q52, lies only 700 bp upstream from J_H1. It is possible that some D_H genes are interspersed with V_H genes and can, therefore, only be joined to V_H genes upstream from them.

The C_H gene cluster

The organization of the murine C_H gene cluster has been elucidated at the DNA level. The exon−intron arrangement of each of the C_H genes, and in most cases the complete sequences, are known. The C_H gene map has been established by cloning overlapping DNA segments. Libraries were made using Charon λ phages, and a walking technique was used to map from one C_H gene to another, starting with C_μ at the 5' end of the cluster. In this way Honjo et al. (1981) mapped about 200 kb, determining the order and spacing of the C_H genes (Figure 7.25). (It should be remembered that classical genetic methods could not map the C_H gene order in mouse since no recombinants have been found in many thousands of meaningful crosses.) Prior to complete mapping, the C_H gene order had been correctly predicted by DNA hybridization kinetic analysis, the interpretation depending upon the hypothesis that C_H genes upstream from the one being expressed in a given cell are deleted from the expressed chromosome.

The map shows that all J_H segments are clustered about 6.5 kb upstream from C_μ. A complete $V_H-D_H-J_H$ unit can be expressed in a μ chain (or a δ chain) by

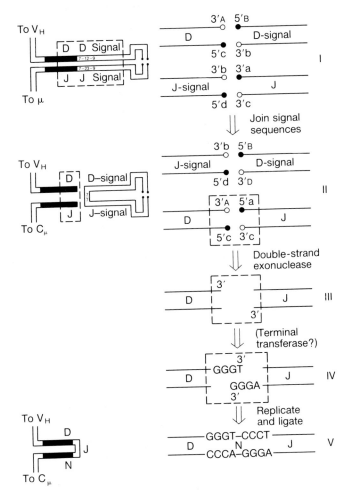

Fig. 7.24 Model for the generation of additional nucleotide sequences (N) at the boundary of joined D and J_H exons. The recognition elements are termed D-signal and J-signal in this model. N sequences are added randomly to free 3' ends of D and J by the action of terminal deoxyribonucleotide transferase. A complementary strand is generated prior to ligation to effect D–J joining. (From Alt & Baltimore, 1982.)

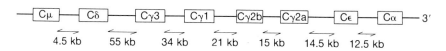

Fig. 7.25 Organization of murine heavy chain constant region genes. This map was constructed by Honjo and co-workers (1981) from clones of genomic DNA using partial overlaps. The J_H cluster lies 6.5 kb upstream from $C\mu$.

transcription through the intron separating J_H and C_μ (and for δ, the intron between C_μ and C_δ). Expression of any $V_H-D_H-J_H$ unit with a C_H gene other than C_μ or C_δ requires a further recombination to put the V region element immediately upstream from the chosen C_H gene. The C_H switch processes are discussed more fully in Chapter 8. The C_H map reveals the presence of S regions (switch regions) upstream from each of the C_H genes except C_δ. Each S region sequence consists of tandem repetitions (of the order of fifty times) of short unit sequences containing common sequences GAGCT and GGGGT. The S regions are sites for general recombination and are involved in the C_H class switch. Many sequences homologous to S exist in the mouse genome other than in the C_H gene cluster, and such sequences are involved in recombinations leading to chromosome translocations (see Chapter 8).

Each domain is encoded by an exon

The idea of independently folded domains within globular proteins stemmed from immunoglobulin chain structures. It was predicted from the repeating sequence homologies and confirmed by three-dimensional structure determination (see Chapter 2). The nucleotide sequences of the C_H genes reveal that each constant region domain is encoded by a separate exon. Each of the exons is bounded by donor (5') and acceptor (3') RNA splice sites that permit the joining of any one exon to its 3' and 5' neighbours, respectively, at the RNA transcript level to generate a contiguous mRNA (see Chapter 8). The hypothesis that each exon corresponds to a functional domain at the protein level has been extended to other proteins, but it is not always as clear as in the immunoglobulin case.

A comparison of the exon−intron arrangement of the C_H genes is shown in Figure 7.26.

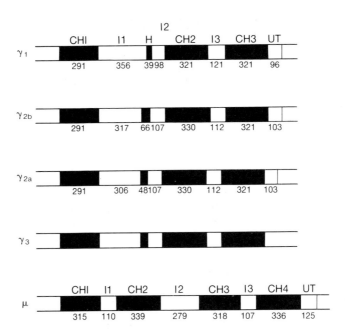

Fig. 7.26 Exon-intron maps of the four murine C_γ genes compared with murine C_μ. (From Honjo *et al.*, 1981.)

The hinge region

The arrangement of the DNA encoding the hinge region differs from one C_H gene to another. In the C_γ genes the hinge region is encoded by a discrete exon located between the C_H1 and C_H2 exons (Figure 7.26). In the C_α gene the hinge region is encoded in a nucleotide sequence contiguous with the 5′ end of the C_H2 exon. The C_μ gene has no H exon but nucleotide sequence comparison suggests that the hinge region in other C_H genes may have evolved from the C_H2 region of a duplicated C_μ gene. The hinge sequence of C_γ genes shows some sequence homology with the 3′ end of the C_H2 exon of the C_μ gene. Also, the C_H2 and C_H3 exons of C_γ genes are more homologous to the C_H3 and C_H4 exons of the C_μ gene than to the C_H2 and C_H3 exons of the latter gene.

The membrane (M) exons

Each class and subclass of immunoglobulin exists in a membrane-bound and a secreted form. The heavy chains of the two forms have distinct C-terminal amino acid sequences and each form is translated from a separate mRNA molecule. The two different mRNA molecules are derived from a transcript of a single C_H gene by using alternate membrane (M) exons. The expression of membrane and secretory heavy chain is described in Chapter 8; the gene structures underlying this process are described here.

Two M exons are downstream from each C_H gene. In each case the membrane domain coding sequence is in the proximal M_1 exon and the intracellular domain of the membrane-bound heavy chain is encoded in the distal M_2 exon, together with the 3'-untranslated portion of the mRNA. The M_1 exon encodes a very hydrophobic 26 residue peptide (highly conserved in sequence between μ and γ) preceded by several acidic amino acid residues. The M_2 exon codes for the intracellular domain that varies in length from 3 amino acid residues on the μ_m chain to 28 residues on γ_m chains (Figure 7.27).

The C-terminal sequences of secreted heavy chains (except δ chains) are encoded in a sequence of nucleotides contiguous with the last domain coding sequence (Figure 7.28). This tailpiece is only 2 codons long for γ_s and ε_s chains but is 20 codons of conserved sequence for μ_s and α_s chains. A consensus donor splice site is located in the coding sequence at the boundary between the terminal domain coding sequence and the tail-

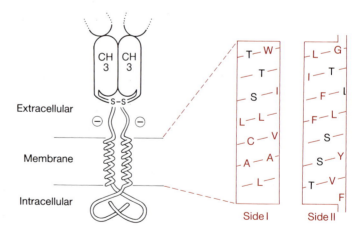

Fig. 7.27 Proposed structure for the transmembrane portion of membrane IgG. The M_1 encoded domains form an α-helix; the opposing sides of this helix are shown here as side I and side II. The M_2 encoded domain is on the cytoplasmic (intracellular) side of the membrane. (From Tyler et al., 1982.)

H chain class	Residue number
	(609)
μ	G K P T L Y N V S L V M S D T A G T C Y
	(477)
α	G K P T H V N V S V V M A G V D G T C Y
	(477)
γ	G (K)
	(609)
ε	G K

Fig. 7.28 Carboxy-terminal amino acid sequences of human immunoglobulin heavy chains. The glycine (G) residue with which each starts is the last residue of the final constant domain of each class of heavy chain. The carboxy-terminal lysine (K) of γ chain is shown in parentheses as it is not always found on the mature polypeptide chain.

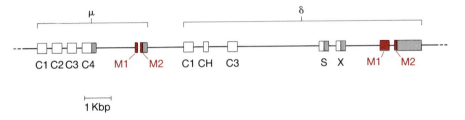

C1 C2 C3 C4   M1   M2   C1 CH   C3   S   X   M1   M2

1 Kbp

Fig. 7.29 Map of the exons encoding C_μ and C_δ genes in the murine germ-line. The positions of M_1 and M_2 for μ and δ and of the δ exon are indicated. (From Tucker *et al.*, 1982.)

piece. The δ_s tailpiece is not homologous to those of α_s and μ_s. The 23 residue tail peptide of δ_s is encoded on an exon located 4.7 kb downstream from the terminal domain sequence and 1.4 kb upstream from δM_1 (see Figure 7.29).

Human immunoglobulin genes

Elucidation of the molecular organization of human immunoglobulin genes followed the molecular mapping of the mouse genome. The major features of the organization of the heavy chain constant region genes resemble those of the murine *Igh* locus (Figure 7.30). In each species the tandem array of C_H genes begins with a cluster of J_H genes close to the C_μ gene and then followed by the C_δ gene. Although there are four C_γ genes encoding the subclasses of γ chain in each species the arrangement of these genes differs relative to the C_α and

Mouse

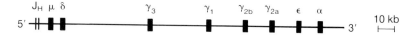

Human

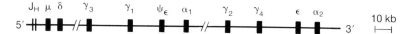

Fig. 7.30 Organization of human (Flanagan & Rabbitts, 1982) and mouse immunoglobulin heavy chain constant (C_H) region genes.

and these are encoded by a $C_\alpha 1$ gene interspersed between the four C_γ genes and a $C_\alpha 2$ gene at the extreme downstream end of the C_H array. The C_ε gene upstream of $C_\alpha 2$ has a counterpart in a pseudo C_ε gene upstream of $C_\alpha 1$. A rational evolutionary explanation of the human C_H array is that a set of C_H genes comprising two C_γ genes, one C_ε and one C_α was duplicated; subsequently the C_γ genes and C_α genes diverged to yield the known subclasses while one of the C_ε genes became a pseudo gene.

Rabbit immunoglobulin genes

Study of the molecular genetics of rabbit immunoglobulin genes should resolve some of the unusual phenomena associated with the *a* and *b* loci (see p. 121 *et seq.*, Chapter 5). Cloning and sequencing of V_H genes differing in allotype (*a* locus) confirm that there are multiple amino acid differences but reveal that many allotype-correlated amino acids are related to each other by single base changes. From available DNA sequence data Mage *et al.* (1984) concluded that the genome of a rabbit of a given V_Ha phenotype contains V_H sequences having unusual combinations of codons at allotype-correlated positions. However, it seems unlikely that all rabbits have a complete set of V_H genes comprising each of the *a* locus allotypes with regulatory control determining the nominal allotypes. It is more likely that latent V_H allotypes are generated by somatic changes, either mutations or gene conversion events, at allotype-correlated positions.

The expression of latent *b* locus allotypes could be due to expression of structural C_κ genes encoding the latent allotype and carried in the genome along with the nominal allotype genes. Rabbits possess multiple C_κ genes, as revealed by Southern blotting, and rabbits of nominal allotypes *b4b5* and *b9* each contain at least two DNA sequences that are highly homologous to a cloned *b4* gene in both coding and flanking regions (Emorine et al., 1983).

References

Alt F.W. & Baltimore D. (1982) Joining of immunoglobulin heavy chain gene segments: implications from a chromosome with evidence of three D—J$_H$ fusions. *Proc. natn. Acad. Sci. USA* **79**, 4118.

Bernard O., Hozumi N. & Tonegawa S. (1978) Sequences of mouse immunoglobulin light chain genes before and after somatic changes. *Cell* **15**, 1133.

Blattner F.R., Williams B.G., Blechl A.E. et al. (1977) Charon phages: safer derivatives of bacteriophage lambda for DNA cloning. *Science* **196**, 161.

Bolivar F., Rodriguez R.L., Greene P.J., Betlach M.C., Heyneker H.L. & Boyer H.W. (1977) Construction and characterization of new cloning vehicles. II A multipurpose cloning system. *Gene* **2**, 95.

Brack C., Hirama M., Lenhard-Schuller R. & Tonegawa S. (1978) A complete immunoglobulin gene is created by somatic recombination. *Cell* **15**, 1.

Emorine L., Drehar K., Kindt T.J. & Max E.E. (1983) Rabbit immunoglobulin κ genes: structure of a germline b4 allotype J—C locus and evidence for several b4-related sequences in the rabbit genome. *Proc. natn. Acad. Sci. USA* **80**, 5709.

Flanagan J.G. & Rabbitts T.H. (1982) Arrangement of human immunoglobulin heavy chain constant region genes implies evolutionary duplication of a segment containing γ, ε and α genes. *Nature* **300**, 709.

Hochtl J., Müller C.R. & Zachau H.G. (1982) Recombined flanks of the variable and joining segments of immunoglobulin genes. *Proc. natn. Acad. Sci. USA* **79**, 1383.

Honjo T., Nakai S., Nishida Y. et al. (1981) Rearrangements of immunoglobulin genes during differentiation and evolution. *Immunol. Rev.* **59**, 33.

Mage R.G., Bernstein K.E., McCartney-Francis N. et al. (1984) The structural and genetic basis for expression of normal and latent V$_H$a allotypes of the rabbit. *Mol. Immunol.* **21**, 1067.

Maxam A.M. & Gilbert W. (1977) A new method for sequencing DNA. *Proc. natn. Acad. Sci. USA* **74**, 560.

Sakano H., Huppi K., Heinrich G. & Tonegawa S. (1979) Sequences at the somatic recombination sites of immunoglobulin light-chain genes. *Nature* **280**, 288.

Sakano H., Kuiosawa Y., Weigart M. & Tonegawa S. (1981) Identification and nucleotide sequence of a diversity DNA segment (D) of immunoglobulin heavy-chain genes. *Nature* **290**, 562.

Sanger F., Nicklen S. & Coulson A.R. (1977) DNA sequencing with chain-terminating inhibitors. *Proc. natn. Acad. Sci. USA* **74**, 5463.

Seidman J.G., Max E.E. & Leder P. (1979) A κ-immunoglobulin gene is formed by site-specific recombination without further somatic mutation. *Nature* **280**, 370.

Southern E.M. (1975) Detection of specific sequences among DNA fragments separated by gel electrophoresis. *J. Mol. Biol.* **98**, 503.

Tucker P.W., Chang H-L., Richards J.E., Mushinski, J.F., Fitzmaurice L. & Blattner F.R. (1982) Genetic aspects of IgD expression: III Functional implications of the sequence and organization of the Cδ gene. *Ann. N. Y. Acad. Sci.* **399**, 26.

Tyler B.M., Cowman A.F., Gerondakis S.D., Adams J.M. & Bernard O. (1982) mRNA for surface immunoglobulin γ chains encodes a highly conserved transmembrane sequence and a 28-residue intracellular domain. *Proc. natn. Acad. Sci. USA* **79**, 2008.

Further reading

Honjo T. (1983) Immunoglobulin genes. *Ann. Rev. Immunol.* **1**, 499.

Chapter 8
Expression of immunoglobulin genes

Introduction

Gene expression in higher organisms involves transcription, modification of each end of the transcript RNA splicing to generate mRNA with a contiguous coding sequence, translation and processing at the polypeptide level. Regulation of expression can occur at any or all of these levels. For immunoglobulin genes the intricacies of expression are greater than for any other known system. The pre-transcriptional step of DNA joining necessary to form a functional light or heavy chain gene is described in Chapter 7. The role of these joining steps in regulating differential immunoglobulin gene expression, e.g. isotype and allotype exclusion, will be dealt with in this chapter.

The finding of split genes, consisting of alternate coding sequences or exons and intervening, non-coding sequences or introns, in higher organisms came as a surprise after the careful elucidation of gene – polypeptide colinearity in bacteria. Many, but not all, genes in higher organisms have an exon – intron structure. For the immunoglobulins and for class I and class II MHC products the exons into which their genes are divided correspond to functional domains. The transcription unit comprises both exons and introns. Transcriptional initiation is regulated by a promoter (a short DNA sequence) upstream from the first exon. The gene can be in a transcriptionally incompetent state, e.g. highly methylated, or in a competent state. Promoters can vary in strength and be subject to other regulating DNA sequences — in particular, enhancer elements.

The initial RNA transcript is processed in three ways (Figure 8.1):

1 Capping — a chemical modification of the 5' end.

2 Polyadenylation — the addition of about 150 adenylate residues at the 3' end.

3 Splicing — the precise removal of intervening sequences with the joining of exons to give a contiguous mRNA.

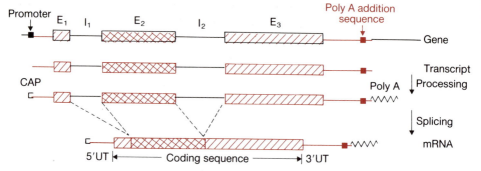

Fig. 8.1 Gene expression. The first step is transcription. The initiation is effected at the promoter and the transcript RNA is synthesized 5′ to 3′ as a continuous molecule comprising exons and introns. Processing of the original transcript commences with capping at the 5′ terminus. The capping terminates RNA with an inverted attachment (3′−5′) of a GTP molecule (see Fig. 8.2). At the 3′ end the poly (A) addition signal sequence is recognized enzymatically and a tail of poly (A) up to 200 residues long is added. Splicing removes the intron sequences (see Fig. 8.2) to generate mature mRNA.

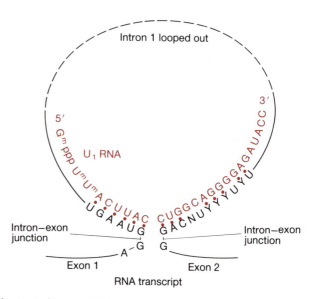

Fig. 8.2 Splicing of RNA transcripts to remove intron sequences involves recognition of intron-exon junctional sequences by a small nuclear RNA molecule U_1. Y designates a pyrimidine and N a variable base in the concensus junctional sequences.

The cap structure and the poly (A) tail can be added prior to splicing. A poly (A) tail is added downstream from the signal sequence AATAAA.

RNA splicing removes introns flanked by a donor sequence at the 5′ end and an acceptor sequence at the 3′ end. These sequences are partially conserved and together show homology with a small nuclear RNA molecule U_1 that is proposed as a recognition unit in the splicing process. The immediate upstream and downstream boundaries of every intron are marked by GT and AG sequences respectively (Figure 8.2).

The immunoglobulin system uses various alternate patterns of gene splicing to generate different mRNA molecules, and ultimately different polypeptides from the same transcription unit. This mechanism allows each heavy chain isotype to occur in two forms, one membrane bound and the other secreted. A similar mechanism allows coexpression of μ and δ chains using the same V region, and coexpression of other C_H genes with μ (and δ) may also use some alternate splicing process.

The sequential expression of C_H genes during clonal development involves a second set of DNA rearrangements, in this case using a more general recombinational mechanism than that used for the $V-D-J$ joining.

All of the above processes are necessary for expression of germ-line genes. The range of V regions expressed is also enhanced by the apparent accumulation of somatic mutations during clonal development.

Transcription of immunoglobulin genes

Competence

The transcriptional competence of DNA can be regulated by the level of methylation of bases. Hypermethylation of DNA causes profound inactivation of transcription. Immunoglobulin V_κ genes have been shown to be transcriptionally silent when in their germline context, even when present in cells expressing other immunoglobulin genes. This transcriptional silence correlates with a high methylation level of the DNA regions containing the V_κ gene family; in this state the V_κ genes are resistant to degradation by DNAase I (transcriptionally competent genes are DNAase I sensitive and this is used as an assay for competence). The other V gene families are probably similarly controlled.

In lymphoid cells C_κ and C_μ genes are transcriptionally competent even in the germ-line (κ°, μ°) state, as evidenced by the presence in the nucleus of B lymphoid cells (both early and late in differentiation state), and even in some T cells, of C_κ and C_μ transcripts. Transcription of germ-line (κ° or μ°) C genes originates at a weak promoter (or pseudopromoter) upstream from the J elements. It is assumed that in non-lymphoid cells C genes are in a non-competent transcriptional state.

C genes appear to become transcriptionally competent at an early stage in lymphoid cell differentiation, prior to V−D−J gene rearrangements. Competence is conferred on a V gene when it is rearranged to form a complete immunoglobulin gene.

DNA sequences regulating transcription

The two important base sequence controls on transcription of immunoglobulin genes are promoters and enhancers. *Promoters* are short DNA sequences that act locally and in *cis* to the gene to be expressed. The primary element of a promoter is a hexameric sequence referred to as a TATA box. A similar but not identical sequence based on TATAAT is located about 30 bp upstream from the 5′ initiation point for the RNA transcript. The TATA box is thought to act as a docking site for RNA polymerase II. A second partially conserved sequence, termed a CAT box is located about 50 bp upstream from the TATA box and also acts in some way as part of the promoter (Figure 8.3). The variation in sequence of the promoter elements may correlate with the efficiency or strength of the promoter (note the weak promoter, referred to above, that allows C_κ expression).

Enhancer, or promoter enhancer, elements were first described in the DNA virus SV40. The prototype enhancer is a 72 bp tandem repeat that increases the rate of transcription by 200-fold. This enhancer acts exclusively *cis* but functions when placed at a considerable distance (up to 3 kb) from the promoter, independent of orientation and either upstream or downstream from the promoter. An enhancer with properties similar to those of viral enhancers, and with some sequence homology, is located in the intron separating J_H and C_μ. Enhancer activity is located in a 140 bp region that lies more than 1 kb upstream from the switch sites (S_μ);

Fig. 8.3 The promoter region involved in initiation of gene transcription consists of two elements termed 'CAT' and 'TATA' boxes upstream from the initiation codon AUG.

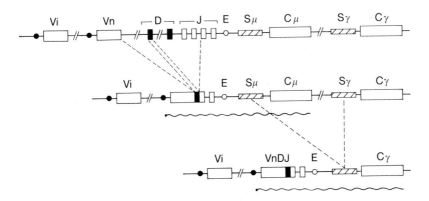

Fig. 8.4 The position of the enhancer, E (○), relative to the switch sequences S_μ (see Fig. 8.11) in the intron separating J_H and C_μ The promoter is shown by ●. Note that after a switch to γ chain expression the function of the enhancer is preserved. (From Gillies *et al.*, 1983.)

thus, the enhancer is retained after class switching and it is probable that the same enhancer is used for the expression of all heavy chain genes (Figure 8.4).

The immunoglobulin enhancer, by contrast with the viral enhancer, is tissue specific, acting very effectively in plasma cells but not at all in fibroblasts. In view of the low-level expression of immunoglobulin genes in pre-B and B lymphocytes relative to the high-rate expression in plasma cells, it is possible that the immunoglobulin enhancer is specific for a particular differentiation stage of B cells.

The mode of action of enhancers is not yet proven. They could act by direct perturbation of DNA tertiary structure or they could serve as binding sites for specific regulatory proteins. The latter mechanism would best explain tissue or differentiation state specificity. A small number of specific enhancers with complementary binding proteins could function to coordinately express sets of genes involved in determining differentiated states.

Immunoglobulin gene expression as a function of differentiation

The expression of immunoglobulin genes is regulated in relation to the differentiation state of lymphoid cells. The following sections discuss the progressive stages in immunoglobulin gene expression during B-cell differentiation. The important topics of allelic exclusion and somatic mutation in expressed immunoglobulin genes are dealt with in more detail at the end of this chapter.

Expression of μ chains in pre-B cells

The first phenotypic change that determines B-lymphocyte differentiation is the expression of intracellular μ chains in pre-B cells. Stem cells that could give rise to either B or T lymphocytes appear to be irrevocably committed to one or the other of these differentiation pathways by the functional rearrangement of $V_H - D - J_H C_\mu$ to yield a pre-B cell, or, by analogy, functional rearrangement of $V - D - JC$ of the T cell receptor gene family (see Chapter 13) to yield a pre-T cell. In view of the similarity of the palindromic sequences involved in immunoglobulin $V - D - J$ recombinations with those involved in T-cell receptor $V - D - J$ recombinations, it is possible that a single enzymic mechanism is involved in the two processes; therefore, determination of B- and T-lymphocyte differentiation from the common stem cell may be a competitive process.

In the formation of a functional μ chain three types of recombination could occur primarily: $V_H - D$, $D - D$ and $D - J_H$. No direct evidence for $D - D$ joining has yet been found; neither have $V - D$ joining events been detected in a wide variety of B-lymphoid cell lines. The evidence points to $D - J$ as being the first joining event to occur en route to a functional μ chain gene in a pre-B cell. Transformed cell lines derived from murine fetal liver or bone marrow exhibit genotypes with rearrangement patterns: DJ/O, DJ/DJ, VDJ^-/DJ, VDJ^+/DJ, VDJ^+/ VDJ^- and VDJ^-/VDJ^- (where $^+$ and $^-$ indicate functional and nonfunctional joining; see p. 201 under 'Allelic exclusion'). The only two genomic patterns seen in plasmacytoma cells are: VDJ^+/DJ and VDJ^+/ VDJ^-. This evidence is consistent with the pathway of early B-cell differentiation shown in Figure 8.5.

Analysis of the genomes of T-cell lines reveals recombined DJ elements but no recombined VDJ elements. The conclusion to be drawn is that DJ joining precedes the determination of a stem cell towards a B- or a T-cell lineage. The correct recombinant joining of V_H to DJ_H allows μ chain expression and is proposed to be the key step determining the pre-B cell differentiated state. Presumably a similar sequence of events at the T-cell re-

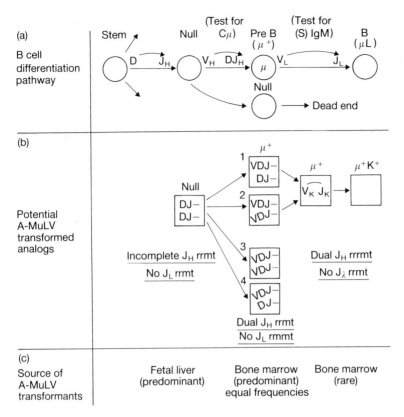

Fig. 8.5 The pathway of B-lymphocyte differentiation involves an ordered rearrangement of V_H and D_H elements (a). (b) shows the alternative rearrangements that can occur in the pathway D_H to J_H followed by V_H to $D_H J_H$. Functionally rearranged genes are written VDJ and aberrantly rearranged genes are written by the skewed VDJ letters. (From Alt *et al.*, 1984.)

ceptor locus (β chain?) determines progression of a stem cell to a pre-T cell. It can be predicted on this hypothesis that DJ_T elements will be found in B-lymphoid cell DNA.

Prior to any recombinant events, C_μ and C_κ are transcriptionally active (see p. 189). After V−D−J recombination the enhancer located between J and C_μ can activate the promoter of the rearranged V gene segment (Figure 8.6). Thus, only that one rearranged V gene will be transcriptionally active, while the hundreds of non-rearranged V genes remain non-transcribed.

Light chain expression: pre-B to B cell transition

Rearrangement of light chain V genes occurs in pre-B

193   GENE EXPRESSION

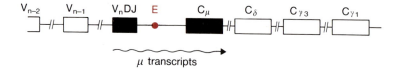

Fig. 8.6 The enhancer (●) activates the promoter of the single V gene that has been somatically recombined with a DJ element.

cells. Functional joining of a V_L gene to a J_L segment allows expression of a light chain in a cell already producing μ chains. With the first appearance of $L_2\mu_2$ molecules and their localization at the cell surface (see below), the pre-B cell becomes an early B cell.

Choice of C_L

Analysis of the genomic patterns of κ-producing B cells with κ probes reveals VJ^+/O and VJ^+/VJ^- but no VJ^+/VJ^+ combinations. When human λ-producing B-cell DNA is analysed with a κ probe neither of the two germ-line copies of C_κ can be detected. By contrast, in κ-producing B cells a germ-line pattern of λ genes can be revealed.

The hypothesis has been advanced that there is a strongly preferential order in which gene rearrangements of light chain loci can be attempted in pre-B cells. The proposed order is shown in Figure 8.7. Rearrangement occurs on one chromosome at a time. Joining at the κ locus precedes joining at the λ locus. The sequence only progresses if previous attempts at VJ joining have been non-productive. Once a productive VJ element is created and light chains are expressed, no further rearrangement events occur at light chain loci.

Choice of V_L

The early B cells, resulting from a single pre-B cell and making indistinguishable μ chains, express physico-chemically distinguishable κ chains and have distinct κ gene rearrangements (Figure 8.8). Within a single pre-B cell clone individual cells can select different V_κ genes for rearrangement. Neither the choice of V_κ set, nor that of the particular V_κ gene, is uniquely determined by the V_H being expressed in the pre-B cell. Moreover, even

Pre-B cells

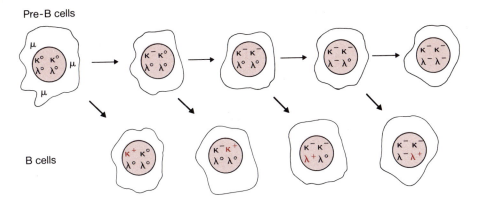

Fig. 8.7 Sequential light chain rearrangements that can occur during pre-B-cell differentiation. In the pre-B cell expressing μ chains, light chain rearrangements begin at one κ locus. Successful rearrangement (κ⁺) generates a B cell synthesizing a μκ product. Non-productive rearrangement events allow further attempts at rearrangement at the second κ locus and failing that each of the λ loci are rearranged. (From Korsmeyer & Waldman, 1984.)

when a particular V_κ gene is functionally rearranged, the κ chains expressed by different B cells may be distinguished electrophoretically. The latter differences may be due to somatic mutation (see section below).

Combinatorial diversity, resulting from the possible combinations of each V_H with each V_L, is theoretically an important element in the generation of antibody diversity (see Chapter 6). The finding that a random selection of V_L regions may be tried with each V_H region within the progeny of a single pre-B cell is practical demonstration that combinatorial diversity occurs during B-cell differentation.

Regulation of μ_m and μ_s expression

Production of membrane and secretory forms of μ chain (μ_m and μ_s respectively) is controlled at the level of processing of the primary transcript of the μ chain gene. Synthesis of μ_m and μ_s occurs by translation of specific μ_m-mRNA and μ_s-mRNA. Generation of these distinct mRNA molecules requires specific processing of the μ gene transcript to include either the twenty codons of the secretory T piece or to splice out the T-piece codons and include the codons of the two M exons (see Chapter 7). The choice of which mRNA is to be expressed could be made by termination of transcription either before or after the M exons. Alternatively, processing of the μ gene transcript could be regulated by the frequency

195 GENE EXPRESSION

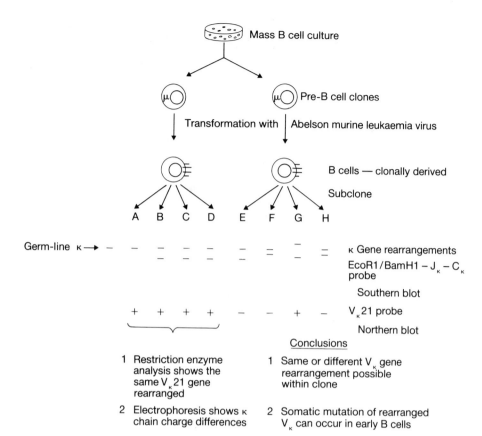

Fig. 8.8 The diversity of κ chain rearrangement within the clonal progeny of a pre-B cell can be demonstrated by culturing pre-B cell clones *in vitro*. The B cell progeny are transformed by Abelson murine leukaemia virus to allow a sufficient number of cells to be grown to yield DNA for Southern blot analysis, RNA for Northern blot analysis and radio-labelled light chain for electrophoretic analysis. The results and the conclusions drawn are from Ziegler *et al.* (1984).

of polyadenylaton at sites I and II (Figure 8.9). Polyadenylation precedes removal of introns. Sequences (AATAAA) specifying polyadenylation occur 3′ to the $C_\mu 4$ exon (site I) and 3′ to the M_2 exon (site II). Polyadenylation at site I precludes processing to generate μ_m-mRNA. The use of site II for polyadenylation could allow splicing to yield either μ_m- or μ_s-mRNA, but use of site II may favour splicing to yield μ_m-mRNA.

Expression of μ_m- and μ_s-mRNA is developmentally regulated. Pre-B cells express both μ_m and μ_s. Early B lymphocytes synthesize predominantly μ_m while plasma cells express mainly μ_s.

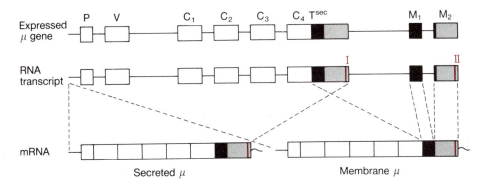

Fig. 8.9 The choice of coding segments to give μ_s or μ_m is made at gene transcription, RNA processing and splicing. Differential use of the polyadenylation signals I and II (red line) regulates the levels of μ_s- and μ_m-mRNA. ▢ 3′ untranslated sequence; ▭ alternative tail sequences T^{sec} or M_1 and M_2; ~ poly (A).

Simultaneous expression of μ and δ genes

IgM and IgD molecules of identical idiotypes are co-expressed on the surface of single B lymphocytes at a certain stage in B-cell development. The idiotypic identity is due to the use in the IgM and IgD molecules of identical V_L and V_H regions; this indicates that the same V_H gene is expressed in conjunction with both C_μ and C_δ genes in an individual cell. Separate mRNA molecules encoding μ and δ chains with identical V_H regions are generated by alternative processing of a transcript of the entire $V_H - C_\mu - C_\delta$ region (Figure 8.10). The mechanism for generation of μ_m- and μ_s-mRNA should also generate δ_m-mRNA (and presumably δ_s-mRNA although secreted IgD is rare). Various mRNA sizes are detected in cells synthesizing both μ and δ chains. However, polyadenylated RNA long enough to contain complete $VDJ - C_\mu - C_\delta$ transcripts have not been found; this is either due to technical deficiencies or to the fact that C_μ gene sequences are spliced out during transcription. Such co-transcriptional splicing has been invoked to explain simultaneous production of another isotype (i.e. γ, α or ϵ) and μ (see below).

Heavy chain isotype switching

The clonal progeny of a single B lymphocyte each express the original idiotype and antigen specificity, but cells of subclones switch from IgM/IgD expression

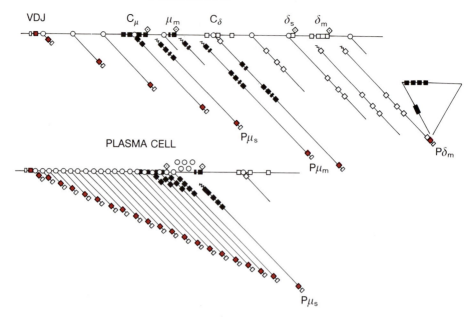

Fig. 8.10 Regulation of expression of the μ–δ locus in B lymphocytes producing membrane IgM and IgD and plasma cells producing and secreting IgM. In B lymphocytes transcription (by RNA polymerase, ○) is continuous through C_μ (■) and Cδ (□) exons. Differential cleavage at poly (A) addition sites (see Fig. 8.9) regulates the levels of μ_s-, μ_m- and δ_m-mRNA. Excision of C_μ exons from transcripts running through δ_m may occur prior to completion of transcription (as indicated by the optional insert shown as a triangle with the $P\delta_m$ transcript). In IgM-secreting plasma cells transcription is initiated more frequently, as indicated by the density of polymerases and is usually terminated at the poly (A) site I (see Fig. 8.9) to yield μ_s-mRNA.

Note that run-on transcripts can be formed after cleavage of the RNA molecule to yield μ_s- or μ_m-mRNA; the run-on molecules lack V regions and are degraded in the nucleus. (From Mather *et al.*, 1984.)

to the expression of another class of immunoglobulin. This switch in expression from one isotype to another involves generation of an mRNA molecule in which the active V−D−J element is joined to a different set of C_H exons. For the coexpression of μ and δ chain mRNA molecules alternate transcription is used. A similar mechanism may explain other coexpression. For all other class switching DNA recombinant events take place sequentially (Figure 8.11).

The isotype switch event is distinct from V−D−J recombination processes; the latter are precise events demanding recognition of heptamer−spacer−nonamer sequences found only in immunoglobulin and T-cell receptor gene families. Switch recombination occurs between two S regions, these being located in the inter-

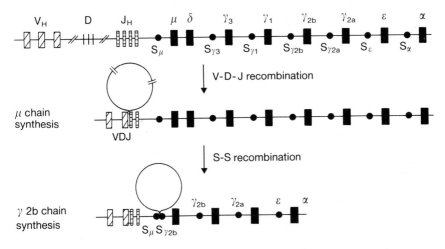

Fig. 8.11 Two different types of rearrangement are necessary to form an expressed γ, ε or α gene. V−D−J recombination allows expression of μ and δ genes. Subsequent S−S recombinant events are necessary to express the other downstream C_H genes, e.g. the synthesis of $γ_1$ as indicated.

genic DNA upstream from each C_H gene (and in the case of $C_μ$ the S region lies between the J cluster and $C_μ$). Each S region consists of multiple (50−100) tandem repeats of short nucleotide sequences sharing dispersed TGAGC and TGGGG runs. The combination of tandem repeats and short common sequences provides a large number of possible switch recombination sites, thus increasing the probability of S−S recombination. The importance of these sequences in S−S recombination is shown by the prevalence of TGAG and TGGG tetramers surrounding the S−S joining sites in plasmacytoma genomes. Class-specific recombinases have been postulated but it is possible that switching is a general homologous recombination process. The latter hypothesis is supported by evidence showing that S−S recombination can be catalysed by an *E. coli* extract (Figure 8.12). The experiment involved S−S recombination between a λ phage carrying the $S_μ−C_μ$ insert and another λ phage carrying the $S_α−C_α$ insert. The phages were designed with different markers on their right and left arms so that S−S recombinants could be selected by growth in appropriate host bacteria. Recombinants catalysed by the *E. coli* enzyme have the tetramers TGAG, TGGG and TGGT surrounding the S−S joining

199 GENE EXPRESSION

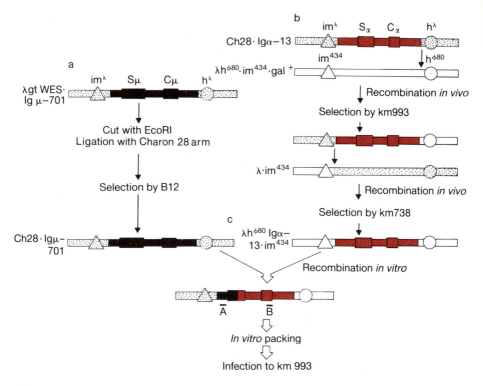

Fig. 8.12 Assay system for S−S recombination *in vitro*. (a) Formation of Ch28.Igμ−701. λgt.WES.Igμ-701 (1) was cleaved with EcoRI and ligated in the presence of Charon 28 arms. The resultant phage DNAs were packaged *in vitro* and permitted to infect *E. coli* B12(sup0). One of the phages grown in B12 was isolated and tested for the presence of the intact 13 kb EcoRI fragment by restriction cleavage. (b) Formation of λh^{ø80}. Igα-13.imm^{434}. Genetic markers h^{ø80} and imm^{434} were introduced into the phage arms of Ch28.Igα-13 by two steps of crosses. (c) Recombination assay *in vitro*. Ch28.Igμ-701 and λh^{ø80}. Igα-13.imm^{434} DNAs are mixed, treated with extracts from B lymphocytes, and packed *in vitro* into phage coats. Recombinants bearing the h^{ø80} and imm^λ markers are then selected. Fragments A and B are the 0.8 kb HindIII fragment of Igμ-701 and the cloned α chain cDNA (pABα-1), respectively. Imm^λ, △, and imm^{434}, △, immunity to wild-type λ phage and to lambdoid phage 434, respectively. (From Kataoka *et al.*, 1983.)

sites. The Chi sequence responsible for stimulating generalized recombination is GCTGGTGG; also S region like sequences have been found in *Drosophila* DNA. These findings suggest a primitive origin for S−S recombination systems.

Switch recombination usually results in a deletion of intervening DNA in a *cis* joining of VDJ to a downstream C$_H$ gene. A simple looping-out model can explain such S−S events. Deletion of C$_H$ genes is clearly the general method of class switching. (Occasionally, chromosomes that have undergone S−S recombination have C$_H$ genes out of their germ-line linear order and retained when they would have been deleted in a looping-out model. It is possible that unequal sister chromatid exchange is responsible for such rare rearrangements.)

Simultaneous expression of μ genes with other C_H genes

The deletion model for isotype switching implies that, with the exception of μ/δ coexpression, simultaneous synthesis of two heavy chain classes on a single cell should not occur. However, cells bearing surface μ together with one other isotype (γ, α or ε) have been observed. Such cells (1) simultaneously synthesize distinct mRNAs for the two isotypes, each expressing an identical V region; and (2) have an expressed heavy chain locus that has not sustained a deletion of any C_H exons.

Read-through transcription, such as has been postulated to explain coexpression of μ and δ chains, would require an initial RNA transcript of about 180 kb to include both μ and ε genes. Splicing of the transcript during the transcription process could generate shorter RNA molecules with VDJ joined to any C_H gene sequence.

The significance of coexpression of μ with either γ, α or ε prior to S−S rearrangement is yet to be established. A role for T cells in directing, or selecting for, particular class switches has been postulated but is unproven.

Allelic exclusion

Each plasma cell synthesizes a single type of antibody molecule, i.e. all molecules produced by a given cell are identical with respect to antibody specificity (controlled by V_H, D, J_H, V_L and J_L gene segments), type of light chain (determined by C_κ or C_λ genes) and class or subclass of heavy chain (determined by the C_H genes expressed at the Igh locus). The limitation to a single type of light chain and a single class of heavy chain per cell is termed isotype exclusion.

In individuals heterozygous at immunoglobulin loci each cell expresses only one of the two alleles. This phenomenon is termed allelic exclusion. In serum both allelic products (allotypes) are found (though not necessarily in equal amounts) without any mixed molecules being detected. This mosaic phenotype exhibited by antibody-producing cells has its only comparable counterpart in X-chromosome inactivation seen in somatic cells of female mammals; the female somatic cell with one of the two X chromosomes (maternal and paternal) expressed is, therefore, functionally equivalent in gene dosage to the male cell, which has only a maternal X chromosome. The mechanisms maintaining X chromosome inactivation and allelic exclusion are probably not similar. The current models for allelic exclusion are discussed below.

Rearrangement on a single chromosome

The simplest mechanism which may be invoked to explain allelic exclusion at the DNA level is to propose that for the chromosome bearing the allele to be expressed, the V_L and J_L or V_H, D_H and J_H elements are somatically recombined to give a rearranged V gene configuration while the chromosome carrying the allele to be excluded remains in the unrearranged, germ-line configuration. Clearly, the rearranged allele can be expressed while the allele remaining in the germ-line configuration cannot. Examples of such events are known but seem not to be the general rule, particularly at the *Igh* locus.

Non-productive rearrangement

Three types of non-productive gene rearrangements have been described for V_L–J_L or V_H–D_H–J_H joining: (1) frameshift rearrangements; (2) incomplete rearrangements; and (3) null rearrangements to a pseudo joining site.

1 DNA rearrangement requires recognition of precise nucleotide sequences, but despite the highly conserved nature of these sequences the somatic recombination event may occur at various points in the recognition sequence. Several examples of frameshift rearrangements have been observed. A frameshift rearrangement mechanism can be invoked to explain the amino acid sequences of BALB/C κ myeloma proteins which cannot be accounted for in terms of known V_κ and J_κ nucleotide sequences. In the case of amino acid 96, which is known to be hypervariable in BALB/C κ proteins, frame-shift rearrangements increase the number of amino acids available at this position (Figure 8.13) and lead to an increase in antibody diversity.

2 In the heavy chain gene system, two distinct recombination events, D_H-J_H and $V_H-D_HJ_H$, are required to generate a functional V_H gene. In some instances only the first of the two DNA joining events, D_H-J_H, will occur on one chromosome and this will be preserved as an incomplete rearrangement (see p. 192).

3 Null rearrangements involve joining of an exon to a pseudo joining site not located adjacent to a J exon (Figure 8.14). In the murine κ chromosome a sequence CACAGTG homologous to the palindromic joining sequence exists between $J_κ$ and $C_κ$. Aberrant rearrangement of $V_κ$ to this pseudo joining site deletes the $J_κ$ elements and leaves the $V_κ$ exon with no RNA splice site 3' to the $V_κ$ sequences; RNA splicing can occur only between the $C_κ$ exon and the hydrophobic leader sequence exon to yield an mRNA containing a leader and $C_κ$ sequences but no $V_κ$ or $J_κ$ sequences.

```
91  92  93  94  95
TyrAlaSerSerPro  - - - - - - -
TATGCTAGTTCTCCTCCCACAGTGATA          Embryonic  VK41
| | | | | | | | | | | | | | |      | |    |
TATGCTAGTTCTCCGTGGACGTTCGGT          Recombined  MOPC41
                | | | | | | | | | | |
GTACTACCACTGTGGTGGACGTTCGGT          Embryonic  J–C
        - - - - - - -  TrpThrPheGly
                       96  97  98  99
                       (J5)
```

```
91  92  93  94  95
TyrAlaSerSerPro  - - - - - - -
TATGCTAGTTCTCCTCCCACAGTGATA          Embryonic  VKM173B
| | | | | | | | | | | | | | |  | |            |
TATGCTAGTTCTCCCCCACGTTGGGAG          Recombined  MOPC173B
| |    | | |          | | | | | | | | | | |
GACACCAGTGTGTGTACACGTTCGGAG          Embryonic  J–C
        - - - - - - -  TyrThrPheGlyG
                       96  97  98  99
                       (J4)
```

Fig. 8.13 Comparison between recombined and embryonic κ genes at the V−J recombination site for MOPC41 and MOPC173B murine myeloma proteins. The embryonic (germ-line) genes have been aligned to illustrate homology with the recombined (expressed) genes. Note the variation of the junction by one nucleotide in MOPC173B; this causes a frame shift in the J sequence leading to mis-sense and termination of translation. (From Max *et al.*, 1980.)

Non-productive rearrangements of types 2 and 3 are very common. In murine plasma cells unexpressed κ chromosomes are non-productively rearranged in about 30% of cases and for unexpressed heavy chain chromosomes the proportion of non-productive rearrangements is 80−100%.

Models for allelic exclusion

The observation that many, though not necessarily all, B lymphocytes exhibit non-productive rearrangements is consistent with a stochastic model for allelic exclusion. The simple model makes three principal predictions:

1 A non-productive rearrangement on one chromosome does not exclude an attempt to form a productive rearrangement on the second chromosome.
2 Multiple rearrangements can occur on a single chromosome.
3 Some cells with two functionally rearranged chromosomes should exist (though possibly at low frequency) in the population.

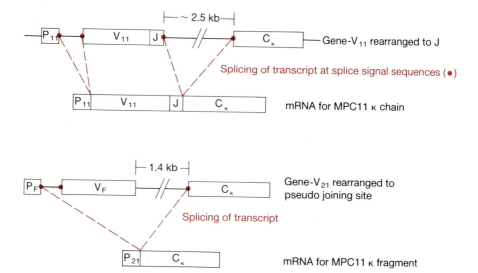

Fig. 8.14 Structures of the two rearranged genes encoding a complete κ chain and a fragment κ chain in MPC11 plasmacytoma cells. The complete MPC11 κ chain is expressed from the V_{11} gene rearranged normally to a $J_κ$. The fragment κ chain consists of a $C_κ$ region with a hydrophobic signal sequence P_F (identical to P_{21} of the MOPC21 κ chain). The gene encoding P_F–$C_κ$ contains a V_F exon but no $J_κ$ exon; consequently there is no V_F–$C_κ$ splicing site and splicing joins P_F to $C_κ$ eliminating V_F. The rearrangement of V_F appears to have occurred by recognition of a pseudo joining site between the J cluster and $C_κ$.

We can represent this model for the κ chromosome by calling the germ-line configuration k°, a productive rearrangement k⁺, and a non-productive rearrangement k⁻ (Figure 8.15).

Allelic exclusion may not be a totally random process as predicted by the stochastic model. The failure to find cells corresponding to the k⁺/k⁺ type raises the possibility that active suppression of further gene rearrangement may occur once a functional gene is formed.

Oncogene translocation to Ig loci

Study of karyotypes of tumour cells has shown an association between particular malignancies and specific chromosomal translocations. These changes are described by the shorthand t(n;m), where t denotes a translocation and n and m are the numbers designating the two chromosomes involved in the genetic exchange. Thus t(9;22) is an exchange of chromatin between chromosomes 9 and 22, and this translocation is associated with chronic myelogenous leukaemia. Translocations t(2;8), t(8;14) and t(8;22) are each associated with Burkitt's lymphoma (see Chapter 15).

Oncogenes were intially observed as transforming genes carried by retroviruses and being the means by which such viruses could convert a normal cell to the malignant state. Immunological methods were used to identify the protein product of the *src* oncogene of the Rous sarcoma virus. Antibodies against *src* protein were raised by inducing tumours in rabbits with the Rous sarcoma virus. These antibodies were shown to combine specifically with the *src* gene product synthesized either in a cell-free translation system or expressed in a transformed cell.

The presence of cellular homologues of oncogenes was discovered by the use of gene probes. These cellular, or c-, oncogenes (contrast the viral or v-oncogenes)

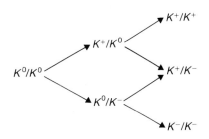

Fig. 8.15 A probabilistic or stochastic model for the generation of κ chain expressing genotypes.

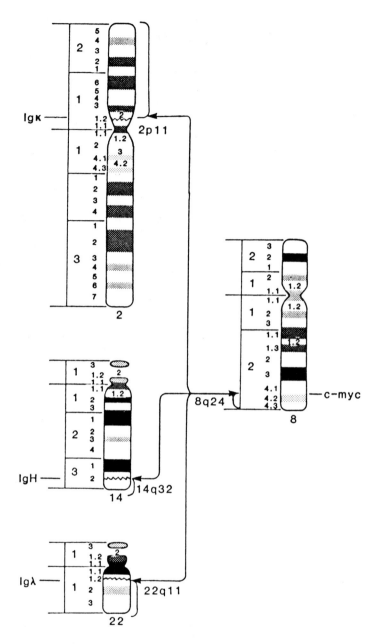

Fig. 8.16 Schematic representation of the human chromosomes involved in specific translocations found in Burkitt lymphoma cells. The shaded banding corresponds to the characteristic Giemsa banding patterns observed in the light microscope. The positions of the κ locus on chromosome 2, the heavy chain locus on chromosome 14, the λ locus on chromosome 22 and the c-myc gene on chromosome 8 are indicated. Arrows point to breakpoints at which chromosomes 2, 14 or 22 reciprocally exchange fragments of chromosome with chromosome 8. (From Leder *et al.*, 1983.)

appear to be expressed in some normal cells and presumably have some role, as yet to be defined, in growth
or differentiation. However, in the transformed cell the
expression of a c-oncogene is enhanced and it can be
shown in some cases that a functional product appears
to be involved in the maintenance of the transformed
state. One of the ways of amplifying c-oncogene expression is by its translocation from one chromosomal location
to another. In Burkitt's lymphoma this translocation
process has been shown to be linked to the rearrangement of immunoglobulin genes. Chromosome 8 is the
normal location of the cellular oncogene c-myc and
chromosomes 2, 14 and 22 are the locations of the Igk,
Igh and Igl loci respectively (see Chapter 1, Table 1.1).
The breakpoints involved in each type of reciprocal
translocation can be identified by observing the Giemsa
banding patterns of normal and rearranged chromosomes. In each case the breakpoints correspond to the
c-myc site in chromosome 8 and to one of the Ig loci in
each of chromosomes 2, 14 or 22 (Figure 8.16).

One such translocation is found in each Burkitt's
lymphoma but the predominant finding (i.e. in about
75% of cases) is a translocation involving chromosomes
8 and 14. In two such cases, where the c-myc–Igh
region of the new chromosome has been cloned and
sequenced, the joining point for the translocated c-myc
gene has been found to be the μ switch region, normally
involved in the recombination events that allow the
switch of class expression in immunoglobulins (see

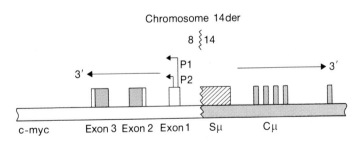

Chromosome 14der

8 ⟨ 14

Fig. 8.17 Map of the arrangement of c-myc and Ig-C_μ genes formed by
translocation between chromosomes 8 and 14 in the Burkitt
lymphoma cell line BL22. Solid boxes indicate exons encoding amino
acid sequences of c-myc or C_μ. The hatched box, S_μ, is the μ switch
region. The arrows indicate the direction of transcription of each
gene. The two c-myc promoters are shown by P1 and P2, with P1
being 5′ from P2. (From Leder et al., 1983.)

p. 197). The structure of exons and introns in the trans-located chromosome of one Burkitt line (BL22) is shown in Figure 8.17.

The c-myc and Igh genes are in opposite transcriptional orientations. Although the reciprocal translocated chromosome region has not been cloned it is clear that the Igh promoter and enhancer sequences have been lost from the new chromsome carrying c-myc. Thus, it is not obvious why expression of c-myc should be greatly increased, as it is, from the translocated allele in the Burkitt cell relative to expression of the normal allele. This question and many others, in particular why increased expression of oncogenes leads to the transformed state, remain to be answered. Immuno-genetic methods will have a key role to play in eluci-dating the function of oncogenes.

Somatic mutations of immunoglobulin genes

An enhanced frequency of, or selection for, somatic mutations in V genes was an early hypothesis to account for the extent of antibody diversity. The pattern of amino acid sequence diversity in murine V_λ regions (see Chapter 6, Figure 6.12) provided the first evidence in favour of somatic mutations as a source of diversi-fication. Comparison of the DNA sequences of V_λ in the mouse germ-line and in various plasma cells can be taken as virtual proof that a single inherited V_λ gene undergoes somatic mutations resulting in expressed λ chain variants.

The major questions concerning somatic mutation of antibody genes can be summarized as When? Where? How? and What?

When do somatic mutations occur in antibody genes?

They accumulate sequentially due to a process opera-ting throughout the proliferation and differentiation of B lymphocytes. Evidence comes from comparison of the V_H and V_κ region sequences of a set of monoclonal antibodies and myeloma proteins that recognize a com-mon hapten and utilize variants of a given germ-line V_H or V_κ gene. In a series of somatic variants of a single V_H gene, those expressed in conjunction with C_γ or C_α regions had a high concentration of somatic mutations. This finding led to the hypothesis that somatic mutation is an event related in time or mechanism to the isotype switch. The occurrence of mutations in V_κ genes of the same cells showed that if there is a switch related mechanism it can act *trans* as well as *cis*. The finding that mutations occur in V_κ regions at the time of their initial expression in pre-B cells differentiating into B cells (see Figure 8.8) is not consistent with a switch related mechanism.

In a given clone the extent of somatic mutation in the V_H region is comparable to the extent in the V_κ region that is coexpressed. The most satisfactory explanation is that a common mechanism of somatic mutation operates throughout the life of B lymphocytes with sequential accumulation of mutations in V_H and V_L (Figure 8.18).

Where are the sites of somatic mutation in B cells?

Somatic mutations occur locally within and surrounding the rearranged V gene. In the series of anti-phosphorylcholine antibodies based on the T15 V_H gene (see Chapter 6) there is a clustered pattern of somatic mutations in a span of approximately 1 kb around the $V_\kappa J_\kappa$ or $V_H D J_H$ genes (Figure 8.19).

The distribution of mutations within the expressed V region gene shows some clustering within the second

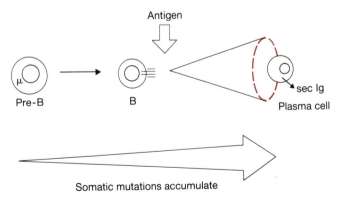

Fig. 8.18 Somatic mutations are preferentially accumulated in rearranged Ig-V genes during the proliferative life span of B lymphocytes. The process introducing somatic mutants remains to be defined. Somatic mutations can be introduced at the pre-B cell level and the process continues during the process of B lymphocyte clonal expansion, a potential of up to 80 cell divisions.

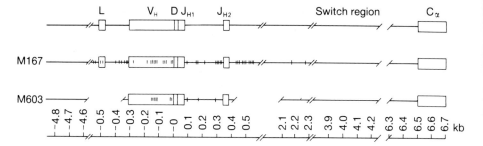

Fig. 8.19 The clustered location of mutations (vertical lines) introduced somatically into the T15 V_H genes of the murine plasmacytomas M167 and M603. Mutations comprise nucleotide substitutions, insertions and deletions. The M167 V_H gene has 44 mutations in 1054 nucleotides analysed (3.8% variation) and M603 V_H has 10 mutations in 714 nucleotides (1.4% variation). Sequencing upstream and downstream from the rearranged V_H shows that variation is concentrated in and around the rearranged V_H gene. (From Kim *et al.*, 1981.)

hypervariable region (Figure 8.20). This could be due to the process of selection either for or against variants.

How are somatic mutations introduced into V genes?

A site-specific mechanism of error-prone DNA replication or repair is the process that best explains the data. Investigation of the mechanism of somatic mutation in V genes promises to be an exciting field. The two major facts to be explained are the high frequency of mutation, estimated at 10^{-3} per base pair per cell generation, and the peculiar clustering of mutations around and within rearranged V genes.

What is the contribution of somatic mutation to antibody diversity?

Given the frequency of mutation in V genes, the large majority of antibodies expressed in any response will be variants of 'germ-line antibodies'. An initial selection of mutants will eliminate those not able to form viable antibody molecules. Antigenic selection can operate on all of those variants that are expressed as receptor antibodies at the B cell surface. The simplest hypothesis claims that antigen will select the highest affinity antibodies from amongst the variants. Evidence for this hypothesis has yet to be found. In the anti-phos-

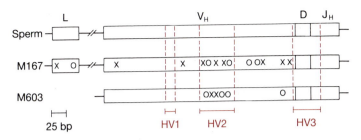

Fig. 8.20 Distribution of silent (O) and amino acid replacement (×) mutations in the coding regions of M167 and M603 V_H genes. Note that somatic mutations cannot be recognized in D until the sequence of germ-line D is known. (From Kim *et al.*, 1981.)

phorylcholine antibody series none of the antibodies using variants of the T15 V_H gene has a significantly higher affinity for the hapten than those antibodies using the germ-line T15 gene.

It is possible that the importance of somatic mutations varies according to the nature of the antigen. It could be that somatic mutations were more important at an earlier stage of evolution of the immune system, or that they are a more recent addition to the system. Evidence is accumulating to show that somatic mutations are used infrequently, if at all, in the V genes of the T cell antigen receptor. The T cell antigen recognition system probably predates the antibody system in evolution, so one concludes that the somatic mutational mechanism is a more recent addition for the diversification of antibody V genes.

Origin of antibody diversity

The origin of the extensive diversity of antibodies described in Chapter 6 does not lie in a single Generator of Diversity (GOD); this once fashionable hypothesis is now replaced by a multifunctional origin for antibody diversity. The nature of these factors has been revealed by molecular genetics though many mechanisms are as yet ill-understood.

Germ-line genes

A multiplicity of V, D and J genetic elements are inherited. In the mouse the number of germ-line genes is estimated as follows:

Kappa	~250 V_κ
	4 J_κ
Lambda	3 V_λ
	3 J_λ
Heavy	~250 V_H
	~10 D_H
	4 J_H

Combinatorial jonining

Any V_κ can join with any J_κ. Thus 250 V_κ × 4 J_κ → 1000 V_κ regions. By contrast each V_λ can be joined only with its own J_λ. At the heavy chain locus any V_H-D-J_H combination can be joined to yield a V_H region gene. Thus 250 V_H × 10 D × 4 J_H → 10 000 V_H regions.

Junctional diversity

This arises because of flexibility in the site of joining. The palindromic sequences align the genetic elements but the cutting and joining can occur at several different positions within the junctional region. The results are: (1) hybrid codons leading to sequence diversity; and (2) junctional regions of differing length. Junctional diversity is particularly important because it occurs in parts of V_L and V_H regions that contribute to the antigen-binding site. Junctional diversity may be estimated to increase the number of V_L regions by a factor of 5 and the number of V_H regions by a factor of 25 (5^2).

Combinatorial association

Each antibody is formed by the combination of a light chain and a heavy chain. Assuming (as seems probable) that each and every combinatorial association of a V_L region with a V_H region can generate a functional antigen-binding site, the number of possible antibody specificities will be the product of the number of V_L and V_H region genes. Simple combinatorial joining of 1000 V_L regions with 10 000 V_H regions generates 10^7 antibodies. Allowing for the increase in V_L and V_H regions generated by junctional diversity, as estimated above, we have 5000 V_L and 250 000 V_H, giving 1.25×10^9 antibodies.

Two other mechanisms contribute additional diversity to the antibody repertoire, one of which is the addition of N regions. The N regions consist of a small number of base pairs, found at the junction between V_H and D, or D and J_H, that cannot be accounted for by any known germ-line genetic elements. It is postulated that N regions are created by the addition (catalyzed by the enzyme terminal deoxyribonucleotidyl transferase) of nucleotides to the ends of V_H, D or J_H elements during the joining process. As with junctional diversity, N regions coincide with parts of V_H regions that contribute to antigen-binding. The quantitative contribution of N regions to the extent of antibody diversity is difficult to estimate.

The other contributory factor in the generation of antibody diversity is a mutational mechanism which appears to operate on rearranged V_L and V_H genes during B cell differentiation. Point mutations, both silent and expressed, accumulate in expressed V genes. This somatic mutational mechanism expands the diversity of the antibody repertoire far beyond the size of approximately 10^9 different molecules quoted above.

Consideration of this list of mechanisms and the estimates of their contribution to the generation of antibody diversity provides a ready explanation of the size of individual repertoires of antibodies elicited by single haptens. In such a large and diverse antibody repertoire degeneracy must be extensive. For a system that must act as a defence against the unexpected it is reassuring to see the extent of antibody diversity and the multiple, flexible mechanisms by which the repertoire is generated.

References

Alt F.W., Yancopoules G.D., Blackwell T.K., Wood C., Thomas E., Boss M., Coffman R., Rosenberg N., Tonegawa S. & Baltimore D. (1984) Ordered rearrangement of immunoglobulin heavy chain variable region segments. EMBO J. **1(6)**, 1209.

Gillies S.D., Morrison S.L., Oi V.T. & Tonegawa S. (1983) A tissue-specific transcription enhancer element is located in the major intron of a rearranged immunoglobulin heavy chain gene. Cell **33**, 717.

Kataoka T., Takeda S. & Honjo T. (1983) Escherichia coli extract-catalyzed recombination in switch regions of mouse immunoglobulin genes. Proc. natn. Acad. Sci. USA **80**, 2666.

Kim S., Davies M.M., Sinn E., Patten P. & Hood L. (1981) Antibody diversity: somatic hypermutation of rearranged V_H genes. Cell **27**, 573.

Korsmeyer S.J. & Waldmann T.A. (1984) Immunoglobulin genes: rearrangement and translocation in human lymphoid malignancy. J. Clin. Immunol. **4**, 1.

Leder P., Battey J., Lenoir G., Moulding C., Murphy W., Potter H., Stewart T. & Taub R. (1983) Translocations among antibody genes in human cancer. *Science* **222**, 765.

Max E.E., Seidman J.G., Miller H. & Leder P. (1980) Variation in the crossover point of kappa immunoglobulin gene V—J recombination: evidence from a cryptic gene. *Cell* **21**, 793.

Mather E.L., Nelson K.J., Haimovich J. & Perry R.P. (1984) Mode of regulation of immunoglobulin μ- and δ-chain expression varies during B-lymphocyte maturation. *Cell* **36**, 329.

Ziegler S.F., Treiman L.J. & Witte O.N. (1984) Kappa gene diversity among the clonal progeny of pre-B lymphocytes. *Proc. natn. Acad. Sci. USA* **81**, 1529.

Further reading

Darnell J.E. Jr. (1978) Implications of RNA—RNA splicing in evolution of eukaryotic cells. *Science* **202**, 1257.

Gearhart P.J. (1982) Generation of immunoglobulin variable gene diversity. *Immunol. Today* **3**, 107.

Gillies S.D., Morrison S.L., Oi V.T. & Tonegawa S. (1983) A tissue-specific transcription enhancer element is located in the major intron of a rearranged immunoglobulin heavy chain gene. *Cell* **33**, 717.

Leder P., Battey J., Lenoir G., Moulding C., Murphey W., Potter H., Stewart T. & Taub R. (1983) Translocations among antibody genes in human cancer. *Science* **222**, 765.

Lewis S., Gifford A. & Baltimore D. (1985) DNA elements are asymmetrically joined during the site-specific recombination of kappa immunoglobulin genes. *Science* **228**, 677.

Perry R.P., Kelley D.E., Coleclough C., Seidman J.G., Leder P., Tonegawa S., Matthyssens G. & Weigert M. (1980) Transcription of mouse kappa chain genes: implications for allelic exclusion. *Proc. natn. Acad. Sci. USA* **77**, 1937.

Shimizu A. & Honjo T. (1984) Immunoglobulin class switching. *Cell* **36**, 801.

Chapter 9
Classical genetics of histocompatibility

The mouse H-2 system

Histocompatibility recognized as under genetic control

When one animal accepts a tissue graft from a second of the same species the two animals are described as being histocompatible with each other. Experiments early this century involving the propagation of tumours by transplantation from one mouse to another revealed that, in general, tumour grafts are not accepted.

The first indication of genetic control of tumour transplantation came from the observation that the rate of successful grafts is higher when the donor and recipient mice are from the same relatively inbred stock. The observation led to the deliberate inbreeding of mice to generate inbred strains.

The genetic theory of tumour transplantation was proposed by Little (1914). The theory postulates that susceptibility to growth of a transplanted tumour is controlled by several dominant genes. This notion explains the earlier results that had baffled geneticists. A tumour (JWA) originating in Japanese waltzing (JW) mice (a partially inbred stock) was shown to grow in most JW mice but not in unrelated mice. The gene controlling susceptibility to the JWA tumour is dominant, as shown by the growth of the tumour in most first generation (F_1) crosses between JW mice and common mice (Figure 9.1). Mendelian concepts of dominant and recessive inheritance did not appear to explain the fact that the JWA tumour did not grow in the F_2 ($F_1 \times F_1$) mice tested. The theory that multiple dominant genes must be co-inherited to confer susceptibility to a tumour graft predicts that the majority of F_2 mice will not accept the tumour but that a proportion will accept the tumour graft. Most important of all is the quantitative prediction of the relationship between the proportion of F_2 mice accepting a tumour graft and the number of dominant genes determining susceptibility. In the original test of this hypothesis only three F_2 mice, out of 183 F_2 mice

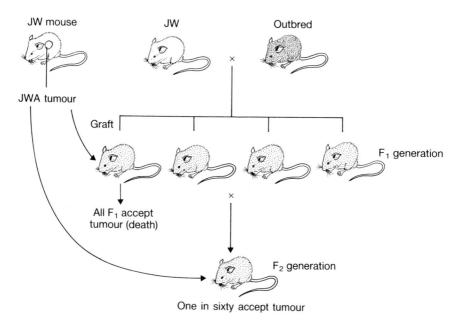

Fig. 9.1 The acceptance of the JWA tumour graft by parental, JW, mice and by most F_1 crosses of JW with outbred mice was explicable by conventional genetic theory. The low number of F_2 generation mice accepting JWA tumour grafts was initially puzzling but is explained by the postulate that transplant acceptance is controlled by several dominant genes.

tested, were susceptibile to the grafted JWA tumour; these numbers indicate, according to the hypothesis, that fourteen to fifteen dominant genes determine susceptibility (see 'The laws of transplantation', p. 220).

Graft rejection: an immunological response to histoincompatibility

The hypothesis that tumour grafts are rejected by an active defence mechanism mounted by the non-susceptible or resistant animal provides a model for histoincompatibility competing with the susceptibility gene hypothesis. For many years both the tumour-specific immunity thought to be necessary to explain the genetics of tumour rejection or acceptance and the nature of susceptibility genes remained unproven, and the two alternative models coexisted.

During the 1930s the alloantigen hypothesis was introduced to explain transplantation phenomena. Landsteiner described the transplantation antigens on human red blood cells, the ABO system, and he suggested the search for other transplantation antigens. The possibility that tumour cells carry alloantigens similar to the blood group antigens was proposed by Haldane. The hypothesis attributed tumour rejection to immune reactions directed against alloantigenic differences distinguishing the tumour cells from the recipient mouse, rather than to immunity against tumour-specific antigens. The alloantigens hypothesis explains the acceptance of tumour grafts by appropriate inbred mice. The search for blood group antigens in mice led Gorer (1936) to define four separate antigens on mouse red cells. Of these antigens, designated I, II, III and IV, antigen II is now known as H-2. This antigen was initially detected by an antiserum raised in a rabbit against strain A red blood cells and absorbed with C57BL red blood cells. The A strain specific H-2 antigen was shown to be present on an A strain tumour. Grafting of this A strain tumour to (A × C57BL)F_1 and F_2 generation crosses showed a pattern of graft rejection compatible with two or three genes being involved, and one of these genes was shown to map with the *H-2* gene. In these observations Gorer demonstrated the importance of alloantigens or histocompatibility antigens in determining graft rejection. Thus the multiple genes counted by the experiments of Little are *not* susceptibility genes that are all required to be simultaneously inherited for acceptance of the tumour; instead the multiple genes can now be understood to be loci determining histocompatibility antigens with inheritance of any one alloantigenic difference being sufficient to determine rejection of the tumour.

In addition to defining histocompatibility antigens, Gorer demonstrated an immune response to these antigens occurring concomitantly with tumour graft rejection. The alloantibodies raised against the tumour cells were shown to recognize the identical A strain H-2 antigen defined by the rabbit antiserum.

The skin-graft model

Grafts of skin provide a simple and convenient test system for mechanisms controlling transplantation. Medawar and his colleagues used skin grafting to establish the immunological characteristics of graft rejection. Autografts and allografts were compared on rabbits. Allograft rejection was shown to be characterized by local inflammation and cellular infiltration during the second week after grafting. Specific immunological

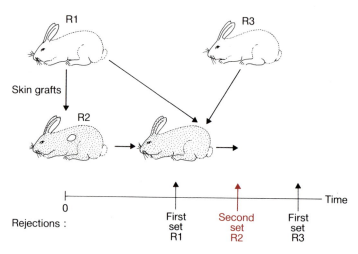

Fig. 9.2 Specific immunological memory in graft rejection was demonstrated by skin grafting between outbred rabbits. A graft of skin from R1 to R2 is rejected by the latter in a given time, as indicated (first set R1). A second set graft from R1 to R2 and a new first set graft R3 to R2 made at the same time, shortly after the first set R1 rejection, demonstrate the immunological memory response; the second set R1 graft is rejected more rapidly than the first set R3 graft that is rejected over a similar time span to the first set R1 graft.

memory was demonstrated by making second grafts of skin from either the original or different allogeneic rabbits (Figure 9.2). A second graft from the same donor is rejected very rapidly relative to the original graft; this accelerated response is termed a second-set rejection. A second graft from a different donor (third party) is rejected at a similar rate to any first-set rejection and with similar pathology. The third party graft demonstrates that it is the specificity of host memory for the histocompatibility type of the original skin graft that triggers the very rapid second-set rejection.

Graft rejection involves cell-mediated immunity

The designated systemic nature of immunological memory persisting after a graft has been rejected is shown by the fact that a second-set rejection is equally rapid no matter where the second graft is located relative to the site of the original graft (Figure 9.3). The cellular nature of immunity and immunological memory specific for grafted tissue was demonstrated by Mitchison (1954) in an adoptive transfer experiment (Figure 9.4). A sensitized or immunologically primed state was found

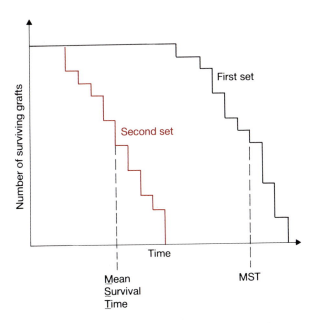

Fig. 9.3 Graphical representation of the results of a transplantation experiment using skin grafts. Considerable biological variation is seen. A useful parameter obtained from these data is the mean survival time (MST). The MST is shorter for second set than for first set grafts. The second set MST is independent of the site of the skin graft.

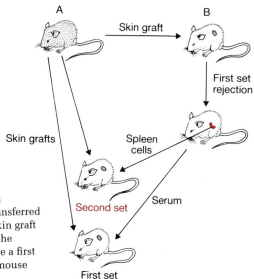

Fig. 9.4 Specific immune memory (a sensitized or primed state) can be transferred to mice that have never received a skin graft by injection of lymphoid cells from the spleen of a mouse that has undergone a first set rejection; serum from that same mouse does not transfer memory.

to be conferred on a mouse by transfer of lymphoid cells from a syngeneic mouse that has rejected a tissue graft. Transfer of serum from the grafted animal does

219 CLASSICAL GENETICS OF HISTOCOMPATIBILITY

not render the recipient sensitive. It was postulated that graft rejection is effected by specific cell-mediated immunity; lymphocytes are now recognized as being the effector cells in graft rejection (see Chapter 13). The adoptive transfer system has proved to be an important tool in the elucidation of the cellular basis of immune responses.

The laws of transplantation

Snell and Stimfling (1966) summarized the evidence in terms of five laws of transplantation:
1 Grafts between syngeneic animals, i.e. within an inbred strain, are retained.
2 Grafts between allogeneic animals, i.e. between inbred strains, are rejected.
3 The F_1 hybrid, resulting from a cross between two inbred strains, will accept a graft from either parental strain; the parental strains reject grafts of F_1 origin.
4 F_1 hybrids accept grafts from F_2 or subsequent intercross generations.
5 Only some members of the F_2 generation accept a graft from one or other of the inbred parental strains; similarly, only certain backcross animals (progeny of mating one parental strain with the F_1 hybrid) will accept grafts from the other parental strain.

The first two laws embody the genetic basis of histocompatibility. The third and fourth laws describe the co-dominant inheritance of histocompatibility genes; the F_1 hybrid expresses both allelic forms of each histocompatibility antigen and therefore accepts either parental allele as self; by the same token either parent sees the non-self allele of the opposite parent expressed in the F_1.

The fifth law describes the independent segregation of multiple histocompatibility genes. The frequency (f) of F_2 animals accepting a parental graft is related to the number (n) of independent loci controlling histocompatibility by the equation:

$$f = \left(\frac{3}{4}\right)^n$$

For the acceptance frequency in backcross animals the formula is:

$$f = \left(\frac{1}{2}\right)^n$$

These formulae were applied by Little to estimate the number of susceptibility genes. The answer obtained was fourteen to fifteen genes and this can be interpreted now as the first estimate of the number of histocompatibility genes.

How many histocompatibility loci are detectable in mice?

Segregation of histocompatibility loci, measured by skin-graft rejection in transplantation from an inbred strain to mice of the F_2 or backcross generation from crosses with a second inbred, strain yields answers for the number of loci ranging from about 10 to over 40.

Bailey (1970) calculated a maximum for the number of histocompatibility loci from an estimate of 5.4×10^{-3} per gamete for the rate of mutation at the sum of all histocompatibility loci. If the average mutation rate for histocompatibility loci is similar to the mutation rate, M, for individual loci, other than histocompatibility loci, then the number of histocompatibility loci, n, can be obtained from: $n \times M$ = overall mutation rate for histocompatibility loci.

In the mouse,

$$M = 7.5 \times 10^{-6} \text{ per gamete}$$

Thus:

$$n = \frac{5.4 \times 10^{-3}}{7.5 \times 10^{-6}} \text{ or } 750$$

This estimate will be high if the average mutation rate at histocompatibility loci exceeds 7.5×10^{-6} per gamete. Detectable mutations at the H-2 locus occur at a rate about two orders of magnitude higher than the average rate (see p. 232). However, the total number of histocompatibility loci is much less important (to immunogeneticists or even to transplantation specialists) than the quantitative and qualitative differences between H-2 and all other histocompatibility loci. These differences emerged from studies with congenic lines of mice.

Development of congenic lines

'The development of congenic lines by Snell was the most significant event in transplantation biology since the introduction of inbred strains. Without the lines, progress in many areas of research would have been impossible or considerably hindered.'

J. Klein (1975)

Congenic lines are produced by backcrossing of the progeny of two inbred strains, one designated as the donor strain, A, and the other as the background strain, B, against strain B with selection for the A strain allele of one locus (Figure 9.5). Successive backcrosses of selected progeny against strain B eventually results in a

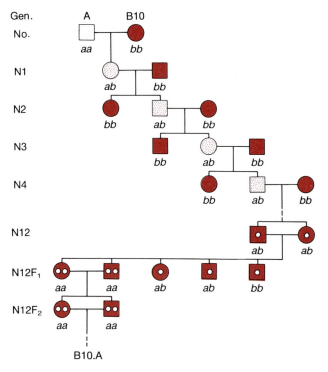

Fig. 9.5 Production of congenic lines of mice by successive backcrosses. The donor strain is A and the gene a (H-2^a) is being bred into the B10 background (H-2^b). Selection of heterozygotes (H-2^a/H-2^b) at each generation (N1, N2, etc.) is made by use of specific anti-H-2 antibodies. After a minimum of 12 generations, inbreeding of aa homozygotes (N12F$_1$, N12F$_2$ generations) yields a congenic line with B10.A that is similar to B10 at all loci except H-2 where H-2^b has been replaced by H-2^a. (Note that congenic lines are not truly coisogenic because of the transfer of loci closely linked to that being assayed and because of other exchanges of genetic material that may have occurred during the backcross procedure.)

line carrying the chosen A strain allele (and an undefined number of closely linked loci) on the B strain genetic background.

Snell constructed congenic lines differing from one to another by one histocompatibility locus. Selection at each generation was made on the basis of resistance to (i.e. rejection of) tumour grafts. Serological tests also provide a convenient selection screen for histocompatibility antigens. A minimum of twelve generations of backcrossing is necessary before a congenic line is established by brother−sister mating. Two lines congenic for an allelic difference at a histocompatibility locus are referred to as congenic resistant because they resist a mutual exchange of tissue grafts. Congenic lines have also been constructed with wild type mice as the donor partner, the selection being for the wild type histocompatibility allele on the inbred background.

Strong and weak, major and minor histocompatibility loci

Skin grafts between congenic lines differing at single histocompatibility loci revealed that differences at the H-2 locus resulted in more rapid rejection of grafts than differences at non-H-2 histocompatibility loci (Figure

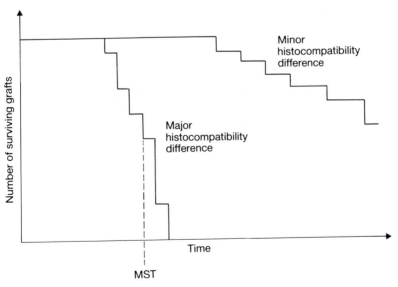

Fig. 9.6 Graphical representation of the results of a skin graft experiment comparing transplantation across major and minor histocompatibility differences. Rejection across a major histocompatibility barrier is rapid and complete. Across a minor histocompatibility difference rejection occurs slowly and only in some animals.

9.6). The concept of strong and weak barriers to transplantation was introduced. All histocompatibility loci discovered have been tested against the criteria for strong and weak loci.

It is now clear that *H-2* is the only strong locus in the mouse and the terminology that has been adopted is to call *H-2* the major locus, or the major histocompatibility complex (MHC) to indicate the multiplicity of genes at *H-2*, and all other non-*H-2* loci are termed minor histocompatibility loci. The differences between the MHC and minor loci are both quantitative and qualitative. The function of cell surface products encoded at the MHC in controlling immune responsiveness (see Chapter 13) explains the extensive difference between MHC and non-*H-2* loci.

Klein (1975) compiled a list of phenomena that distinguish the properties of allelic differences at the MHC

Table 9.1 Phenomena distinguishing properties of MHC allelic differences from effects of minor histocompatibility differences (Klein, 1975)

Phenomenon	MHC allelic difference	Minor histocompatibility difference
Graft rejection	Acute, rapid invariable rejection	Chronic, slow, some grafts retained
Genetic complexity	Many genes map to H-2, pleiotropic effects	Simple, single loci
Polymorphism	Extreme polymorphism	Two or three alleles per locus
Control of immune responsiveness to other antigens	Linked to Ir genes	Not associated with Ir genes
MLR, GVH	Strong reactions	Little or no reaction
Proportion of lymphocytes specifically involved in MLR or GVH		
(a) primary	High percentage (~1% T lymphocytes)	Low percentage
(b) secondary	Similar percentage to primary	Increased percentage
Nature of immunity		
(a) humoral	+	±
(b) cellular	+++	+
Cooperation between B and T lymphocytes	Identity necessary	No effect
Tolerance	Difficult to establish	Relatively easier to establish
Immunosuppression of response	Difficult	Relatively easier

from the effects of allelic differences at non-*H-2* histocompatibility loci (Table 9.1).

Probably most alloantigenic differences expressed on the cell surface of transplanted tissues can be detected as minor histocompatibility antigens. To the list of phenomena in Table 9.1 can be added MHC restriction of responsiveness to antigens (see Chapter 13); the restriction phenomenon puts minor antigens in a similar category to viral antigens.

H-2: the murine major histocompatibility complex

H-2 is in linkage group IX

Location of *H-2* in the same linkage group (IX) as *fused tail* (*Fu*) was first indicated by the observation that the two loci were transferred together during construction of a congenic line. Backcross linkage tests established linkage between *H-2*, *Fu*, *T* (Brachary) and *Ki* (Kinky).

Linkage group IX is on chromosome 17

Individual chromosomes are arranged by morphological criteria in the karyotype. Each chromosome can be identified by a specific banding pattern revealed by differential staining (Figure 9.7). Linkage groups are assigned to particular chromosomes by correlating translocations of chromosome bands with translocations of linked markers.

Translocations and banding pattern correlation indicate the approximate location of *H-2* in band c.

Serology of H-2: the two-locus model

The mapping of *H-2* has been one of the more involved projects in classical immunogenetics. Serological methods have been used extensively in analysing the *H-2* complex. Alloantisera are raised by immunization of mice of one inbred strain with cells from another. Inbred strains are usually found to differ from each other by multiple antigenic differences mapping at *H-2*. The major problems in the interpretation of the data have lain in the complex nature of both the antigenic differences between strains and the complex nature (i.e. multispecificity) of antisera raised in these immunizations.

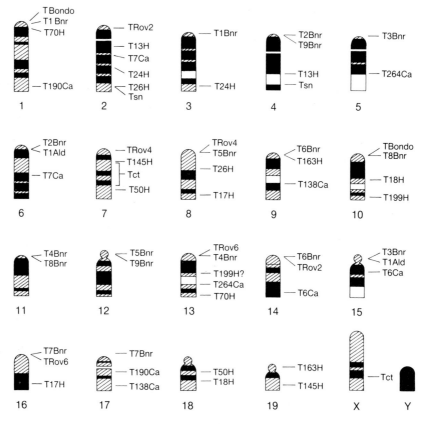

Fig. 9.7 Idiogram showing the quinacrine fluorescent banding patterns of mouse chromosomes, arranged and numbered systematically. Chromosomes involved in a translocation can be identified by virtue of the fact that the banding pattern of each segment of a translocation chromosome is similar to the banding pattern of the original segment in the normal chromosome. Cytologic breakpoints (T, followed by an alphamerical code) are shown for some of the translocations that have been analysed by fluorescent banding. (From Miller & Miller, 1972)

Initially, analysis was attempted on the assumption that each antiserum defined a single antigen. Later, recognition of the multispecific nature of antisera with extensive cross-reactions with individual antigens led to the two-locus model for the *H-2* complex.

In the two-locus model all H-2 antigenic differences are mapped to either the *K* or the *D* regions of *H-2* or to both. Detailed recombinant analysis has mapped *K* and *D* at a spacing of 0.33 map units apart. This is a considerable stretch of DNA estimated to be approximately 2×10^6 base pairs (bp). This segment of genome from *K* to *D* inclusive is designated the *H-2* complex.

H-2 is a complex locus

Prior to the advent of recombinant DNA techniques there were approximately ten genes identified within the H-2 complex. The known H-2 genes, in addition to K and D genes (now termed class I genes), were the I region (class II) genes (mapping in the subregions I-A, I-B, I-J, I-E and I-C), the SS/S1p locus, and the L locus. The L locus is a D region class I histocompatibility gene apparently analogous to the K and D genes in function. (The presence of a third major histocompatibility locus at H-2 makes the system more analogous with the HLA complex comprising the A, B and C genes.) The other known H-2 genes are discussed at more length below.

It has been proposed that the length of DNA ($\sim$ 2000 kb) designated as H-2 could contain up to 1000 structural genes. Present evidence indicates that structural genes may be separated by extensive (> 10 kb) non-coding sequences and the stretches of DNA containing structural genes themselves are much longer than was predicted on the basis of a gene being colinear with a polypeptide, due to the presence of intervening sequences (introns) that split genes into several coding segments (exons). Although further genes have been map-pated at H-2 by classical techniques (and some have not been confirmed, e.g. I-B, I-C, I-J and G), it has been the application of molecular genetics that has answered the question of the number (and nature) of the genes at H-2 (see Chapter 10).

Serological analysis of H-2: antigenic specificities

The difficulties encountered in the interpretation of serological evidence, prior to the two-locus model, is illustrated by reviewing, in the light of our accumulated knowledge of antigenic specificities of H-2 K and D products, the cross-reactions seen in the raising and the testing of antisera between four common inbred mouse strains: A, BALB/C, C3H and C57BL.

Three antisera: (1) BALB/C anti-C3H; (2) C57BL anti-BALB/C; and (3) C3H anti-C57BL, can, with appropriate absorption define six different antigenic specificities. Three of these six antigens (2, 4 and 11) are limited to, and characteristic of, one strain each, while three are shared between strains thus:

C57BL   2   —   —   5   6   —

BALB/C	–	3	4	–	6	–
C3H	–	3	–	5	–	11

A fourth antiserum, C57BL anti-A, tested before and after absorption on C3H cells, can be used in conjunction with the first three antisera to show that A strain cells share the antigens detected on BALB/C cells, in addition to sharing those found on C3H cells, thus:

A		3	4	5	6	11

This H-2 genotype for A strain mice can be explained as a recombinant between the H-2 genotypes of BALB/C and C3H strains. The latter simple explanation requires the verification that the antigens defined by each of the antisera are each determined by a gene or genes mapping at H-2, and this evidence is provided by linkage analysis in backcrosses. The A strain thus appears to have arisen by a recombinant event during the inbreeding of mice to yield the present day laboratory strains.

Public and private H-2 antigens

More than fifty antigens of H-2 have been defined to date and the conclusions drawn from the simple, illustrative analysis shown above continue to hold good. Certain antigens are found only in one inbred strain (or derivatives of that strain) and these are referred to as *private antigens*; e.g. 2, 4 and 11 are private antigens for strains C57BL, BALB/C and C3H in the above example. (NB In a fuller antigenic analysis of these three strains, antigens 2 and 4 are still found to be private antigens.)

The antigens common to more than one inbred strain are referred to as *public antigens*. Antigens 3, 5 and 6 are thus public antigens, each being shared between two of the three strains tested. Further analysis involving other strains places antigen 11 in the public category also. It should be noted that the finding of antigens 2 and 4 in strain A does not invalidate their classification as private antigens because of the assumption that strain A could have arisen by recombination.

The definition of antigen 11 as private on the basis of a limited analysis, but public on the basis of a fuller strain survey, should indicate the arbitrary nature of the private and public classification. Nevertheless, the H-2 antigens may be arranged in a spectrum ranging from 'strictly' private to 'broadly' public. If this spectrum has a common genetic basis with a variety of antigens present on a single structural gene product, then there is a possible evolutionary significance: the range public to private may correspond to the results of mutations that occurred long ago or more recently, respectively.

H-2 haplotypes

A linked set of allelic forms of the genes comprising the H-2 complex is generally termed a haplotype. Each complete genotype (diploid set of genes per cell) has two haplotypes. In an inbred strain the two haplotypes are identical since the animals are homozygous. In the heterozygote the two haplotypes are different and are transmitted independently (except for rare recombinant events) to the next generation.

Each haplotype can be described by listing all known alloantigenic markers encoded at the particular H-2 locus. For convenience a particular lower case letter is used as a shorthand to designate each of the haplotypes known in standard laboratory strains of mice. Among the standard strains nine unrelated haplotypes have been recognized (Table 9.2).

Since, by definition, public antigens are those shared between two or more strains, it is useful to classify these haplotypes according to the private specificities (Table 9.2) at K or D (or as yet unassigned, e.g. 18).

A full description of each H-2 haplotype of independent origin listing all known antigens, is given in Klein et al. (1983). Certain conclusions can be drawn:
1 Each haplotype can be divided into two series of antigens, one defining the K locus and the other the D locus.

Table 9.2 H-2 haplotypes

Haplotype	Prototype strain	Private antigens		
		K	?	D
H-2^b	C57BL/10	33		2
H-2^d	DBA/2J	31		4
H-2^f	A−CA	?		9
H-L^k	CBA	23		32
H-2^p	P/Sn	16		?
H-2^q	DBA/1	12		30
H-2^r	RIII/Wy		18	
H-2^s	A-SW	19		12
H-2^z	NZW	?		?

2   Each K and D locus is defined by a particular pattern of public antigens (in addition to the characteristic private antigen).

3   Public antigens are often common to K and D locus products.

H-2 recombinant haplotypes

The majority of H-2 haplotypes determined for inbred strains of mice can be explained by invoking recombinant events between K and D regions of two of the unrelated H-2 haplotypes. The classical example is the $H\text{-}2^a$ haplotype (A strain mice) that could have arisen by recombination between $H\text{-}2^d$ and $H\text{-}2^k$ haplotypes (Figure 9.8).

The occurrence of that recombination can only be surmised. A critical test establishing that the $H\text{-}2^a$ haplotype is made up of K^k and D^d comes from the acceptance of A strain grafts by $(H\text{-}2^k \times H\text{-}2^d)$ F_1 hybrids. Other natural 'putative' recombinant haplotypes are known. Moreover, recombinant events are seen to occur during crosses between inbred strains and new recombinant haplotypes appear. Recombination is the major event used to map the H-2 locus, not only the two major transplantation antigen loci K and D but also all the regions in between.

H-2 mutants

The number of known H-2 alleles make the H-2 locus the most polymorphic locus of the mouse. Each allelic difference represents an inherited genetic alteration arising by a presumed mutation from an ancestral gene. The occurrence of these mutations is inferred from the recognized antigens but the nature of each event cannot be ascertained since we have no knowledge of the gene

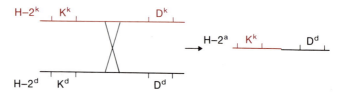

Fig. 9.8 The $H\text{-}2^a$ haplotype is a putative recombinant between $H\text{-}2^k$ and $H\text{-}2^d$.

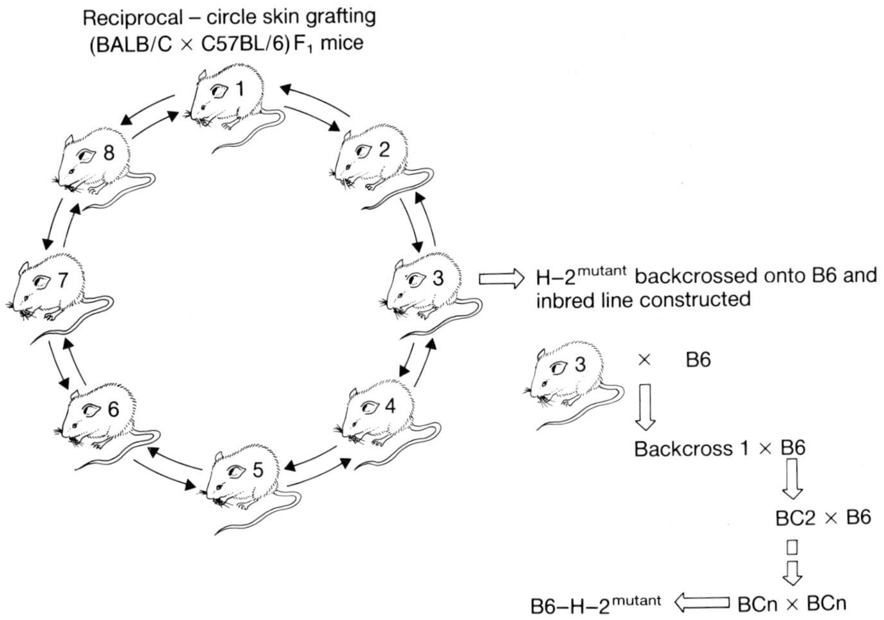

Reciprocal – circle skin grafting
(BALB/C × C57BL/6) F₁ mice

H−2^{mutant} backcrossed onto B6 and
inbred line constructed

3   ×   B6

⇓

Backcross 1 × B6

⇓

BC2 × B6

⇓

B6−H−2^{mutant} ⟸ BCn × BCn

Fig. 9.9 Reciprocal-circle skin graft experiment performed to select mutations at major histocompatibility loci (see text for explanation). Each new mutant is backcrossed onto the B6 background to yield an H-2^{mutant} congenic line (see Fig. 9.5). Method due to Bailey & Usama (1960) and employed by Bailey, Egorov, Kohn and Melvold.

prior to each mutation. Nor can the frequency of mutation be estimated from the extent of polymorphism without a time-scale for the evolution of H-2.

To understand more about mutations and the mutation rate for H-2, mutants must be identified experimentally as they occur. The type of mutation which has been sought is that affecting antigenicity of H-2 products. The change can be the loss of an antigen, the creation of an antigen or both simultaneously. The favoured method of screening for H-2 mutants is to exchange skin grafts in a reciprocal-circle manner amongst a group of up to twenty mice (Figure 9.9). The mice studied are F₁ hybrids of two inbred strains (preferably two H-2 congenic lines). Each mouse in the circle donates two grafts of skin and each receives two grafts with the neighbours in the circle as recipients and donors respectively.

In the reciprocal-circle test using F₁ hybrids of H-2 congenic mice, a new antigen arising in any animal is registered by rejection of both of the skin grafts donated by the mutant mouse. Proof that the mutation occurred at H-2 can then be obtained by linkage analysis. A mutation leading to the loss of an antigen is registered by rejection of both of the skin grafts received by the mutant mouse. Because the F_1 hybrids are homozygous at all loci except H-2, any detectable antigenic loss must map at H-2. Reciprocal rejection of grafts received and donated indicates a mutation simultaneously eliminating an antigen and creating a new antigen. Each putative mutation indicated by graft rejection is subjected to progeny testing. If the new pattern of graft rejection is inherited by the progeny of the putative mutant the mutation is established as a homozygous line. This new line will be truly coisogenic with the original parental inbred strain. Each mutant haplotype is given a genetic symbol derived from the parental symbol thus:

$$H\text{-}2^b \xrightarrow{\text{mutation}} H\text{-}2^{bm1}$$

Subsequent independent mutations of the H-2^b haplotype are then H-2^{bm2}, H-2^{bm3}, etc.

Mutations are mapped to regions of H-2 by using intra-H-2 recombinants. Serological and ultimately biochemical evidence (see p. 57 and p. 264) locates the mutation in a particular locus. Of the first fourteen mutations mapped twelve occurred at the K locus, one at the D locus and one at the L locus; none mapped at the H-2 I locus (which can code for transplantation antigens causing rapid graft rejection and therefore mutants at H-2 I should have been detected in the reciprocal-circle assay).

A spontaneous mutation frequency for a single H-2 locus has been calculated, on the basis of some 50 000 mice grafted, as 4×10^{-5} per gamete. This is to be compared with mutation rates two or three orders of magnitude lower for recessive or dominant genes in the mouse. The H-2 loci are apparently highly mutable. Of the individual loci tested the highest mutation rate determined is 2.2×10^{-4} per locus per gamete for the H-2 K^b gene, higher than for any other mammalian gene. One explanation that has been offered for the high mutation rate of H-2 K and D (L) loci is that the screening assay for antigenic differences is more sensitive than assays for detection of mutants at other loci. However, the absence of detectable mutants at the H-2 I loci and the differential frequency of observed mutation at individual alleles of K and D suggest that one must seek an intrinsic biological reason for the high mutation rate. (Molecular genetics show that gene conversion is operative in the *H-2* region; see Chapter 10.) It is probable that:

1 A high mutation rate at *H-2* loci has been selected positively during evolution.

2 The complex nature of *H-2* allelism, with multiple differences between allelic polypeptide sequences (see Chapter 3), is due to the high mutation rate.

3 The large number of *H-2* alleles, making *H-2* the most polymorphic locus in the mouse, are the result of the high mutation rate.

H-2 I region: immune response genes (Ir)

Genes controlling immune responsiveness have been mapped within *H-2*, nearer to *K* than to *D*. These immune response (*Ir*) genes are in a region termed I located between the K and S regions.

   Ir genes have been demonstrated to control the production of antibody against defined antigens such as immunoglobulin alloantigens or synthetic polypeptides of limited structural heterogeneity. Particularly useful is a series of polypeptides constructed on a poly-L-lysine backbone with side chains of poly-D, L-alanine terminating in oligopeptides containing L-glutamine with either L-tyrosine, or L-histidine, or L-phenylalanine. It is now known that it is the terminal oligopeptides that constitute the antigenic determinant in each case. Differences in the response of two strains of rabbits to (T,G)-A-L were noted by Humphrey and McDevitt (1962) but the genetics of responsiveness were studied successfully only when differences between inbred strains of mice were observed.

   Immunization of different strains of mice with (T,G)-A-L under standard conditions results in a good antibody response in certain strains and a low or undetectable response in other strains. Although strains are often, for convenient shorthand, referred to as responder and non-responder strains, a better terminology is high and low responder strains respectively. Difference in response is not absolute. Antibody production is dependent in all strains on the dose of antigen. However, even comparing optimal responses in different strains allows high and low responder strains to be defined.

   Mice showing a high response to (T,G)-A-L and

(H,G)-A-L were found to have their responder status determined by an autosomal dominant gene that was termed the *immune response-1* or *Ir-1* (now *Ir-1A*; see below) gene. In the two inbred strains C57 and CBA the responsiveness to (T,G)-A-L and (H,G)-A-L is reciprocally related (Table 9.3).

The F_1 hybrid reveals the dominance of high responder status towards each antigen. Backcross analysis shows that high responsiveness segregates with *H-2* type. Backcrossing to low responder parent is the informative cross (Table 9.4).

This type of linkage analysis confirmed the observation made in *H-2* congenic strains that *Ir-1* responder status is associated with *H-2* haplotype. Intra-*H-2* recombinant strains allow *Ir-1* to be mapped just to the right of the K region. A new region designated I was defined by this recombinant mapping of *Ir-1*.

The region I was divided into two subregions I-A and I-B by the mapping of responsiveness to allotype antigens of murine IgA and IgG. Responsiveness to IgG_{2a} (murine myeloma protein MOPC173) was shown to be controlled by a single, autosomal dominant gene mapping within the *H-2* complex and distinct from *Ir-1A*. This new locus is termed *Ir-1B* and it defines the I-B subregion. The I-B subregion is distinguished from the I-A subregion and the S region by the assay of two recombinant inbred strains, B10.A(4R) and B10.A(5R). The genetic constitution of the parental and recombinant haplotypes can be written in shorthand form, as shown in Table 9.5.

Control of immune response to the random polypeptide GLT, composed of L-glutamic acid, L-lysine and L-

Table 9.3 Dominance of responder status

	C57	CBA	(C57 × CBA)F$_1$
(T,G)-A-L	High	Low	High
(H,G)-A-L	Low	High	High

Table 9.4 Responder status segregrates with H-2. From McDevitt *et al.* (1972)

Either:	(H-2^b × H-2^k)F$_1$ × H-2^b →		H-2$^{b/b}$ + H-2$^{b/k}$	
(H,G)-A-L	High	Low	Low	High
Or:	(H-2^b × H-2^k)F$_1$ × H-2^k →		H-2$^{k/k}$ + H-2$^{b/k}$	
(T,G)-A-L	High	Low	Low	High

Table 9.5 Responder status maps to the I-A region. From Lieberman et al. (1972)

	K	I-A	I-B	S	D	Response to IgG allotype
B10	b	b	b	b	b	Low
B10.A	k	k	k	d	d	High
B10.A(5R)	b	b	k	d	d	High
B10.A(4R)	k	k	k	b	b	High

tyrosine (57:38:5), maps to a third I subregion designated as I-C.

Control of the immune response to more than forty different antigens has been mapped to the I region. Certain common factors can be seen:

1 All antigens under Ir gene control are T lymphocyte dependent (see Chapter 13).

2 Many antigens under Ir gene control are structurally 'simple'.

3 Ir gene control of response to 'complex' antigens (e.g. ovalbumin or bovine gammaglobulin) is demonstrable at low doses of antigen.

Ir genes can be detected by antigen-specific stimulation of T lymphocyte proliferation

An assay involving presentation of antigen to T lymphocytes *in vitro* with measurement of the degree of proliferation induced has proved most useful both for mapping Ir genes and for elucidating their function (see below and Part 3). One source of T cells is the fluid of the peritoneal cavity; the acronym PETLES, standing for peritoneal exudate, T lymphocyte-enriched subpopulation, has been devised for the assay. The T lymphocyte is presented with antigen on the surface of an 'antigen-presenting cell' (see Part 3). Proliferation is assayed by measuring incorporation of [³H]-thymidine into DNA. The assay was first devised for studies on guinea pigs but is now also used to detect Ir genes in mice (Schwartz et al. 1976).

Ir gene complementation

Immune responsiveness to certain antigens requires the simultaneous presence of correct alleles at two independent loci mapping within the I region. The requirement for two genes has been shown by genetic complementation. Complementation can be either *cis* or *trans*. Bringing together the two required genes on the same chromosome by recombination can effect *cis* complementation. A heterozygote with one gene on one parental chromosome and the complementary gene on the opposite parental chromosome effects *trans* complementation.

Two complementary Ir genes determining the response to a synthetic polypeptide were first discovered in a *trans*-complementation cross. The two *H-2* congenic strains B10.A (*H-2*^{a}) and B10.S (*H-2*^{s}) are both low responders to the random linear polymer composed of L-glutamic acid, L-lysine and L-phenylalanine (this polymer is termed GLΦ for brevity). An F_1 hybrid, produced by mating the two low responder strains, is able to make antibody to GLΦ and shows GLΦ-specific stimulation of T cell proliferation in a PETLES assay, although in both assays the F_1 hybrid is less efficient than a responder inbred mouse.

Complementation between Ir genes arranged in *cis* configuration is seen with intra-*H-2* recombinant haplotypes. The appropriate recombinant inbred strains bring together the alleles of the I subregions necessary for the high responder phenotype (Table 9.6).

Other recombinant haplotypes limit the effects to the I region. Therefore the two complementing genes map at I-A/I-B and I-E/I-C respectively.

Complementation between genes in *trans* configuration raises the question as to whether the complementary genes need to be expressed in the same cell. The answer is that they do and the evidence comes from experimentally constructed chimeric and allophenic mice. Bone-marrow chimeras were made by reconstituting lethally irradiated {B10.A(2R)×B10}F_1 mice (high responders) with either syngeneic bone-marrow cells or a mixture of 2R and B10 bone-marrow cells (low responder parents). The 2R↔B10 chimera PETLES did not

Table 9.6 Complementation (*cis*) between Ir genes. From Dorf *et al.* (1979)

	K	I-A	I-B	I-J	I-E	I-C	D	Response to GLΦ
B10.A	k	k	k	k	k	d	d	Low
B10.S	s	s	s	s	s	s	s	Low
B10.S(9R)	s	s	s	k	k	d	d	High
B10.HTT	s	s	s	s	k	k	d	High

proliferate when challenged with GLΦ, although control responses to other antigens showed both parental cell populations to be active.

Allophenic mice were constructed by fusing blastocysts from two mouse strains with complementing low responder haplotypes. The resulting mosaic mice, having lymphoid cells of both genetic types, were also low responders to GLΦ.

Co-dominant Ir gene expression: gene dosage effects

The F_1 hybrid between a high responder and a low responder mounts a good antibody response and shows a positive antigen-specific T cell proliferation in the PETLES assay. This pattern was originally described as showing that responsiveness is dominant. However, quantitative analysis of several Ir gene systems shows that the F_1 hybrid is intermediate, in both antibody production levels and T cell proliferative index, between the high and low responder parents. These quantitative results indicate co-dominant inheritance of Ir genes.

The quantitative results obtained in studying complementation between Ir genes show a pronounced gene dosage effect also indicative of co-dominant inheritance. The appropriate crosses using recombinant haplotypes allow comparison of homozygous high responders with heterozygotes involving either *cis* or *trans* complementation (Table 9.7).

Table 9.7 Complementing alleles. From Schwartz *et al.* (1979)

	I-A	I-E	I-C		Response to GLT
B10.A	k k	k k	d d	Homozygote	Low
B10	b b	b b	b b	Homozygote	Low
B10.A(4R)	k k	b b	b b	Homozygote	Low
B10.A(5R)	b b	k k	d d	Homozygote	High
(B10 × B10.A)F₁	b k	b k	b d	Heterozygote (*trans*)	Intermediate
{B10(4R) × B10(5R)}F₁	k b	b k	b d	Heterozygote (*cis*)	Intermediate

Mixed lymphocyte reaction (MLR)

The co-culture *in vitro* of allogeneic leucocytes results in a mutual proliferative stimulation. The proliferating cells are T lymphocytes and the most efficient stimulator cells are allogeneic B lymphocytes. The standard method of detecting and measuring cell proliferation is to assay the incorporation into DNA of [³H]-thymidine during a period of several hours. In the simplest MLR involving lymphocytes from two different inbred strains there is a two-way stimulation with T cells of both haplotypes proliferating. For ease of interpretation a one-way MLR is now normally used. A one-way stimulation is ensured in one of three ways:

1 Stimulator cells are from an F_1 hybrid, e.g. CBA cells stimulated by (CBA × C57)F_1 cells.

2 Mitomycin C treatment of stimulator cells blocks their DNA synthetic capacity.

3 X-irradiation of stimulator cells also prevents DNA synthesis.

In the one-way MLR the nature of the antigenic difference required for stimulation can be determined.

Differences at the *H-2* complex were initially recorded as leading to stimulation. The most effective differences were mapped (using lymphocytes from *H-2* congenic mice and intra-*H-2* recombinant mice) to the K end of *H-2* and to loci within the I region. The antigens involved had not, at the time of this mapping (1965 – 1972), been seen serologically; hence, the stimulating antigens were termed 'lymphocyte determined', or LD antigens, by contrast with the 'serologically determined', or SD antigens, of the *K* and *D* loci. Also, the LD antigens were designated as the products of loci coding for 'lymphocyte activating determinants' and hence these loci were termed *Lad*-1 *Lad*-2, etc. It now seems certain that Lad antigens are synonymous with Ia antigens (see below).

The importance of Ia antigens in stimulating the MLR is demonstrated by the ability of anti-Ia sera to inhibit the proliferative stimulus.

Non-*H-2* differences are also able to stimulate T cell proliferation in the MLR. The degree of stimulation seen when the stimulus is a single minor histocompatibility antigen is usually either low or insignificant. One non-*H-2* linked locus determining a strong MLR has been designated *M*. Four alleles of the *M* locus have been described. Also *H-2* mutants mapping at *K*, *D* or *L* stimulate T cell proliferation in the MLR.

Lymphocytes give a positive MLR when stimulated by xenogeneic cells in place of allogeneic cells. It is possible that Ia antigens on the xenogeneic cells are important but this remains to be proved. Serological and lymphocyte-defined antigenic cross-reactions are detectable between certain species.

Graft-versus-host (GVH) reaction

When lymphoid cells are the allograft given to an animal not capable of mounting an immune response, a one-way response of the grafted cells stimulated by host antigens occurs. In the standard assays the recipients are either neonatal mice or adult mice lethally irradiated to render them immunologically incompetent; alternatively, an F_1 hybrid provides a non-responsive host for injection of lymphoid cells from either of the parental strains.

The response to the host by the grafted cells is manifested in a multiplicity of ways as lymphocytes proliferate and differentiate, giving rise to specific cytotoxic effector cells. Any of the resultant effects of GVH reaction can be assayed. The common assay in mice is measurement of splenomegaly, i.e. increase in spleen size, made by weighing the spleen excised about 10 days after grafting.

The genetic differences predominantly controlling the GVH response map at the *H-2* complex with I region differences determining the strongest responses.

Immunological specificity of MLR and GVH reactions

Both reactions are due to a cellular immune response to histocompatibility antigens. The MLR being carried out *in vitro* is a more tractable system and the genetics are consequently better mapped. The events occurring during a GVH reaction are complex and involve non-specific host mechanisms. In the MLR the initially proliferating cells can be shown to be T lymphocytes. Late in the course of the MLR, after the peak of cell proliferation, cytotoxic effector T cells appear in the culture. Although the initial proliferation can be blocked by antisera directed against the Ia antigens of the stimulating cells, the cytotoxic cells generated later are specifically directed against K and D antigens of the stimulating cells. The simplest explanation is as follows: (1) the initially proliferating T cells are Ia specific, or I region antigen restricted in their response to other alloantigens (see Chapter 13); (2) the effector T cells are K or D antigen specific, or K/D antigen restricted; and (3) the I region antigen restricted T cells are necessary as helper cells for the generation of effector T cells.

Serologically detected I region products: immune associated (Ia) antigens (class II MHC antigens)

Immunological recognition of I region differences is not restricted to cell-mediated reactions. Alloantibodies can be produced against I region products, now termed Ia, or class II, antigens (earlier names were Ir-1 or Lna antigens). The nature of Ia antigens is described in Chapter 3. Polymorphism of Ia antigens is not as pronounced as that of class I H-2 antigens, but is still higher than at most other loci. Mapping of genes encoding Ia antigens has been facilitated by the use of recombinant inbred strains such as ATL and ATH (Figure 9.10). These strains differ only at the I and S regions. The sub-specificities of the anti-Ia sera raised between these strains are revealed by testing against a range of inbred strains.

ATH anti-ATL antisera
   against H-2^s negative reaction
   against H-2^b; H-2^d; H-2^f; H-2^q and H-2^p
      positive cross-reaction
(ATH × B10)F$_1$ anti-ATL
   against H-2^b and H-2^q negative reaction
   against H-2^d; H-2^f; H-2^p positive cross-reaction

Each positive reaction defines a particular Ia specificity.

ATH anti-ATL against B10 targets defines Ia.3
ATH anti-ATL against ACA (H-2^f) targets defines Ia.1

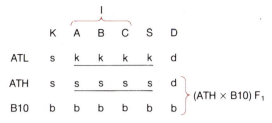

	K	A	B	C	S	D	
ATL	s	k	k	k	k	d	
ATH	s	s	s	s	s	d	(ATH × B10) F₁
B10	b	b	b	b	b	b	

Fig. 9.10 Mapping of genes encoding Ia antigens using recombinant inbred strains ATL and ATH.

The relationship between *Ir* genes and *Ia* genes requires explanation and was, for a while, controversial. The molecular genetic analysis of the I region has simplified the genetic map (see Chapter 10). The I-B, I-C and I-J regions are not definable at the DNA level.

The mapping of Ir function onto cell−cell interaction (see Chapter 13) goes some way towards relating Ia products to *Ir* genes but a wholly satisfactory explanation is still awaited.

The human HLA system

The human HLA system is located on the short arm of chromosome 6, and a provisional map of some of the major gene loci is shown in Figure 9.11. It is convenient to group the products of these loci into three separate categories called class I, class II and class III antigens.

Class I antigens

Class I antigens are found on the surfaces of most

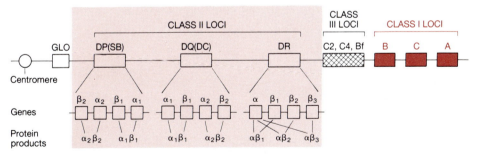

Fig. 9.11 The human HLA map showing relative locations of various gene loci on the short arm of chromosome 6. Class I loci code for the serologically typed HLA-A, -B and -C antigens. Class II loci code for the family of polymorphic antigens involved in the regulation of the immune response (previously known as the D locus — SB and DC are the original names used for the DP and DQ loci). Class III loci code for the complement components C2, C4 and Factor B. GLO = glyoxalase gene.

Table 9.8 Polymorphism in the human MHC

HLA-A	HLA-B	HLA-C	HLA-DP (SB)	HLA-DQ (DC)	HLA-DR	
A1	B5	Cw1	DPw1	DQw1	DR1	Dw1
A2	B7	Cw2	DPw2	DQw2	DR2	Dw2
A3	B8	Cw3	DPw3	DQw3	DR3	Dw3
A9	B12	Cw4	DPw4		DR4	Dw4
A10	B13	Cw5	DPw5		DR5	
A11	B14	Cw6	DPw6		DRw6	
Aw19	B15	Cw7			DR7	Dw7
A23(9)	B16	Cw8			DRw8	Dw8
A24(9)	B17				DRw9	
A25(10)	B18				DRw10	
A26(10)	B21				DRw11	Dw5
A28	Bw22				DRw12	
A29(w19)	B27				DRw13	Dw6
A30(w19)	B35				DRw14	Dw9
A31(w19)	B37					
A32(w19)	B38(16)				DRw52	
Aw33(w19)	B39(16)				DRw53	
Aw34(10)	B40					
Aw36	Bw41					
Aw43	Bw42					
Aw66(10)	B44(12)					
Aw68(28)	B45(12)					
Aw69(28)	Bw46					
	Bw47					
	Bw48					
	B49(21)					
	Bw50(21)					
	B51(5)					
	Bw52(5)					
	Bw53					
	Bw54(w22)					
	Bw55(w22)					
	Bw56(w22)					
	Bw57(17)					
	Bw58(17)					
	Bw59					
	Bw60(40)					
	Bw61(40)					
	Bw62(15)					
	Bw63(15)					
	Bw64(14)					
	Bw65(14)					
	Bw67					
	Bw70					
	Bw71(w70)					
	Bw72(w70)					
	Bw73					

w = a provisional 'workshop' classification.
Where an antigen is followed by a bracketed number there is evidence that the antigen in question is a 'split' of a previously recognized antigen. For example, Aw23(9) and Aw24(9) are 'splits' of A9.

Table 9.9 Gene frequencies of various HLA-A locus alleles in different ethnic groups

	Allele	British Caucasian (%)	N. American Caucasian (%)	N. American Negroes (%)	Japanese (%)
	A1	18.5	16.1	8.1	1.2
	A2	29.1	28.0	16.3	25.3
	A3	14.4	14.1	7.0	0.7
	A9	9.5	9.2	15.7	37.2
A9	Aw23	2.0	1.9	10.6	
	Aw24	6.7	7.3	5.1	
	A10	4.4	5.9	2.7	12.7
A10	A25	1.4	2.6	0.4	
	A26	2.2	3.4	2.3	
	A11	9.2	5.1	2.8	6.7
	A28	4.0	4.2	5.8	0.0
	Aw19	11.1	14.9	25.1	11.9
	A29	4.4	3.6	2.3	0.2
	Aw30	1.6	2.9	13.0	0.5
Aw19	Aw31	1.0	4.5	2.8	8.7
	Aw32	2.4	3.7	1.9	0.5
	Aw33	0.2	1.2	5.1	2.0
	Blank	0.2	1.3	16.5	4.2

Adapted with permission from Batchelor & Welsh (1982). Population numbers ranged from 128 to 290. Data in pink is for an antigen which is split into two or more subcomponents, e.g. A9 subdivides into Aw23 and Aw24.

nucleated cells in the body and are usually detected by lymphocytotoxic tests. This involves the separation of the whole lymphocyte fraction from a blood sample, followed by incubation of the cells with appropriate typing sera and a source of complement. Cell death indicates a positive reaction and is usually evaluated by the uptake of an appropriate dye such as trypan blue. The class I antigens which are typed in this manner are the HLA-A, -B and -C antigens. The expression of these antigens is controlled by three loci and many different allelic variants are possible.

The major alleles recognized at the 1984 HLA workshop are listed in Table 9.8, and the occurrence of some of these in different ethnic groups are listed in Tables 9.9–9.12. The antigens listed in these tables are constantly reviewed at a series of WHO workshops on HLA nomenclature. Antigens which are internationally accepted receive a letter and a number, e.g. HLA-A1 and HLA-B8. Those antigens which have received only provisional recognition also carry a 'w' to denote a qualified workshop acceptance, e.g. Aw-23. In the case

Table 9.10 Gene frequencies of various HLA-B locus alleles in different ethnic groups

Allele	British Caucasian (%)	N. American Caucasian (%)	N. American Negroes (%)	Japanese (%)
B5	4.6	5.9	4.9	20.9
B7	12.5	10.5	12.6	7.1
B8	13.4	10.4	5.5	0.2
B12	20.8	13.8	14.0	6.5
B13	2.0	2.6	0.4	0.8
B14	2.8	5.1	4.6	0.5
B15	6.5	5.9	4.7	9.3
Bw16	3.0	3.9	0.8	6.5
Bw16 [Bw38	0.8	2.5	0.4	1.8
Bw39	1.6	1.4	0.4	4.7
B17	6.2	1.9	4.3	0.2
B18	4.4	3.1	3.6	0.0
Bw21	2.2	3.8	4.4	1.5
Bw22	2.4	2.3	3.9	6.5
B27	4.3	5.6	0.8	0.3
Bw35	7.6	8.6	12.5	9.4
B37	0.8	1.7	1.2	0.8
B40	5.4	9.2	3.9	21.8
Bw41	0.2	0.0	0.0	0.0
Bw42	0.0	0.0	3.2	0.0
Blank	0.9	5.7	11.0	7.6

Adapted with permission from Batchelor & Welsh (1982). Population numbers ranged from 128 to 290.

Table 9.11 Gene frequencies of various HLA-C locus alleles in different ethnic groups

Allele	British Caucasian (%)	N. American Caucasian (%)	N. American Negroes (%)	Japanese (%)
Cw1	5.2	3.5	2.8	12.6
Cw2	5.9	6.0	12.3	0.6
Cw3	14.2	9.6	12.3	32.9
Cw4	8.8	9.9	16.1	3.9
Cw5	7.5	3.9	3.2	0.9
Cw6	6.8	7.3	4.5	1.9
Blank	51.6	59.8	48.8	47.2

Adapted with permission from Batchelor & Welsh (1982). Population numbers ranged from 128 to 290.

of the A and B loci the numbers are always different — this avoids changing long established numbers which were assigned before the A and B loci were recognized as distinct. C loci are, however, numbered sequentially.

Chromosome 6, on which are located the A, B and C loci, is one of 45 pairs of autosomal chromosomes found

Table 9.12 Gene frequencies of various HLA-DR locus alleles in different ethnic groups

Allele	British Caucasian (%)	N. American Caucasian (%)	N. American Negroes (%)	Japanese (%)
DR1	6.0	5.5	8.5	5.0
DR2	13.0	13.0	13.5	16.5
DR3	13.5	11.0	12.5	1.0
DR4	20.0	16.0	13.5	13.0
DR5	9.0	12.0	16.0	13.0
DR6	9.4	12.0	17.5	6.0
DR7	14.0	11.0	15.0	0.5
Blank	15.1	19.5	3.5	45.0

Adapted with permission from Batchelor & Welsh (1982). Population numbers ranged from 128 to 290.

in all cells. Each individual inherits one paternal chromosome 6, plus one maternal chromosome 6, and the A, B and C loci function as codominants. This means that usually two antigens of each series are detected in the cytotoxicity assay — one set of A, B and C antigens controlled by the paternal chromosome 6 (the *paternal haplotype*) and another set of A, B and C antigens controlled by the maternal chromosome 6 (the *maternal haplotype*). A typical inheritance of HLA antigens is illustrated in Figure 9.12. Occasionally less than two antigens of a series will be detected and this indicates either an incomplete panel of identifying antisera or homozygosity at that locus. The A, B and C loci are very closely linked (especially B and C) and genetic recombinants accordingly occur very rarely. (The frequencies usually quoted are 0.8% between A and B loci and 0.2% between B and C loci.) This accounts for the tendency of a linked set of A, B and C antigens to be inherited as a group or haplotype.

As can be seen in Figure 9.12 there is a one-in-four chance that two sibling children would be HLA identical, inheriting the same haplotypes from both parents.

Class II antigens

A fourth locus within the HLA region was originally called the D locus and was recognized because its gene products were able to stimulate a proliferative response in mixed lymphocyte cultures (so-called MLC). The D locus was mapped between the centromere and the

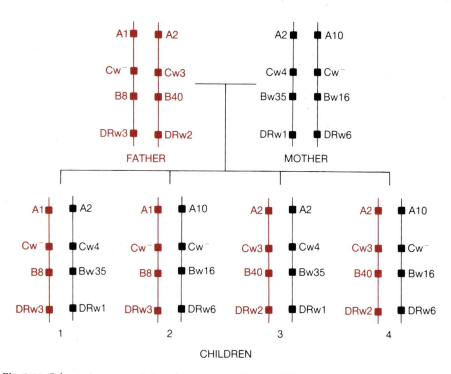

Fig. 9.12 Schematic representation showing inheritance of HLA antigens encoded at the A, B, C and D loci of chromosome 6. Each individual family member inherits both a maternal and a paternal chromosome 6 and will express the gene products of both chromosomes on cell surfaces. Four different combinations of the parental haplotypes are possible and are illustrated in the children. Note that child 3 inherits A2 from both parents and is therefore homozygous at this locus.

B locus, and was found to have a recombination frequency of 0.7% with the latter. In MLC cultures (typically of 5−7 days duration) homozygous stimulator cells are used in an attempt to induce proliferation in the untyped lymphocytes. Failure of stimulation suggests that the unknown cell has the same D allele as that present on the surface of the homozygous typing lymphocyte. D locus typing is time-consuming and technically difficult — not least because of the problems associated with compiling an appropriate library of homozygous typing cells. The most fruitful source of the latter are the offspring of first cousin marriages — as illustrated in Figure 9.13.

The problems of D locus typing led to a great deal of effort to identify the same antigens by serological techniques and the eventual description of a series of cell surface antigens now designated as the DR antigens. These were distinct from the A, B and C series antigens

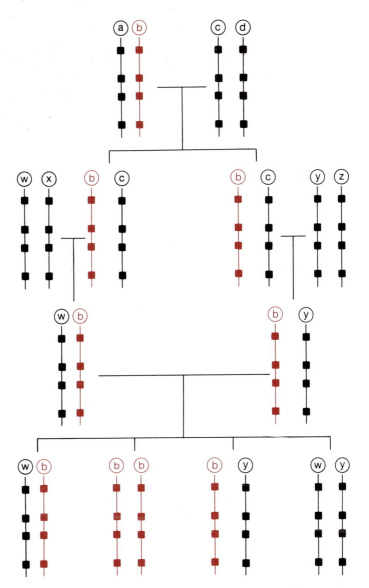

Fig. 9.13 MHC haplotype inheritance pattern involving a first cousin marriage where both partners have inherited the same haplotype (shown in red) from one of the grandparents. There is a one-in-four chance that one of the children of such a marriage will be homozygous for that haplotype. The lymphocytes of these individuals have been extensively used as stimulator cells in mixed lymphocyte cultures for D locus typing.

and were shown to be present on endothelial cells, B cells, macrophages and a subset of T cells. This cellular distribution, and the fact that no cross-over between D and DR has yet been demonstrated, suggests that they are probably one and the same. The International HLA Nomenclature Committee was sufficiently confident that this was so, that DR antigens were given numbers corresponding to the D locus antigens with which they showed an association. Thus DR1 was thought to be equivalent to Dw1 and so on. Nevertheless, a measure of caution was also evident in the choice of 'DR' to symbolize D-related. DR typing is performed in a manner resembling A, B, C typing after first separating out a preparation of DR-positive lymphocytes. The known DR antigens and their population frequencies are shown in Tables 9.8 and 9.12.

It is now clear that the HLA-D region contains a number of closely related genes, specified by at least three distinct loci (see Figure 9.11). These are:

1   The DR locus. This appears to be homologous to the I-E locus in the mouse H-2 complex and its products are identified by alloantisera, T cell responses and monoclonal antibodies. As shown in Figure 9.11, there is a single DR_α gene, the product of which appears to associate with the products of the DR_β genes.

2   The DQ locus. This locus (formerly designated DC, DS and MB) encodes proteins which are identified by alloantisera, T cell responses and monoclonal antibodies. The DQ gene family includes two α genes, $DQ_{\alpha1}$ and $DQ_{\alpha2}$, and two β genes, $DQ_{\beta1}$, and $DQ_{\beta2}$. The gene products of the DQ locus share sequence homology with the mouse I-A α chains.

3   The DP locus. This locus (formerly designated SB) encodes proteins which are identified by T cell responses and monoclonal antibodies. There are two α genes and two β genes in the DP family — $DP_{\alpha1}$, $DP_{\alpha2}$, $DP_{\beta1}$ and $DP_{\beta2}$.

A further Class II locus (DZ/DO) is also recognized. Located between the DP and DQ loci it comprises at least two genes — DZ_α and DO_β.

It seems probable that the β chains of the families associate exclusively with the α chain(s) within their own family, and that the various DR specificities are encoded on the differing β chains. The latter is suggested by the high degree of polymorphism observed in the β_2 structural domains of β chains encoded by all the loci. The delineation of the relationships between the genes, the protein products and the defined Ia specificities is one of the most active areas of current research in human immunogenetics.

Class III antigens

The class III antigens which are encoded within the MHC include three components of the complement system, namely C2, C4 and Factor B. Aspects of the genetics of these important proteins and several other complement components are considered in a later section (see p. 273).

Linkage disequilibrium

The frequencies of the HLA antigens have been established in various population groups, as shown in Tables 9.8—9.12. If one assumes totally random associations between the antigens then a particular combination of alleles should occur at a frequency given by the product of the respective gene frequencies. However, this is clearly not the case for several combinations of antigen. In Caucasian populations the haplotype combinations A1—B8, A2—B12 and A3—B7 are all much commoner than would be expected from random assortment. Similarly, Dw3 and DR3 are also in linkage disequilibrium with B8. There may be selective advantages for individuals having certain haplotypes, but a totally convincing explanation for the mechanism is still awaited.

HLA-disease associations

It has been known for more than 25 years that malignant disease of the gastrointestinal tract is associated with certain ABO blood groups. However, the link is so weak as to be virtually useless for either diagnosis or prognosis. When Lilly et al. (1964) reported an association between certain antigens determined by genes in the MHC of the mouse and susceptibility to leukaemogenic viruses, interest in this field was re-awakened. Immediately, investigators began comparing the frequencies of individual HLA antigens in patients suffering from particular diseases with the frequencies of the same antigens in appropriate controls. Initially attention was focussed on malignant disease but the early studies in acute lymphoblastic leukaemia and Hodgkin's disease gave conflicting results. It seems plausible that the reason for this is that in any disease with a significant early mortality the surviving patients do not constitute the most appropriate study group. In such cases associations should ideally be sought using only newly diagnosed individuals.

It is now recognized that considerable care is required at the planning stage of such studies if all the possible pitfalls are to be avoided. The selection of both the patients and the controls is crucial. Obviously, both should be ethnically homogeneous, and strict pre-defined clinical criteria should be met before acceptance into either study group. In the evaluation of results it is now accepted that allowance should be made for the number of comparisons by multiplying the uncorrected probability value by the number of antigens investigated (see Svejgaard et al., 1974, 1975). It is also important to consider the possible role of linkage disequilibrium (see above) when associations are found. Many disease associations with HLA-B8 have been reported in recent years, but this antigen is in strong linkage disequilibrium with DR3, and it is now believed that the primary association is with some component of the D locus.

The literature on HLA and disease is now extensive and the interested reader who wishes to go further into the subject should consult some of the key references and review articles listed at the end of this section. There are numerous associations with autoimmune disorders and this area is discussed in more detail in Part 4. Some of the associations are tabulated in Table 9.13 together with the appropriate relative risk. The latter is a commonly used expression of the probable risk of contracting the particular disease if one has that HLA antigen.

Mechanisms of HLA-disease associations

It is generally accepted that the biological functions of the HLA gene products are likely to be similar to those of the mouse H-2 gene products. This would lead us to expect the glycoproteins specified by the HLA-A, -B and -C loci to play a key role in effective T cell recognition of specific target antigens, as shown in the mouse by Zinkernagel and Doherty (1977). In contrast, the gene products of the D locus, expressed mainly on the surface of B lymphocytes and macrophages, would be expected to be involved in interactions between those cells and T helper cells. In the mouse the immune response genes of the H-2 I region are believed to

Table 9.13 Selected diseases showing associations with HLA antigens

Disorders	Antigen	Relative risk
Rheumatological disorders		
Rheumatoid arthritis	Dw4	4.2
	DR4	5.8
Juvenile rheumatoid arthritis	B27	4.5
Ankylosing spondylitis	B27	90.0
Reactive arthropathies:		
Reiter's disease	B27	33.0
Post-salmonella arthritis	B27	17.6
Post-shigella arthritis	B27	20.7
Endocrine disorders		
Graves' disease	B8	2.3
	Dw3	3.6
	DR3	3.5
Hashimoto's disease	DR3	2.6
Juvenile onset diabetes	B8	2.6
	DR3	5.7
	DR4	2.8
Neurological disorders		
Myasthenia gravis (late onset disease in females)	B8	12.7
Multiple sclerosis	B7	1.8
	Dw2	4.05
	DR2	4.8
	Dw3	2.7
Dermatological disorders		
Psoriasis vulgaris	B37	6.4
	B17	4.7
	B13	4.7
	Cw6	13.3
Dermatitis herpetiformis	B8	8.7
	Dw3	13.5
	DR3	56.4
Pemphigus:		
Caucasian	A10	2.8
Japanese	A10	6.0
Renal disorders		
Goodpasture's syndrome	DR2	13.1
Berger's disease	Bw35	2.5
Gastrointestinal and hepatic disorders		
Chronic active autoimmune hepatitis	B8	9.0
	DR3	13.9
Coeliac disease	B8	8.3
	Dw3	10.9
Idiopathic haemochromatosis	A3	8.2
	B7	3.0
	B14	4.7

modulate the function of T suppressor cells, and a human counterpart would be anticipated in the region of the HLA-D locus. This immediately suggests a possible mechanism for some of the observed HLA-disease associations.

Although it is conceivable that macrophages possessing different D locus products might differ in the efficiency with which they process antigens, it has been suggested that the associations are not directly with HLA-D locus products but with closely linked genes analogous to the Ir genes of the mouse. There is, however, no direct evidence for such genes in man and some of the best circumstantial evidence is that from Marsh and his colleagues (1977). These workers initially showed a highly significant association between an allergic individual's ability to develop marked skin sensitivity to allergen Ra 5 (following exposure to ragweed pollen) and the possession of the HLA antigen B7. Subsequently it became clear that the strongest associations between a specific IgE response and a particular HLA type were seen in the individuals with lowest total IgE levels (see Figure 9.14). This was seen with both ragweed and Rye grass responsive individuals. Alternative interpretations of these data are possible but it was hypothesized that certain Ir locus alleles controlling responsiveness to the Ra 3 and Ra 5 antigens are in linkage disequilibrium with the alleles HLA-A2 and HLA-Dw2 respectively (see Figure 9.15).

Subsequently both qualitative and quantitative IgG and IgE responses to Ra 5 were found to be strongly associated with HLA-Dw2 (Marsh et al., 1982a). Further studies showed that the primary association of the Ra 5 response was with Dw2 rather than DR2, and that various combinations of A3, B7 and Dw2 were less strongly associated than Dw2 alone. Furthermore, after artificial immunization with ragweed antigens the specific IgG antibody response to Ra 5 was significantly associated with Dw2 ($P < 0.0001$) (Marsh et al., 1982b).

These Dw2−Ra 5 associations are the strongest HLA associations observed with a defined immune response in humans and provide strong evidence for the role of an HLA-D specificity in determining the immune response in outbred, highly polymorphic human populations. Nevertheless, the variable expression of the response suggests the involvement of a further genetic locus.

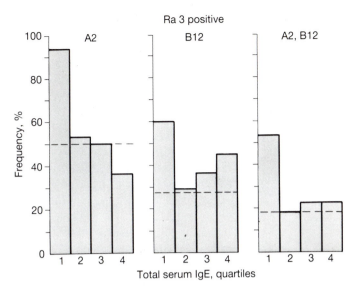

Fig. 9.14 Distribution frequencies of HLA-A2, B12 and the A2, B12 phenotype in 76 North American Caucasian individuals responding to the ragweed antigen Ra 3. Groups subdivided into quartiles based on the total IgE levels of a parent group of 126 allergic individuals. Quartile divisions were at 127, 202 and 430 U ml^{-1}. (Adapted with permission from Marsh *et al.*, 1977.)

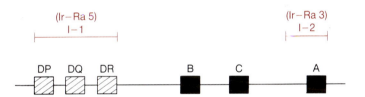

Fig. 9.15 Gene maps of the HLA region showing possible locations of hypothetical immune response subregions. The I-1 region contains the postulated Ir-Ra 5 locus (known to be strongly associated with HLA-Dw2) and the I-2 region contains the Ir-Ra 3 locus which appears to be linked to HLA-A2. (Adapted with permission from Marsh *et al.*, 1977.)

Although the Ir gene hypothesis may explain many of the DR associations there will remain a number of HLA-disease associations requiring an alternative explanation. One of the earliest suggestions was Snell's 'molecular mimicry' hypothesis, and this is believed by many workers to provide a satisfactory explanation of some of the B27 associations. According to this view an invading microorganism shares a close antigenic resemblance to a particular HLA antigen and individuals with the antigen fail to recognize the organism as foreign. It has been claimed that such cross-reactivity exists between B27 and certain *Klebsiella* species.

A third group of hypotheses suggests that HLA antigens may act as receptors and bind viruses, hormones and/or other biological messengers. The associations of Cw6 with psoriasis and A3 with haemochromatosis may be explicable by such mechanisms.

The close proximity of the structural genes for the complement components C2, C4 and Factor B to the HLA-B and HLA-D loci may provide yet another mechanism for some of the observed associations. In immune complex disease with associated C2 deficiency there is known to be defective clearance of antigen— antibody complexes, and C2 deficiency has been closely linked with Dw2.

To date none of the HLA associations with disease has been shown to be absolute, suggesting that possession of a particular tissue type is only one of several aetiological factors involved. The diagnostic and therapeutic relevance of the associations has, accordingly, been rather limited thus far, although most workers remain convinced that there is enormous potential awaiting appropriate exploitation.

References

Bailey D.W. (1970) Four approaches to estimating number of histocompatibility loci. *Transplant. Proc.* **2**, 32.

Bailey D.W. & Usama B. (1960) A rapid method of grafting skin on tails of mice. *Transplant. Bull.* **7**, 424.

Batchelor J.R. & Welsh K.I. (1982) Association of HLA antigens with disease. In *Clinical Aspects of Immunology*, 4th edn, (eds. Lachmann P.J. & Peters D.K.), p. 283. Blackwell Scientific Publications, Oxford.

Dorf M.E., Stimpfling J.H. & Benacerraf B. (1979) Gene dose effects in *Ir* gene-controlled systems. *J. Immunol.* **123**, 269.

Gorer P.A. (1936) The detection of antigenic differences in mouse erythrocytes by the employment of immune sera. *Brit. J. exp. Path.* **17**, 42.

Klein J. (1975) *Mouse Major Histocompatibility Complex*. Springer-Verlag, New York.

Klein J., Figueroa F. & David C.S. (1983) H-2 haplotypes, genes and antigens: second listing. II The H-2 complex. *Immunogenetics* **17**, 553.

Lieberman R., Paul W.E., Humphrey W. Jr & Stimpfling J.H. (1972) H-2 linked immune response (*Ir*) genes: independent loci for Ir-IgG and Ir-IgA genes. *J. exp. Med.* **136**, 1231.

Lilly F., Boyse E.A. & Old L.J. (1964) Genetic basis of susceptibility to viral leukaemogenesis. *Lancet* **ii**, 1207.

Little C.C. (1914) A possible Mendelian explanation for a type of inheritance apparently non-Mendelian in nature. *Science* **40**, 904.

Marsh D.G., Chase G.A., Goodfriend L. & Bias W.B. (1977) Mapping of postulated Ir genes within HLA by studies in allergic populations. *Monogr. Allergy* **11**, 106.

Marsh D.G., Hsu S.H., Roebber M., Ehrlich-Kautzky E., Freidhoff L.R., Meyers D.A., Pollard M.K. & Bias W.B. (1982a). HLA-Dw2: a genetic marker for human immune response to short ragweed pollen allergen Ra 5. I. Response resulting primarily from natural antigenic exposure. *J. exp. Med.* **155**, 1439.

Marsh D.G., Meyers D.A., Freidhoff L.R., Ehrlich-Kautzky E., Roebber M., Norman P.S., Hsu S.H. & Bias W.B. (1982b). HLA-Dw2: a genetic marker for human immune response to short ragweed pollen allergen Ra 5. II. Response after ragweed immunotherapy. *J. exp. Med.* **155**, 1452.

McDevitt H.O., Deak B.D., Shreffler D.C., Klein J., Stimpfling J.H. & Snell G.D. (1972) Genetic control of the immune response: mapping of the Ir-1 locus. *J. exp. Med.* **135**, 1259.

Miller D.A. & Miller O.J. (1972) Chromosome mapping in the mouse. *Science* **178**, 950.

Mitchison N.A. (1954) Passive transfer of transplantation immunity. *Proc. Roy. Soc. Lond. B*, **142**, 72.

Schwartz R.H., Jackon L. & Paul W.E. (1976) T lymphocyte-enriched murine peritoneal exudate cells. I. A reliable assay for antigen-induced T lymphocyte proliferation. *J. Immunol.* **115**, 1330.

Schwartz R.H., Merryman C.F. & Maurer P.H. (1979) Gene complementation in the T lymphocyte proliferative response to poly $(Glu^{57}Lys^{38}Tyr^{5})$: evidence for effects of polymer handling and gene dosage. *J. Immunol.* **123**, 272.

Snell G.D. & Stimfling J.F. (1966) Genetics of tissue transplantation. In *Biology of the Laboratory Mouse*, 2nd edn, (ed. Green E.L.), p. 457. McGraw-Hill, New York.

Svejgaard A., Jersild C., Staud Nielsen L. & Bodmer W.F. (1974) HL-A antigens and disease. Statistical and genetical considerations. *Tissue Antigens* **4**, 95.

Svejgaard A., Platz P., Ryder L.P., Staud Nielsen L. & Thomsen M. (1975) HL-A and disease associations — a survey. *Transplant. Rev.* **22**, 3.

Zinkernagel R.M. & Doherty P.C. (1977) Major transplantation antigens, viruses and specificity of surveillance T cells. *Contemp. Top. Immunobiol.* **7**, 179.

Further reading

Bach F.H. (1985) The HLA class II genes and products: the HLA-D region. *Immunol. Today* **6**, 89.

Benacerraf B. (1981) Role of MHC gene products in immune regulation. *Science* **212**, 1229.

Benacerraf B. & McDevitt H.O. (1972) Histocompatibility linked immune response genes. *Science* **175**, 273.

Bodmer W.F. (1978) The HLA system. *Brit. Med. Bull.* **34(3)**, 324.

Bodmer W.F. (1980) The HLA system and disease. *J. Roy. Coll. Phys. (Lond.)* **14**, 43.

Klein J. (1978) H-2 mutations: their genetics and effect on immune functions. *Adv. Immunol.* **26**, 55.

Klein J. (1986) *Natural History of the Major Histocompatibility Complex*. John Wiley, New York.

Moller G. (1983) HLA and disease susceptibility. *Immunol. Rev.* **70**, 218.

Chapter 10
Molecular genetics of the major histocompatibility complex

Introduction

The application of molecular genetic methods to analysis of the major histocompatibility complex (MHC) of mouse and man began later than the molecular analysis of immunoglobulin genes. The delay reflects the relative difficulty in obtaining cDNA probes for MHC products. However, once an entry was effected into the MHC region, knowledge of the molecular organization of the genes accrued rapidly and about 30% of the murine MHC had been cloned within a couple of years. DNA sequence data continues to accrue. Mouse and human probes have been used interchangeably and the data are discussed in parallel in this chapter.

The molecular genetic map of the murine MHC (Figure 10.1) is strikingly different to the classical genetic map in two major respects. First, many more class I genes have been found than expected: five genes map to polymorphic K and D regions and a further thirty-one genes map in the conserved Qa and TL regions. Second, fewer class II genes have been found than classical genetics predicted; only α and β chain genes corresponding to I-A and I-E subregions have been found and no genes for I-B or I-J products appear to be present in the expected location between the I-A and I-E subregions. The I-B subregion (like I-C) may not exist. The I-J subregion and its products are better documented; possible explanations of the I-J genetics are discussed in the light of the molecular data.

DNA sequences and mapping of restriction enzyme polymorphisms leads to new ideas on the generation of the extensive polymorphism in some MHC genes of both class I and class II while other genes of each class are relatively conserved.

The gene encoding β_2-microglobulin (essential for the expression of class I gene products) has been cloned and sequenced. The intron−exon maps of class I, class II and β_2-microglobulin genes reinforce the evidence for homology (first noted from sequence data) between these genes and those for immunoglobulins.

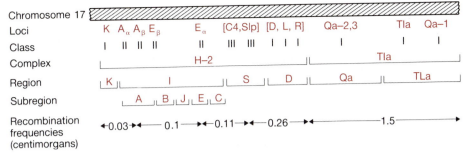

Fig. 10.1 Genetic map of the murine major histocompatibility complex (H-2) located on chromosome 17. The adjacent *Tla* complex encodes many class I genes homologous to the class I genes encoded in the K and D regions of H-2. (From Steinmetz *et al.*, 1982.)

Class I genes

Class I cDNA

The low abundance of class I mRNA (~1.0%) has been circumvented by two strategies:

1 cDNA libraries were screened by positive mRNA selection, cell-free translation and specific immunoprecipitation to identify class I cDNA (see Chapter 7).

2 Synthetic oligonucleotides, with sequences corresponding to class I polypeptide sequences, were used either to screen cDNA libraries or to prime the synthesis of cDNA by reverse transcription from heterogeneous mRNA.

Human class I cDNAs were cloned and selected by the above methods. A human class I cDNA clone was then used to probe murine cDNA libraries. Sequences of class I cDNA clones reveal the presence in the 3′ untranslated region of a repeated element termed the Alu repeat (or the Alu-like repeat in mouse) that is present in 300 000 copies in the human genome. Alu sequences are similar to transposons in having terminal direct repeats. That Alu sequences may function as transposons in mammalian evolution is evidenced by comparing two class I genes, a Qa gene and L^d; the Qa gene contains an insertion of approximately 1000 bp bounded by Alu-like repeat sequences at each end.

Structure of class I genes

A pattern of similarity emerges from the first three murine class I genes to be cloned and sequenced and a single human class I gene. The murine class I genes each have eight exons correlating exactly with the functional domains of the class I product (Figure 10.2). The major difference between the murine and human class I gene structure is that the cytoplasmic domain and 3′ untranslated region are spread over three exons in the murine gene and between two exons in the human gene.

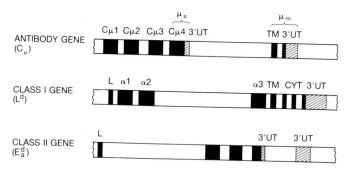

Fig. 10.2 The arrangement of exons and introns in a murine class I gene compared with an antibody (C_μ) and a class II gene. The α_1, α_2 and α_3 exons correspond to the eponomous domains of the protein product. TM encodes the transmembrane domain and three short exons (CYT) encode the cytoplasmic domain and the 3′ untranslated portion of the mRNA. L encodes the signal (or leader) peptide. (From Hood et al., 1983.)

The multigene family encoding class I polypeptides

A single class I cDNA probe used to develop a Southern blot of genomic DNA (mouse or human) reveals about fifteen bands, some of high intensity, implying multiple gene copies per band. Clearly there are at least fifteen class I genes in either the murine or human genomes. Knowledge of the products and in particular amino acid sequence data is sparse, so that assignment of individual class I genes to particular products requires special techniques such as DNA-mediated gene transfer. Indeed, even identifying the polypeptides encoded by a cloned class I cDNA can be contentious due to lack of amino acid sequence data.

One approach to identifying the products of human class I restriction fragments is to analyse, both by Southern blotting and by serology, HLA loss variants generated by γ-radiation of a human cell line. Precise assignment of DNA fragments to HLA products will require analysis of many loss mutants.

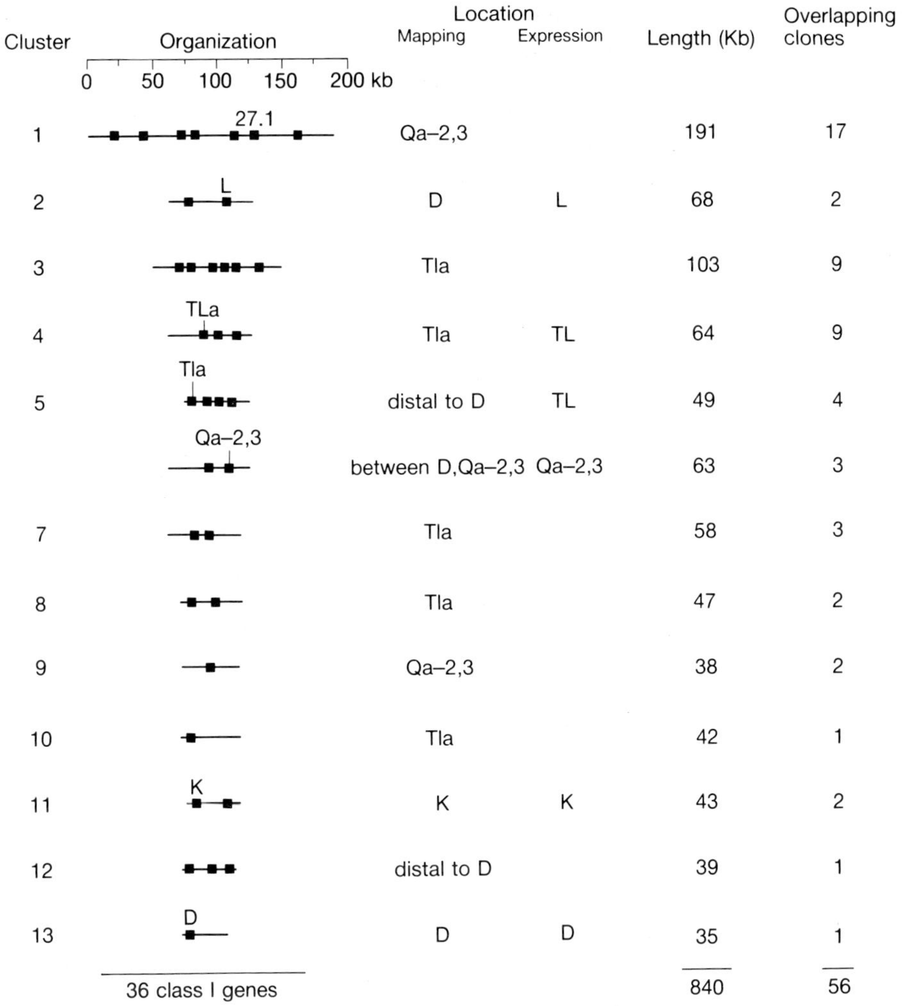

Cluster	Organization	Location Mapping	Location Expression	Length (Kb)	Overlapping clones
	0 50 100 150 200 kb				
1	27.1	Qa–2,3		191	17
2	L	D	L	68	2
3		Tla		103	9
4	TLa	Tla	TL	64	9
5	Tla	distal to D	TL	49	4
	Qa–2,3	between D,Qa–2,3	Qa–2,3	63	3
7		Tla		58	3
8		Tla		47	2
9		Qa–2,3		38	2
10		Tla		42	1
11	K	K	K	43	2
12		distal to D		39	1
13	D	D	D	35	1
	36 class I genes			840	56

Fig. 10.3 Class I genes of the mouse have been isolated in 13 cosmids as shown here. The 36 class I genes (■) were located by genetic mapping using restriction enzyme site polymorphisms (see Fig. 10.5) and by expression upon DNA-mediated gene transfer to suitable cultured mammalian cells.

A more direct approach to ordering the class I genes of the mouse involves cosmid cloning. This method, applied to BALB/C sperm DNA, yielded fifty-six cosmid clones hybridizing to a class I probe. Fine mapping with restriction enzymes was used to identify overlapping DNA inserts and the clones were ordered into thirteen gene clusters (Figure 10.3). In this way a total of thirty-six class I genes have been identified in a stretch of 840 kb of DNA. Most of the class I genes of the BALB/C mouse appear to have been cloned. Cosmid cloning from B10 mouse DNA has yielded seventeen class I genes on seven clusters. It is not known yet whether these numbers reflect a true difference between mouse strains. However, the number of class I genes may well differ between mouse strains. B6 and AKR mice lack expression of the H-2L polypeptide, and Southern blot analysis using specific probe corresponding to the 5'-flanking region of L^d indicates a deletion of that sequence in B6 and AKR DNA.

A similar deletion has occurred in the BALB/C mutants dm1 and dm2, each of which fails to express the L^d gene.

The large number of class I genes had not been predicted by classical genetics. Some pseudogenes may be present in the MHC, but by DNA-mediated gene transfer twenty-one of the thirty-six BALB/C genes can be expressed. Only five genes map in the K, D and L regions and of these only three are expressed after DNA-mediated gene transfer. The expressible genes presumably correspond to the serologically detected K, D and L products and to the genes identified by extensive polymorphism. The remaining thirty-one genes map in the Qa-2, 3 and TL regions. The fact that products of these genes went undetected by classical genetics can be attributed to the limited serological polymorphism of Qa-2, 3 and TL polypeptides.

That the genes in Qa-2, 3 and TL are class I is evidenced by: (1) detection with a class I specific DNA probe (the α_3 domain exon); (2) similarity of exon−intron structure; (3) overall size of the polypeptide product; and (4) expression as a mixed dimer with β_2-microglobulin.

In addition, the sequence data for one Qa-2, 3 gene shows about 80% homology with the genes for classical transplantation antigens (K, D and L). However, TL gene products show many differences when compared to transplantation antigens in peptide maps; it may be that only the α_3 domain is conserved.

Expression of class I genes in Qa-2, 3 and TL regions

Of the thirty-one class I genes mapping in the Qa-2, 3 and TL regions, eighteen are expressed when they are used to transfect cultured murine cells. Expression is measured by a serologically detected increase in β_2-microglobulin on the surface of transformed cells. The increase in β_2-microglobulin is attributed to the presence of a novel, additional class I gene product requiring association with β_2-microglobulin for plasma membrane localization. The products of three of the expressed genes have been characterized serologically as a Qa-2, 3 antigen and two Tla antigens. The other expressed genes appear to code for previously undetected class I antigens; that this is so has been shown for one such cloned gene using a method that could be used for each of them. The first gene on cosmid cluster 1 was used to transform a cell line derived from C3H (H-2^k haplotype) mice and the transformed cells were used to immunize C3H mice. The resultant antiserum reacted with the novel gene product and immunoprecipitation yielded a polypeptide with Mr 45 000. Examination of the tissues of BALB/C mice using the C3H antiserum revealed the presence of the novel class I antigen in low amounts on spleen cells but it was undetectable on other tissues.

It is possible (indeed probable) that individual unassigned class I genes expressed in BALB/C mice (or in the present context expressable when transferred to C3H cells in culture) are not expressed in other mouse strains. Of the known antigens, expression of L, Qa and TL is restricted to certain mouse strains.

Mapping class I genes by restriction enzyme site polymorphisms

The map of class I genes shown in Figure 10.4 was derived by mapping restriction enzyme site polymorphisms. The method is outlined in Figure 10.5. The probes (ii) prepared from each cosmid (i) are selected to hybridize with one or a few genomic restriction fragments. Genomic DNA from parental (P_1 and P_2) and

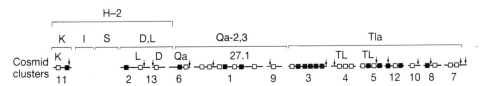

Fig. 10.4 Map of the locations of the 13 class I gene clusters (see Fig. 10.3) in the H-2 and T1a regions of the murine MHC. Genes expressed (■) by DNA-mediated gene transfer are differentiated from non-expressed genes (□). Arrows indicate the locations of probes used for mapping. (Clusters 5 and 12 map either to μa^{-2}, 3 or to T1a). (From Hood *et al.*, 1983.)

recombinant inbred mouse strains (R_1, R_2, R_3, etc.) are digested with various restriction enzymes and Southern blots developed successively with each radioactive probe (iii). In the example, the enzyme $EcoR_1$ generates differently sized fragments specific for each of the two parental strains and for each of the two probes. Finally (iv), the restriction site information is correlated with the serological genetic map to define more exactly the point of crossing-over. From serological evidence recombination is known to have occurred between genes A and B, a distance that might be of the order of 10 kb; restriction enzyme site mapping determines the cross-over point much more exactly. Cloning from recombinant strains and sequencing around the cross-over site could define the site exactly.

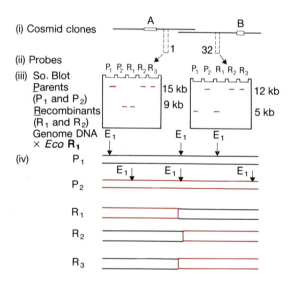

Fig. 10.5 Molecular mapping of recombinants by use of restriction-site polymorphisms. See text for explanation.

Polymorphism and mutation in transplantation antigens could be generated by gene conversion

The major transplantation antigens (K, D and L in the mouse and B, C and A in man) exhibit extreme polymorphism (Chapter 9). Experiments with inbred mice show that mutant transplantation antigens occur with a very high frequency (4×10^{-5}/cell/generation; see Chapter 9). It seems reasonable that these two phenomena might be interrelated. Molecular genetics has shown

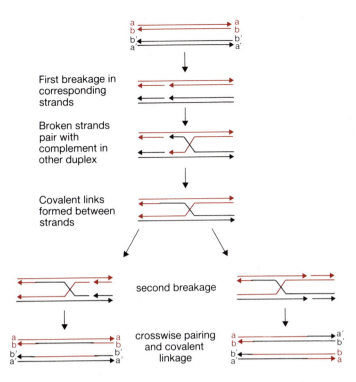

First breakage in corresponding strands

Broken strands pair with complement in other duplex

Covalent links formed between strands

second breakage

crosswise pairing and covalent linkage

Fig. 10.6 Holliday model (1964) for reciprocal recombination between double-helical DNA molecules (genes) with formation of a hybrid DNA. Nicks are introduced into single strands of the same polarity (←,←). These strands can then form new pairings with complementary sequences in the opposite DNA strand and ligation yields new covalent molecules. Repetition of the breakage, pairing and joining as shown on the right effects recombination. If the second breakage occurs in the same strands involved in the first exchange (left hand side) then formal recombination of flanking genetic markers does not occur (although a short, local segment of DNA has been exchanged).

how such a relationship might happen. Nucleotide sequences of wild-type and mutant genes support the hypothesis that the extensive polymorphism and the apparent high rate of mutation in transplantation antigens are predominantly due to non-allelic gene conversion. This hypothesis can also explain two other paradoxes described in Chapter 9: (1) the similarity of the amino acid sequencs of non-allelic transplantation antigens; and (2) the complex nature of allelism with multiple amino acid changes between allelic transplantation antigens and even between wild-type and mutant antigens.

Gene conversion was described by Holliday (1964) as resulting from heteroduplex formation between allelic genes with subsequent mismatch repair by excision of one strand (W) of the heteroduplex and resynthesis of the W strand using the intact C strand as template (Figure 10.6).

This mechanism was invoked to explain allelic gene conversion in fungi in which conversion can occur either at meiosis or mitosis. Also, conversion can occur in either direction.

A similar principle is assumed in the hypothesis of intergenic conversion between related genes in a multigene family.

To demonstrate the possible involvement of intergenic conversion it is desirable to be able to compare both genes participating in the recombinant event before and after the putative conversion. Knowledge of nucleotide sequences of genes before and after conversion would be ideal.

The best evidence for the occurrence of gene conversion amongst class I genes in the MHC comes from cloning and sequencing of a gene for a mutant transplantation antigen (bml, a mutant of the K^b gene). Comparison of the bml sequence with the wild-type allelic sequence and the known sequences of other non-allelic class I genes identified the gene L as a possible partner for an intergenic conversion event changing K^b into K^{bm1}. The sequences are compared in Figure 10.7.

A cluster of seven base differences distinguish the K^{bm1} sequence from the K^b sequence; these seven changes occur within a stretch of 13 bp in exon 3. Five changes alter the codons for amino acids 155 and 156 from Arg.Leu to Tyr.Tyr; these amino acid changes

	145					150		152			155	156				160			165	
K^b	His	Lys	Trp	Glu	Gln	Ala	Gly	Glu	Ala	Glu	Arg	Leu	Arg	Ala	Tyr	Leu	Glu	Gly	Thr	K^b
K^b	CAC	AAG	TGG	GAG	CAG	GCT	GGT	GAA	GCA	GAG	AGA	CTC	AGG	GCC	TAC	CTG	GAG	GGC	ACG	K^b
K^{bm-1}								CT			TAT	TA								K^{bm-1}
L^d	G							CT			TAT	TA							GA	L^d
L^d	Arg							Ala			Tyr	Tyr							Glu	L^d
Gene A	A	A	CT		A	T		TT		A		GAG							G	gene A
Gene A	Glu		Trp			Val		Val				Glu							Ala	gene A

Fig. 10.7 Sequence comparison between K^b, K^{bm-1} and L^d genes between codons 145 and 165; amino acid sequence differences between K^b and K^{bm-1} α chains had been found in this region (see Fig. 3.11). Only nucleotide and amino acid sequence differences from K^b are shown in the K^{bm-1} and L^d genes. The genes can be differentiated by restriction-site mapping using the enzymes Pst I and Hinf I, the sites for which are underlined. Data from Weiss *et al.* (1983).

had been determined in the K^{bm1} polypeptide (see Chapter 3). The remaining two base changes are in the codon for amino acid 152; the resultant change from Glu to Ala had not been observed at the polypeptide level. The possibilities that seven independent point mutations have occurred to generate the K^{bm1} gene, or that a double recombinational event has occurred, are considered unlikely. The data are most easily explained by a gene conversion and any one of the many other class I genes could be the donor of the novel nucleotide sequence. Of the six known H-2 class I sequences only the L^d gene contains the exact sequence found in the K^{bm1} gene (see Figure 10.7).

Since the bm1 mutation was initially detected in a heterozygous (B6 x BALB/C) F_1 mouse (H-2$^{b/d}$) it is possible that the L^d gene could have been the donor for a gene conversion but only if that event occurred in the zygote. It is also possible (and more likely) that the donor was a b haplotype class I gene identical in sequence to the L^d gene in the crucial region. Other mutant class I genes also contain clustered changes in amino acid sequence and apparently identical mutations have been isolated independently. Both of these points support the hypothesis that intergenic conversion accounts for most of the observed mutations. The disparate rates of mutation measured for different class I genes presumably reflect variations in the efficiency of gene conversion; one factor would be the availability of suitable donor genes in a given haplotype.

Restriction enzyme site mapping shows that the extensive polymorphism of K, D and I-A region exons extends through introns and flanking regions (Figure 10.8). Likewise, the limited polymorphism of Qa-2, 3, TL, S and I-E region genes is also a feature of introns and flanking sequences in those regions. Many questions concerning this intriguing regulation of polymorphism are left to be answered. There appears to be a selective advantage for extensive polymorphism in certain class I

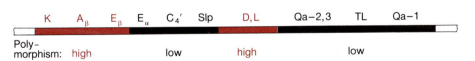

Fig. 10.8 Map showing the regions of high and low polymorphism in the murine MHC as determined by restriction enzyme cleavage sites.

and class II genes. Most of this polymorphism appears to be due to intergenic conversion. Is the extension of this polymorphism to flanking regions due to the same mechanism? The number of suitable donor genes will be a major factor in determining the extent of polymorphism. The number and nature of the genes in the multigene family can be regulated by expansion and contraction involving unequal sister chromatid exchange. The finding that different H-2 haplotypes contain different numbers of class I genes is consistent with the operation of an expansion and contraction mechanism.

The β₂-microglobulin gene

The invariant light chain (β₂-microglobulin) of class I molecules has a strong homology at the amino acid sequence level with an immunoglobulin domain and to the related domains of class I transplantation antigens. The murine gene for β₂-microglobulin has been cloned and the exon−intron organization determined by DNA sequencing (Figure 10.9). The arrangement of the four introns shows a strong similarity to the exons of immunoglobulin genes; one exon encodes primarily the hydrophobic signal sequence; the bulk of the β₂-microglobulin sequence (amino acid residues 3−95) is encoded in the second exon (similar in size to a V-region exon); the third short exon codes for the remainder of the β₂-microglobulin polypeptide and some of the 3′-untranslated region, the bulk of the latter being in the fourth exon.

The single gene for β₂-microglobulin is located on chromosome 2 in the mouse and on chromosome 15 in man; in both cases the β₂-microglobulin gene is syngeneic with the sorbitol dehydrogenase locus, indicating a conservation of autosomal linkage groups.

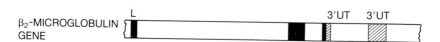

β₂-MICROGLOBULIN GENE

Fig. 10.9 The arrangement of exons and introns in the murine β₂-microglobulin gene. See text for explanation and compare with Fig. 10.2.

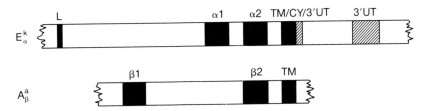

Fig. 10.10 The arrangement of exons and introns in the class II genes E_α^k and A_β^d. (From Hood et al., 1983.)

Class II genes

Class II cDNA

Full length cDNA clones have been prepared and sequenced for several murine and human class II genes. DNA sequences confirm the homologies indicated from amino acid sequences. The homologous murine and human class II genes are as follows:

Mouse: I-Aα, I-Aβ, I-Eα, I-Eβ.

Man: DQα, DQβ, DRα, DRβ.

There appear to be no murine counterparts of human DPα and DPβ genes.

Structure of class II genes

The pattern of exons and introns for the class II genes shows the close interrelationship between these genes and reveals similarities of organization in common with class I genes and the β_2-microglobulin gene (Figure 10.10). The exons of the class II genes generally correspond to the domains of the respective polypeptides. The major external domain exons α_1 and α_2 (or β_1 and β_2) are very similar in size to the α_1 and α_2 exons of class I genes. The Eα and Aβ genes differ slightly in organization. In the Eα gene the transmembrane and cytoplasmic domains are located in a single exon and the 3′-untranslated region is split by an intron; in the latter regard the β_2-microglobulin gene is similarly arranged. The Aβ gene has separate exons for the transmembrane and cytoplasmic domains, a pattern similar to the class I genes.

The human DRα gene has a pattern of exons and introns similar to that of Eα.

The most striking homology is found in a comparison of the DNA sequences of class I α_3 exons with class II α_2 or β_2 exons and with the major exon of β_2-microglobulin.

The class II gene family: the I region molecular map

A molecular map of 230 kb of the I region of the BALB/C mouse has been derived by Steinmetz and colleagues. A chromosome walk was initiated with a human DRα-chain cDNA probe; this probe selected, from a cosmid library of BALB/C sperm DNA, four overlapping clones (shown in red in Figure 10.11) covering about 60 kb around the Eα gene. Two single copy probes (also shown in red in Figure 10.11) were prepared, one from each end of the defined 60 kb region. Using these probes, two cosmid clones lying upstream from Eα and eight lying downstream were selected. Walking along the chromosome in this way generated an overlapping set of twenty cosmid clones that include the downstream boundary of the I region as defined by the C4 complement component gene but not the upstream boundary of the I region as defined by the K gene. Within the cloned limits five class II genes are identified either by known sequences, comparing with cloned cDNA or by cross-hybridization with human cDNA probes. The four genes coding for the serologically defined products Aβ, Aα, Eβ, and Eα occur in that order. The fifth class II gene Eβ$_2$ cross-hybridizes with a human DCβ chain cDNA cloned probe. No product corresponding to the Eβ$_2$ gene is known and the Eβ$_2$ might be a pseudogene. Screening of a cosmid library prepared from AKR mouse DNA (haplotype H-2^k) revealed the presence of an analogous set of five class II genes in the I region and their arrangement is essentially similar to that seen in the H-2^k I region.

The availability of the two cloned I regions of differing haplotype allows comparative restriction enzyme site mapping to determine the extent of polymorphism. That part of the I region (I-A) proximal to the Eα gene is highly polymorphic, as is the K region on the upstream border of I-A. The Eα gene and the DNA downstream, up to and including the C4 complement component gene, are highly conserved between the two haplotypes d and k. This finding correlates with the serological data showing only two alleles each at the Eα, C4 and S1p loci, in contrast to the extensive polymorphism of Aβ and Eβ products. The restriction enzyme site mapping indicates that the extent of polymorphism, either high or low, extends to the sequences flanking the respective class II and class III genes. This finding is directly analogous to that made for class

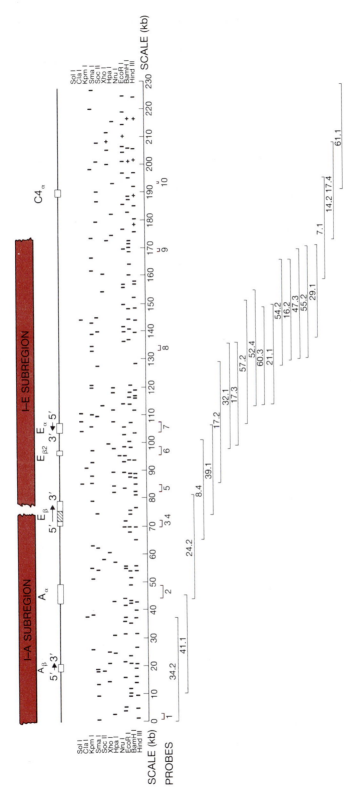

Fig. 10.11 Molecular map of the I region of the MHC of the BALB/C mouse. The locations of the five class II genes identified by probes are shown in black (□). The 20 overlapping cosmid clones used to define this region are listed at the bottom. The map positions of the 10 single-copy DNA probes used to define the location of the I-A and I-E subregions are shown in red. The restriction sites for 11 different enzymes are shown (**1**). (From Hood et al., 1983.)

I genes (see Figure 10.8). The murine MHC divides into two highly polymorphic regions and two conserved regions.

The missing I-B and I-J genes

An almost complete molecular map of the I-A and I-E regions shows that no more than about 1 kb of DNA separates the two regions. If indeed there is any space between I-A and I-E, it is insufficient to accommodate Ia genes in the I-B and I-J regions that appear to map there (see Figure 10.11). Comparison of the molecular map (based on restriction enzyme site polymorphisms in the I region of recombinant congenic mice) with the classical genetic map reveals that recombinant events defining the I-A and I-E subregions have, in all nine cases examined, occurred in approximately (and perhaps exactly) the same place at the Eβ gene (Figure 10.12). Four structural class II genes corresponding to two serologically defined Ia products are located on the molecular map. No gene coding for a unique I-J polypeptide has been found even though I-J products have been identified serologically with both alloantisera and monoclonal antibodies. No I-B subregion product has even been identified serologically. It is most probable that the Ir gene that defined the I-B subregion is explicable as a complementation between the I-A and I-E subregions.

RNA prepared from suppressor T-cell lines, that should contain mRNA for the I-J product, fails to hybridize to DNA sequences located at the I-A/I-E boundary region. In the absence of a definable structural gene corresponding to and mapping in the I-J subregion it has been suggested that either the classical genetic map is incorrect (this could be so if the recombinational events used to define I-J were multiple rather than single events), or the I-J subregion encodes a regulatory element that controls the expression of structural genes located elsewhere.

How many class II genes are present in the murine and human genomes?

This question is relevant to the search for I-J structural genes and to correlations between human and mouse genetics.

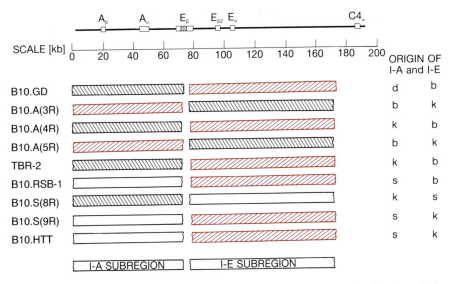

Fig. 10.12 Recombinant events defining I-A and I-E subregions in nine inbred strains of mice map, by restriction enzyme site location (see Fig. 10.11), to within a single stretch of DNA 8 kb or less in length. (From Hood *et al.*, 1983.)

Two class II α genes are detected by Southern blot analysis of BALB/C genomic DNA using the Eα gene as probe. These two α genes are Aα and Eα. Contrary to the evidence from limited N-terminal amino acid sequences (see Chapter 3), the cloned Aα and Eα cDNA sequences reveal about 50% amino acid sequence homology between Aα and Eα. No other class II α genes have been detected at low stringency conditions of probing with Eα.

From the evidence of molecular mapping and Southern blotting there appear to be two functional murine β chain genes, Aβ and Eβ$_1$, with an additional pseudo Eβ$_2$ gene. Thus, there are fewer class II genes in the mouse genome than might have been expected from classical genetic mapping.

The human D region, encoding class II genes, initially seemed less complex than the murine H-2 I region. However, molecular analysis has revealed that the human genome has a greater number of class II genes than the mouse genome. Human class II genes have been classified into three families, DP, DQ and DR. A provisional gene map for the human D region is shown in Figure 10.13. The DP, DQ and DR families contain five α genes and seven β genes; an additional α gene, designated DZα, has not been assigned to one of the families and is not shown in Figure 10.13. Molecular

271 MOLECULAR GENETICS OF THE MHC

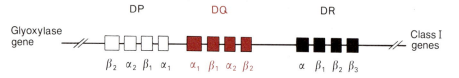

Fig. 10.3 Provisional map of the human HLA-D region encoding class II α and β genes. The order of the genes within the DP family has been ascertained, but the order within the DQ and DR families is not definitely known, nor is the order of the DQ and DR families relative to the class I genes. Based on Bach (1985).

mapping of the D region is yet to be completed, so the order of the families and of the genes within each family is not yet certain.

The products of DR and DQ have been identified by alloantisera and monoclonal antibodies; the DP products have been identified by monoclonal antibodies. At least nine distinct human class II proteins, each consisting of a different α-β chain combination, have been observed.

References

Bach F.H. (1985) The HLA class II genes and products: the HLA-D region. *Immunol. Today* **6**, 89.

Holliday R. (1964) A mechanism for gene conversion in fungi. *Genetic Res.* **5**, 282.

Hood L., Steinmetz M. & Malissen B. (1983) Genes of the major histocompatibility complex of the mouse. *Ann. Rev. Immunol.* **1**, 529.

Steinmetz M., Minard K., Horvath S. *et al.* (1982) A molecular map of the immune response region from the major histocompatibility complex of the mouse. *Nature* **300**, 35.

Weiss E.H., Mellor A., Golden L., Fahrner K., Simpson E., Hurst J. & Flavell R.A. (1983) The structure of a mutant H-2 gene suggests that the generation of polymorphism in H-2 genes may occur by gene conversion-like events. *Nature* **301**, 671.

Chapter 11
Genetic aspects of the complement system

Introduction

Although polymorphism of various complement components was reported in the late 1960s, a major stimulus to contemporary work in this area was the association of the gene coding for mouse C4 with the murine H-2 region. The demonstration that Factor B and C2 were also closely linked to the major histocompatibility complex (MHC) helped to focus the interest of many workers on such problems as the mechanisms underlying the association of various diseases with certain HLA types and the significance of immune complex disease in association with certain complement deficiencies.

Two major types of genetic variation have been described for the proteins of the complement system. These are electrophoretic polymorphism and isolated deficiency states. The latter will be discussed in detail in Chapter 14, so this chapter will largely be confined to a description of known complement polymorphism. Most of this work has been done in man, and data on other animals is still very incomplete. The detection of complement allotypes almost always involves the demonstration of charge differences in the particular component after high voltage electrophoresis or isoelectric focusing. Different techniques are used to detect the component of interest. Simple protein staining will suffice for C3 allotypes but immunofixation using appropriate specific antisera is required for many of the components present at lower concentrations (e.g. Factor B, C6, etc.). Alternatively, highly sensitive haemolytic overlay techniques may be used in which the electrophoretic plate is covered with a gel containing red cells and all the reagents necessary for haemolysis except the component in question. Some of these techniques are described in greater detail in Lachmann and Hobart (1978).

Some of the recognized human complement allotypes are listed in Table 11.1, together with data of allelic

Table 11.1 Human complement allotypes.

Component	Principal alleles	Allelic frequencies		
		Caucasian	Negroid	Mongoloid
C4	C4*A3	0.71		
	C4*B1	0.75		
C2	C2*C	0.96	0.97	0.97
	C2*B	0.04	0.03	0.03
	C2*A	<0.01	—	—
C3	C3*S	0.77−0.81	0.93	0.99
	C3*F	0.19−0.22	0.06	0.01
C5	C5*1	0.93		
	C5*2	0.07		
C6	C6*A	0.6	0.56	0.59
	C6*B	0.37	0.38	0.35
C7	C7*1	0.98		
	C7*2	0.01		
	C7*3	0.01		
C8	C8*A	0.67	0.67	0.79
	C8*B	0.34	0.19	0.21
	Bf*S	0.71−0.83	0.24−0.44	0.82−0.89
B	Bf*F	0.15−0.28	0.51−0.65	0.11−0.18
	Bf*S1	0.01	—	—
	Bf*F1	—	0.05	—
D	Df*1	1.00	0.98	
	Df*2	0	0.02	

Modified from Lachmann and Peters (1982).

frequencies in Caucasian, Negroid and Mongoloid populations.

C4

C4 is one of three complement components known to be coded within the MHC. The close linkage of the C4 gene to the HLA-B locus is shown by the absence of recombinants between C4 and HLA-B (Table 11.2) There is also clear evidence of linkage disequilibrium between the C4 alleles and various HLA haplotypes. The study of C4 polymorphism has been particularly difficult and until recently there were two concepts on the genetics of this component. One of these strongly favoured the existence of two closely linked loci (possibly as a result of a gene duplication) which were thought not to be allelic to each other. The other view remained uncommitted with respect to the number of loci and the formal genetics.

Table 11.2a Linkage of human C4 to various chromosome 6 loci

	HLA loci				Other loci	
	A	C	B	D	Bf	GLO*
Ratio of non-recombinants to recombinants	155:3	48:0	165:0	7:0	53:0	35:2

*GLO = glyoxalase.

Table 11.2b MHC associations of C4 alleles.

Ascertained for	Associated allele
C4*A6	HLA-B17 (Bw57)
C4*A6	HLA-B37
C4*A6	HLA-B27
C4*A2	HLA-Bw50
C4*A2	Bf*S1
C4*B3	HLA-Bw47
C2 deficiency	C4*A4, C4*B2

Modified from Lachmann and Hobart (1978).

In 1982 an agreement was reached at an international workshop to adopt a common nomenclature for C4 allotypes. This is based on agarose gel electrophoretic separation of neuraminidase-treated C4 followed by appropriate immunofixation with an anti-C4 antiserum. This technique reveals a range of allotypic variants with different electrophoretic mobilities (Figure 11.1). The protein of which most variants have a more acidic (anodal) migration — previously termed F (fast) — is now to be called C4A, whereas the protein of which the majority of variants have a more basic (cathodal) migration — previously termed S (slow) — is now to be called C4B. The different C4A and C4B variants will now be classified according to electrophoretic mobility, functional haemolytic activity and, whenever possible, by α chain typing.

In general, each C4 allotype pattern consists of at least one major anodal band and two minor cathodal bands. The most common allotypes have been numbered A1−A6 and B1−B7 for allotypes migrating from cathode to anode. In the case of those allotypes migrating more slowly than A1 or B1 the convention will be to number sequentially from anode to cathode with a number 9 prefix, e.g. A91, A92, A93, etc. At least thirteen alleles of C4A and twenty-two of C4B are known to exist (Mauff et al., 1983).

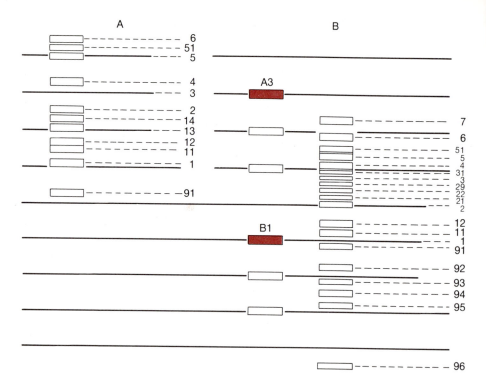

A   B

6
51
5

4   A3
3

2
14
13
12
11
1

—91

7
6
51
5
4
31
3
29
22
21
2

B1

12
11
1
91

92
93
94
95

96

Fig. 11.1 Relative electrophoretic positions of human C4 allotypes. (Except for C4A 3 and C4B 1, only the major bands are shown.)

The symbols adopted for genes, alleles and phenotypes of C4 are those embodied in the International System for Human Gene Nomenclature (ISGN). For example, the most common C4A allotype is C4A 3 and the most common C4B allotype is C4B 1 (see Figure 11.1). The respective alleles are *C4A*3* and *C4B*1*.

Provision exists for the designation of newly discovered allotypes and this should not interfere with the nomenclature for existing variants.

A totally unexpected aspect of C4 polymorphism was reported by O'Neill *et al.* in 1978. These workers were able to show that the Chido and Rodgers blood group antigens are in fact related to the C4B and C4A variants respectively (called S and F at that time). It appears that C4b fragments are readily deposited on red cell surfaces possibly after activation of the critical thioester bond discussed in Chapter 4 and are thus readily detected by standard blood grouping procedures.

Recently, further remarkable findings concerning C4 polymorphism have been published by Law *et al.* (1984). It appears that C4B binds to antibody-coated red cells twice as effectively as C4A, whereas the opposite is found with immune aggregates containing the protein antigen BSA. At the molecular level it could be shown that C4B forms covalent (ester) bonds with hydroxyl groups much more effectively than does C4A. In contrast, the latter is better able to form covalent (amide) bonds with amines than is C4B. The extent of these differences is astonishing if one considers the small number of differences in the amino acid sequences between the two forms of C4 molecule. Belt *et al.* (1984) have detected differences in only thirteen residues out of 1722 and only four of these are certainly ascribed to class differences between C4A and C4B (see Figure 11.2). It is suggested by Law *et al.* (1984) that the duplication of the C4 gene has been biologically advantageous because it facilitates complement activation with a wide range of structurally different antigens and antibodies. It is also tempting to speculate that these differences may be a factor in the susceptibility to systemic lupus erythematosus and other autoimmune diseases, which correlate with the presence of particular tissue type haplotypes flanking the C4 loci.

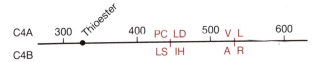

Fig. 11.2 Part of the α chain of C4 showing differences in the amino acid sequences between C4A and C4B. Residue numbering is from the N terminus. (Data derived from Belt *et al.*, 1984.)

C2

The immunogenetics of C2 are rather confusing. Allotypes occur in man and are clearly in close linkage with various MHC genes (Table 11.3). This much is generally accepted. More controversial is the interpretation of C2 deficiencies in various population studies. High frequencies of low C2 and low C4 concentrations, particularly in association with B18, Dw2, have been interpreted by some authors as evidence of heterozygous C2 deficiency. However, there is a bimodal distribution of both C2 and C4 concentrations in healthy Dw2-positive blood donors which suggests that homozygous deficiencies of both C2 and C4 should occur at much higher frequencies than is the case. A possible explanation of this

Table 11.3a Linkage of human C2 to various chromosome 6 loci

	HLA loci				Other loci	
	A	C	B	D	Bf	C4
Ratio of non-recombinants to recombinants	55:1	35:0	83:0	29:0	12:0	10:0

*GLO = glyoxalase.

Table 11.3b MHC associations of C2 alleles

Ascertained for	Associated allele
C2*B	HLA-B15
C2*B	Bf*S
C2⁰	HLA−B18

Modified from Lachmann and Hobart (1978).

discrepancy can be reached by invoking duplication of both the C4 and the C2 loci. The peak of low C2 concentrations would then be explained by the presence of a null allele at only one of the duplicated loci.

However, recent data suggests that the C2 and Factor B genes are single copies less than 1 kb apart. Moreover, as might be predicted for two such structurally similar proteins, cDNA sequence analysis has revealed the presence of internal homology regions similar to those recently reported for Factor B (see p. 281). Nevertheless, overall there is only 35% sequence homology between the two proteins and it is, as printed out by Reid (1985), surprising that the serum concentration of Factor B is some fifteen-fold greater than that of C2.

C3

C3 is quantitatively the predominant complement component and as such is readily visualized after most electrophoretic separations. Polymorphism was first described in 1968 and in most populations two major variants — designated F (fast) and S (slow) — have been demonstrated following high voltage agarose gel electrophoresis. C3*S is the most common allele and migrates relatively slowly on prolonged electrophoresis (see Figure 11.3). It has a gene frequency of 0.77−0.80 in Caucasians and a much higher frequency in Negroids

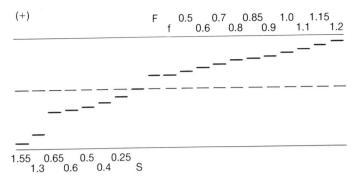

Fig. 11.3 Relative electrophoretic positions of the known variants of C3 in man.

and Mongoloids. The other common or fast allele, C3*F, has a gene frequency of 0.19–0.22 in Caucasians. The F and S variants are inherited as autosomal codominant alleles. A further sixteen less common variants have also been described and all have normal C3 haemolytic activity. Their accurate typing demands critical attention to the Ca^{2+} ion concentration since many of them have altered Ca^{2+} binding properties. All of the structural determinants of these allotypes are confined to either the α or β peptide chain. After earlier studies had shown that human C3 allotypes were not apparently linked to the MHC locus somatic cell hybrids and a molecular probe isolated with the aid of a mouse C3 cDNA clone were used to localize the C3 gene on chromosome 19. In contrast the mouse C3 locus appears to map within 12 centimorgans of the H2 histocompatibility complex on chromosome 17.

The mouse cDNA library of C3-specific clones has been used to demonstrate the existence of a nucleotide sequence coding for a signal peptide at the 5′ end of the pro-C3 coding sequence. This suggests that the secretion of C3 involves a process resembling that described for immunoglobulin light chains and many other molecules. The 'leader' peptide sequence is lost during the process of secretion—presumably as a result of proteolytic cleavage.

The complete cDNA sequence of human C3 has been reported by de Bruijn and Fey (1985) and confirms that the precursor C3 molecule comprises a 22-residue signal peptide, a region of 645 residues corresponding to the β chain and a region of 992 residues corresponding to the α chain. The α and β chains are separated by four basic residues which are believed to be those excised during conversion of pro-C3 to the native two-chain form.

An increased frequency of the C3F variant has been noted in patients with mesangiocapillary glomerulone-phritis and partial lipodystrophy. These individuals show accelerated destruction of their C3 and often have evidence of a circulating IgG antibody called nephritic factor which binds to the alternative pathway conver-tase, thereby protecting it from the action of the regula-tor proteins.

Factor B

Factor B polymorphism was established in 1969 by a combination of high voltage electrophoresis and immunofixation. Two common alleles, now designated Bf*S and Bf*F, and two rare alleles Bf*S1 and Bf*F1 were described. At least three other rare alleles are now known to exist. The product of each allele separates into four or five bands on electrophoresis (see Figure 11.4).

Whereas the common allelic variants F and S appear to be associated with structural features located on the Ba fragment of Factor B (mol. wt. 30 000), the minor variants F1 and S1 are defined by structural differences in the larger Bb fragment (mol. wt. 60 000).

cDNA clones corresponding to Factor B and ranging in size from 1.0 to 2.3 kb have now been isolated from a human adult liver library. Preliminary studies using the largest clone showed that it was able to hybridize with both mouse and guinea pig Bf sequences, spanning some 90% of the nueclotide sequence. Recent detailed study of the Factor B gene (see Figure 11.5) reveals eighteen exons of which three appear to encode internal homology regions of approximately 60 amino acid res-idues. All are present in the Ba fragment and appear to resemble similar homology regions in C2 and C4b-binding protein. The carboxy-terminal half of the cataly-tic Bb fragment is similarly homologous to the catalytic chains of other serine proteases. In each case there are distinct exons encoding the functionally important regions but in addition there is one exon in Factor B which appears to be unique. Whether this will also be a feature of the exon structure for C2, and its precise significance, remain to be investigated.

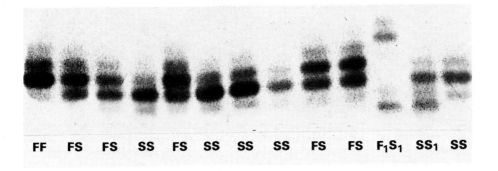

FF FS FS SS FS SS SS SS FS FS F₁S₁ SS₁ SS

Fig. 11.4 Electrophoretic patterns of the common genotypes of human Factor B. (Kindly provided by Susan Cross, Genetics Department, University of Oxford.)

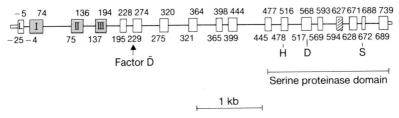

Fig. 11.5 Structure of the gene coding for human Factor B. The boxes represent exons and the numbers refer to the amino acid residue number in the protein product. The Factor D cleavage site is indicated. L is the exon encoding the leader peptide. I, II and III are three homologous regions in the Ba fragment. In Factor B the serine proteinase domain is located in the C-terminal half of the Bb region. Codons for the active site residues h (histidine), D (aspartic acid) and S (serine) are distributed as shown. The shaded exon has no known counterpart in other serine proteinases. (Adapted from Campbell *et al.*, 1984.)

It was established at a very early stage that the Factor B locus is very close to the MHC region in man, the mouse and the guinea pig. In man genetic recombinations between the Factor B locus and the HLA-B, C2 and C4 loci are exceedingly rare (see Table 11.4a). Furthermore, there is clear evidence of linkage disequilibrium between HLA-B antigens and C2 complement allotypes on the one hand and Factor B alleles on the other (Table 11.4b).

cDNA probes specific for C4, Factor B and C2 have recently been used to map the genes for these three complement proteins in the HLA class III region of chromosome 6, by overlapping cloned fragments from a cosmid library of human genomic DNA (Carroll *et al.*, 1984). In 98 kb of genomic DNA four class III genes were identified. Two C4 loci (corresponding to C4A and C4B) were separated from each other by approximately 10 kb. Some 30 kb further towards the centromere were

Table 11.4a Linkage of human Factor B to various chromosome 6 loci

	HLA loci				Other loci		
	A	C	B	D	C2	C4	G
Ratio of non-recombinants to recombinants	315:6	58:0	321:0	24:3	12:0	53:0	9

*GLO = glyoxalase.

Table 11.4b MHC associations of Bf alleles

Ascertained for	Associated Factor B allele
HLA-B8	Bf*S
HLA-Bw35	Bf*F
C2*B	Bf*S

Modified from Lachmann Hobart (1978).

single loci for C2 and Factor B less than 2 kb apart (see Figure 11.6).

Other complement components

Polymorphism of both C6 and C7 is well documented in man and the frequencies of the principal alleles are shown in Table 11.1. It is known that these two complement components, which are structurally similar (see Chapter 4), show strong genetic linkage. However, to date, it has not been possible to assign these genes within the genome other than to exclude a close linkage with the MHC.

C8 is a three-chain protein in which the α and γ chains are covalently linked and associated non-covalently with the β chain. The cases of C8 deficiency described to date (see Chapter 14) have been characterized as functionally defective molecules lacking either the α and γ chains or the β chain. This suggests that two distinct structural genes code for these chains and that the functional molecule is assembled post-synthetically. Neither of these genes appear to be HLA linked, and recent work suggests that the α and β genes of C8 are encoded on chromosome number 1.

The cDNA-derived amino acid sequence of human C9 is now available (Di Scipio et al., 1984) and is remarkable for the contrast between the largely hydrophilic character of the N-terminal half and the essen-

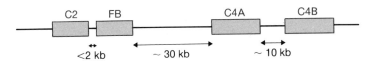

Fig. 11.6 Map of 98 kb section of the short arm of the human chromosome 6 showing the location of the four complement genes coding for C2, Factor B (FB) and the two C4 genes C4A and C4B. These so-called class III genes map between the Class I HLA-B locus and the Class II DR locus. (Data extracted from Carroll *et al.*, 1984.)

tially hydrophobic nature of the C-terminal half. It is tempting to equate this structural feature with the insertion of the C-terminal portion of C9 into phospholipid membranes and the subsequent development of membrane lesions.

The C4b-binding protein is one of the major control proteins of the complement system and the isolation of a human cDNA clone has now permitted the prediction of the complete amino acid sequence of the 549 residues comprising each subunit peptide chain. There appear to be eight internal homology regions of approximately 60 residues and three of these resemble similar homology regions in both Factor B and C2.

The application of recombinant DNA technology to studies of complement proteins has yielded impressive advances in a very short time. The human and mouse complement proteins for which genomic or cDNA clones had been isolated (up to March, 1985) are listed in Table 11.5. It seems likely that many more proteins will be added to the list in the very near future.

Complement receptors

At the time of writing four allotypic variants of CR1 have been described (Table 11.6) and are known to be expressed co-dominantly. Despite the large differences in mol. wt. no functional differences have been ascribed to the variants.

There is growing evidence to suggest that the CR1 membrane protein, Factor H and C4-binding protein are encoded by a closely linked family of genes, not itself linked to the HLA complex.

Immunogenetics of complement in other animals

There is evidence of polymorphic genes controlling C2,

Table 11.5 Complement proteins for which genomic or cDNA clones have been isolated

Protein	Species	Type of clone	Coding sequence
C1q (B chain)	Human	cDNA	Partial
	Human	Genomic	Complete
C2	Human	cDNA	Partial
C4	Human	cDNA	Complete
	Human	Genomic	Partial
	Mouse	cDNA	Partial
	Mouse	Genomic	None
Factor B	Human	cDNA	Partial
	Human	Genomic	Complete
	Mouse	cDNA	Partial
	Mouse	Genomic	None
C3	Human	cDNA	Complete
	Mouse	cDNA	Complete
	Mouse	Genomic	Partial
C5	Human	cDNA	Partial
C9	Human	cDNA	Complete
C4b-BP	Human	cDNA	Complete
C1-INH	Human	cDNA	None
Factor H	Human	cDNA	None

Adapted with permission from Reid (1985).

Table 11.6 Polymorphism of CR1 receptors

Allotypic variant	Mol. wt.	Gene frequency
A	190 000	0.83
B	220 000	0.16
C	160 000	0.01
D	250 000	<0.01

C4 and Factor B encoded within the MHC of the mouse, Rhesus monkey and guinea pig. In fact, mouse C4 polymorphism was the first example of a genetically determined variation of a complement component to be described (Schreffler & Owen, 1963). At that time the protein was designated Ss and neither its true identity nor the chromosomal location of its structural gene between H-2K and H-2D of chromosome 17 were established until much later.

The C4 gene in the mouse is also duplicated, but the protein product of one of the genes (Slp) is not haemolytically active. The thiolester site of this variant differs from that of active C4 and is not readily cleaved by C1.

In the mouse the C3 locus is located on the same chromosome as the H2 complex (chromosome 17), but the loci are clearly distinct. As in man the protein is coded by a single gene.

The MHC-linked complement genes of the mouse are coming under increasing investigation and the sequence has been found to mirror that in man, i.e. $5'$-C2-B$_f$---C4-$3'$. The Slp gene (see above) appears to be closer to B$_f$ than the gene for C4.

The cDNA derived amino acid sequence of the catalytic Bb fragment of mouse Factor B has a remarkable 83% homology with the human Bb sequence, suggesting that the structure of the protein has changed little during evolution.

Another mouse complement protein which appears to be encoded by a gene in the MHC region is C4BP. This locus has been placed between H-2D and Qa-2 on chromosome 17 by Takahasi et al. (1984) but, interestingly, in man it is not linked to the HLA locus.

References

Belt K.T., Carroll M.C. & Porter R.R. (1984) The structural basis of the multiple forms of human complement component C4. *Cell* **36**, 907.

Campbell R.D., Bentley D.R. & Morley B.J. (1984) The Factor B and C2 genes. *Phil. Trans. Roy. Soc. Lond. B* **306**, 367.

Carroll M.C., Campbell R.D., Bentley D.R. & Porter R.R. (1984) A molecular map of the human major histocompatibility complex Class III region linking complement genes C4, C2 and Factor B. *Nature* **307**, 237.

De Bruijn M.H.L. & Fey G. (1985) Human complement component C3: cDNA coding sequence and derived primary sequence. *Proc. natn. Acad. Sci. USA* **82**, 708.

Di Scipio R.G., Gehring M.R., Podack E.R., Kan C.C., Hugli T.E. & Fey G.H. (1984) Nucleotide sequence of cDNA and derived amino acid sequence of human complement component C9. *Proc. natn. Acad. Sci. USA* **81**, 7298.

Lachmann P.J. & Hobart M.J. (1978) Complement technology. In *Handbook of Experimental Immunology*, 3rd edn., (ed. Weir D.M.), p. 5A1. Blackwell Scientific Publications, Oxford.

Lachmann P.J. & Peters D.K. (1982) Complement. In *Clinical Aspects of Immunology*, Vol. 1, 4th edn., (eds. Lachmann P.J. & Peters D.K.), p. 18. Blackwell Scientific Publications, Oxford.

Law S.K.A., Dodds A.W. & Porter R.R. (1984) A comparison of the properties of two classes, C4A and C4B, of the human complement component C4. *EMBO J.* **3**, 1819.

Mauff G., Alper C.A., Awdeh Z., Batchelor J.R., Bertrams J., Bruun-Petersen G., Dawkins R.L., Demant P., Edwards J., Grosse-Wilde H., Hauptmann G., Klouda P., Lamm L., Mollenhauer E., Nerl C., Olaisen B., O'Neill G., Rittner C., Roose M.H., Skanes V., Teisberg P. & Wells L. (1983) Statement on the nomenclature of human C4 allotypes. *Immunobiology* **164**, 184.

O'Neill G.J., Young Yang S., Tegoli J., Berger R. & Dupont B. (1978) Chido and Rodgers blood groups are distinct antigenic components of human complement C4. *Nature* **273**, 668.

Reid K.B.M. (1985) Application of molecular cloning to studies on the complement system. *Immunology* **55**, 185.

Schreffler D.C. & Owen R.D. (1963) A serologically detected variant in mouse serum: inheritance and association with the histocompatibility-2 locus. *Genetics* **48**, 9.

Takahasi S., Takahasi M., Kaidoh T., Natsuume-Saki S. & Takahasi T. (1984) Genetic mapping of mouse C4-BF locus to the H-2D-Qa interval. *J. Immunol.* **132**, 6.

Further reading

Alper C.A. (1981) Complement and the MHC. In *The Role of the Major Histocompatibility Complex in Immunobiology*, (ed. Dorf M.E.), p. 173. Garland Press Publishing Inc., New York.

Alper C.A. & Nathan D.G. (1981) Serum proteins and other genetic markers of the blood. In *Hematology of Infancy and Childhood*, Vol. 2, (eds. Nathan D.G. & Oski F.A.), p. 1459. Saunders,

Colten H.R. (1983) The complement genes. *Immunol. Today* **4**, 151.

Colten H.R., Alper C.A. & Rosen F.S. (1981) Current topics in immunology. Genetics and biosynthesis of complement proteins. *New Engl. J. Med.* **304**, 653.

Hauptmann G. (1979) Genetic polymorphism of human complement proteins. *Blood Trans. Immunohaem.* **22**, 587.

Lachmann P.J. & Hobart M.J. (1978) Complement genetics in relation to HLA. *Brit. Med. Bull.* **34**, 247.

Lachmann P.J. & Rosen F.S. (1978) Genetic defects of complement in man. *Springer Semin. Immunopath.* **1**, 339.

O'Neill G.J. (1985) Complement component polymorphism. In *Methods in Complement for Clinical Immunologists*, (ed. Whaley K.), p. 266. Churchill Livingstone, Edinburgh.

Porter R.R. (1984) The complement components of the major histocompatibility complex. *CRC Crit. Rev. Biochem.* **16**, 1.

Reid K.B.M. (1985) Application of molecular cloning to studies on the complement system. *Immunology* **55**, 185.

Part 3
Cells of the
Immune System

Chapter 12
Differentiation antigens

Introduction

Cellular differentiation is a function of a pattern of gene expression characteristic of the functional state of the cell. The expression of genes encoding cell surface polypeptides or encoding enzymes involved in the biosynthesis of glycoproteins or glycolipids can be viewed as a microcosm of the differentiated state. Cell surface molecules recognized by specific antibodies as characteristic of a particular cell type or differentiation state are termed differentiation antigens. That such antigens are expressed on cells at a given stage of differentiation (or through a series of stages) does not necessarily imply that the antigens are functionally related to the differentiated function of the cell.

Differentiation antigens are powerful markers for genetic studies, particularly somatic cell genetics. The use of differentiation antigens in the characterization and functional dissection of cell types that are important in the immune system is an excellent illustration of the power of the methodology. Initially, conventional allo-antisera were used to define differentiation antigens of cells involved in the murine immune system. More recently, monoclonal antibodies have proved invaluable in defining human differentiation antigens. Identification and isolation of cell surface proteins displaying differentiation antigens has been facilitated by the use of monoclonal antibodies. Now the cloning of genes encoding differentiation antigens is adding to our knowledge of these molecules.

Prior to describing differentiation antigens of cells of the immune system, the major cell types are introduced in the following paragraphs.

Immunocompetent cells

Lymphocytes and their progeny are the cells which exhibit or control immunologically specific functions, i.e. immunocompetent cells. A variety of lymphocytic cells distinguished by simple histological techniques are found in peripheral blood and in lymphoid organs; *primary*: thymus and bursa of Fabricius (*aves*) or bone marrow (*mammalia*); *secondary*: lymph nodes and spleen. The quiescent or resting nature of small lymphocytes (7 − 12 µm diameter) is indicated by the deeply basophilic nucleus, consisting of densely packed chromatin, and by the very small volume of cytoplasm (Figure 12.1). Small lymphocytes respond to immunogenic stimuli by proliferation and differentiation, giving rise to effector cells which are responsible for immune functions, i.e. the humoral and cellular responses.

Stem cells

Lymphocytes arise from stem cells of the haemopoietic system. Stem cells arise in the yolk sac, proliferating in fetal liver and adult bone marrow. Proliferating stem cells can undergo a variety of differentiating influences which take the daughter cells into either lymphoid or myeloid compartments (Figure 12.2). Murine stem cells are assayed by the ability of each cell to form a macroscopic colony of proliferating myeloid cells in the spleens of irradiated mice used as adoptive hosts. An

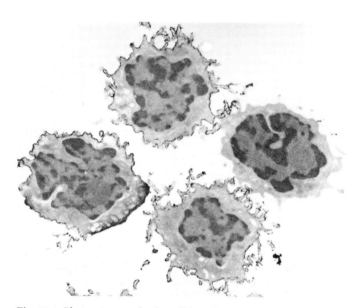

Fig. 12.1 Photomicrograph of small lymphocytes stained with peroxidase conjugated anti-IgM with diamino benzidine-osmium detection. Some lymphocytes have surface IgM (B cells) and others do not (T cells). Photograph kindly supplied by Prof. B.D. Lake, © Prof. B.D. Lake.

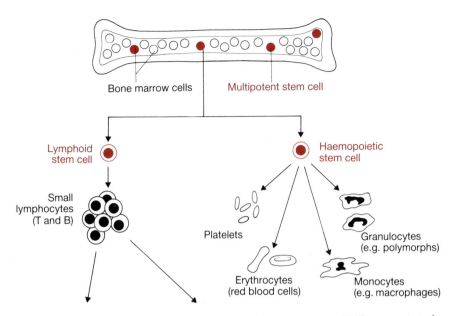

Fig. 12.2 Pluripotential stem cells proliferate in adult bone marrow and differentiate into the red and white blood cells that populate the periphery.

experiment performed by Trentin and colleagues to demonstrate the pluripotentiality of stem cells is illustrated in Figure 12.3. The number of discrete colonies in the spleens of secondary hosts indicates the number of stem cells repopulating the host. As few as four stem cells generate 'complete immunocompetence', as judged by the response to three randomly chosen test antigens. This result supports the view that each stem cell has most, if not all, of the genetic information needed to produce the complete spectrum of antibody specificities (see Chapter 6).

A modified form of the stem cell repopulation experiment demonstrates that there are stem cells that have undergone various degrees of commitment, i.e. they are progressively restricted with respect to the cell types into which they can differentiate. Irradiation of the stem cell population prior to transfer generates random chromosome damage yielding characteristic chromosomal markers in individual cells. The distribution of an individual marker among differentiated cell populations in the reconstituted animal allows the precommitment and differentiation potential of the stem cell originally sustaining the chromosomal damage to be assessed (Figure 12.4).

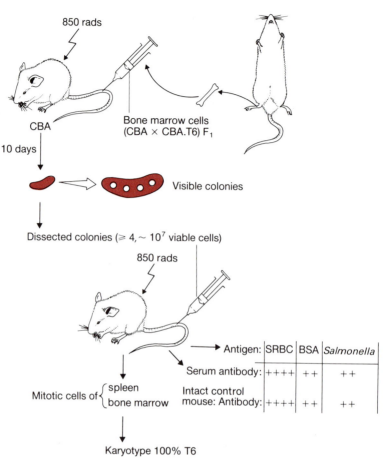

Fig. 12.3 Lethally-irradiated mice repopulated with as few as four clones of lymphoid cell precursors produce antibodies to diverse antigens at levels comparable to those produced in normal mice. The precursor cells are derived from a T6 chromosomally marked mouse, and in the test animal 100% of mitotic figures in spleen and bone marrow show the T6 karyotype. (From Trentin *et al.*, 1967.)

T and B cells

Two separate lineages of lymphocytes are generated according to the environmental influence exerted on the stem cell. In the thymus, epithelial cells exert a hormonal influence on stem cells to generate T lymphocytes (thymus-derived lymphocytes) that pass into the circulation and colonize the secondary lymphoid organs. The other class of lymphocytic cells found in the circulation and in secondary lymphoid organs are

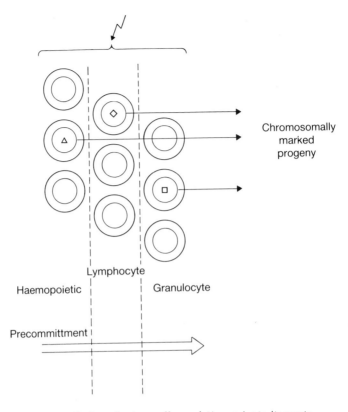

Fig. 12.4 Irradiation of a stem cell population, prior to its use to reconstitute a lethally-irradiated inbred animal, generates chromosomally-marked cells at different stages of commitment. The pattern of markers in committed cells in the host allows differentiation pathways of stem cells to be elucidated.

named B lymphocytes. In birds, stem cells undergo differentiation into B lymphocytes in a gut-associated organ named the bursa of Fabricius (Figure 12.5). The role of thymus and bursa in generating lymphocytes can be shown by the effects of removal of these organs. The functions of peripheral T or B lymphocytes can also be deduced from the type of immunodeficiency resulting after thymectomy (or congenital absence of thymus) or bursectomy. A mammalian equivalent of the avian bursa cannot be so clearly defined; fetal liver and bone marrow are the most probable sites for generation of B lymphocytes in mammals.

Newly produced lymphocytes, prior to antigenic encounter, are referred to as virgin lymphocytes. B lymphocytes synthesize small amounts of immunoglobulin

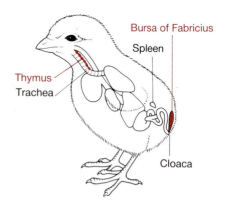

Fig. 12.5 The locations of the thymus and bursa of Fabricius in the bird.

which is displayed on the cell surface at a density of $10^4 - 10^5$ molecules per cell (see Chapter 8). The B lymphocyte lineage embraces the antibody forming cells. The high-rate antibody secreting cell, of which the mature form is recognized as a plasma cell, is derived from a B lymphocyte by a differentiation process, usually progressive and coupled with cell proliferation (Figure 12.6).

The presence of immunoglobulin on the surface of T lymphocytes was for many years a controversial issue. In mouse and man standard tests using fluorescent labelled anti-immunoglobulin antibodies to detect surface immunoglobulin have been used frequently to distinguish B lymphocytes (positive fluorescence) from T lymphocytes (non-fluorescent) (see Chapter 13). Immunoglobulin genes are not expressed in T cells. The nature of the antigen receptor molecule on T cells is discussed in Chapter 13.

The T lymphocytic lineage produces a variety of effector cells (Figure 12.7). These effector cells have been defined functionally and include cells active in delayed type hypersensitivity, helper T cells, suppressor

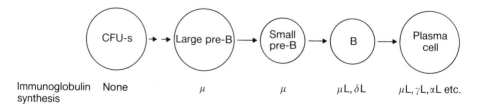

Fig. 12.6 Defined stages in the differentiation of the B cell lineage. Stem cells are shown as colony forming units (CFUs).

294   CHAPTER 12

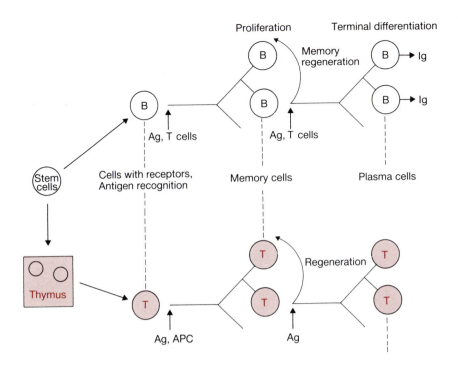

Fig. 12.7 The expansion of clones of B and T lymphocytes involves antigen driven proliferation and differentiation. Both memory cells and effector cells can be generated within each clone of either B or T cells.

T cells and cytotoxic T cells. Antigenic markers (differentiation antigens) allowing these individual populations to be recognized and separated are necessary because easy morphological identification of T cell subsets is not available (see below). The functional interactions of different subclasses of lymphocytes with each other and with non-lymphoid cells (e.g. macrophages) are discussed later in Chapter 13.

In both B and T lymphocyte lineages functional memory is a feature of responses to antigen. Memory cells are not morphologically distinct from non-stimulated virgin lymphocytes. Memory cells are, however, much longer lived than virgin lymphocytes.

Antigen presenting cells

Macrophages were the first cells to be implicated in immune responses in a non-antigen specific manner.

Antigen can be taken up by phagocytic cells and subsequently presented at the cell surface (Figure 12.8). Macrophages are the mature differentiated cells derived from monocytes of bone marrow origin. It is now recognized that dendritic cells of non-bone-marrow origin can process and present antigens to lymphoid cells. The nature of the processing and presentation steps in the antigen presenting cell (APC) remains a mystery. The presence of Ia antigens on APCs seems to be an essential element for successful presentation of antigen (see Chapter 13). The ability to produce the T-cell activating factor interleukin-1 (IL-1) is common to macrophages and certain epidermal dendritic cells; IL-1 production may be a necessary function for all APCs.

Thy-1 (θ, theta): a marker for murine T cells

Thy-1 is a differentiation antigen expressed on murine T lymphocytes but *not* on murine B lymphocytes. Murine nerve cells, especially brain cells, and some other non-lymphoid cells also express Thy-1, but amongst cells of the immune system in the mouse Thy-1 is a marker of the T lymphocyte lineage. The production, assay and use of antisera specific for

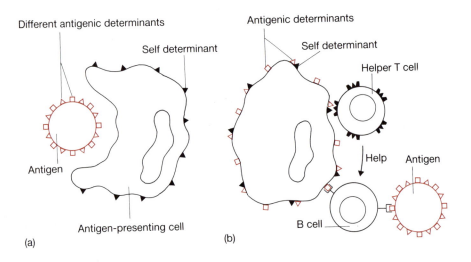

Fig. 12.8 Antigen-presenting cells (APC) digest antigen (a) and display processed fragments on the surface (b). Processed antigenic fragments are associated with self-determinant molecules on the cell surface; helper T cells recognize antigen in association with class II MHC molecules. B cells recognize antigen directly without association with MHC antigens. B cells have the ability to process antigens and to display fragments for presentation to helper T cells.

Thy-1 is typical of the methods used for the definition of differentiation antigens by alloimmunization, an approach used extensively in the mouse.

Cytotoxic alloantisera are raised by injecting C3H thymocytes into AKR mice (or vice versa) (Figure 12.9); both strains are H-2^k but they differ at a number of non-MHC loci (Lyt-1, Lyt-3 and Lyt-6 in particular) so antisera need to be absorbed to improve their specificity for Thy-1. The AKR and C3H strains express Thy-1^a and Thy-1^b alleles respectively, the corresponding antigens being Thy-1.1 and Thy-1.2. The specificity of antisera raised by alloimmunization can be improved, in principle, by using congenic mouse strains, e.g. A.Thy-1^a and A.Thy-1^b. The problems of antiviral antibodies, directed against cell surface viral glycoproteins, and autoantibodies are not removed by using congenic strains. Moreover, the immune response to Thy-1 in alloimmunizations between some congenic strains is poor, apparently because of the involvement of immune response (Ir) genes. The definitive reagents are now

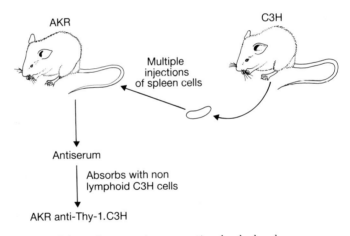

Fig. 12.9 Scheme for preparing conventional polyclonal alloantiserum specific for the Thy-1 (θ) antigen.

cytotoxic monoclonal antibodies specific for each of the Thy-1 specificities.

The Thy-1 alloantigen was first described in 1964 by Reif and Allen but it was five years later that Raff showed that the Thy-1 antigen is a marker for thymus-derived lymphocytes and that murine peripheral lymphocytes can be divided into B cells bearing surface immunoglobulin but not Thy-1, and T cells expressing

Thy-1 but not immunoglobulin. Elucidation of the ontogeny and functions of murine T cells (see Chapter 13) has been greatly facilitated by the use of antisera to Thy-1.

The function of the Thy-1 molecule is not understood. The presence of Thy-1 on murine T lymphocytes and not on murine B lymphocytes benefits the murine immunologist enormously. By contrast, the Thy-1 antigen is present on rat lymphocytes but is not a specific differentiation antigen of rat T cells. Thus Thy-1 does not appear to have a T lymphocyte specific function. This is underlined by the fact that Thy-1 antigen is not detectable on human lymphocytes. Definition of differentiation antigens on human T cells has required production of monoclonal antibodies (see below).

The Thy-1 glycoprotein has been isolated and sequenced. The 110 amino acid chain constitutes a single structural domain with distant homology to an immunoglobulin V region domain. Thy-1 protein is held on the cell surface by a lipid molecule covalently attached to the C-terminal end of Thy-1 and inserted into the lipid bilayer (Figure 12.10); this contrasts with the use of transmembrane hydrophobic peptides that secure immunoglobulins and MHC antigens to the cell surface.

Thymus leukaemia antigens (TL antigens): murine thymocyte differentiation antigens

The murine alloantigens encoded at the *Tla* locus were discovered as antigens characteristic of leukaemic cells in certain mouse strains. The alloantisera specific for TL antigens are cytotoxic for leukaemic cells, even for leukaemias syngeneic with the mouse strain in which the alloantiserum was raised. Absorption of antisera

Fig. 12.10 Representations of the structure of the Thy-1 cell surface antigen. The protein is folded as a single domain and is found on the plasma membrane attached via a covalently bound lipid (on right, marked old). The gene encoding Thy-1 includes a typical transmembrane hydrophobic peptide sequence so when originally synthesized Thy-1 molecules are presumably attached to the endoplasmic reticulum via this peptide as shown on the left (marked new).

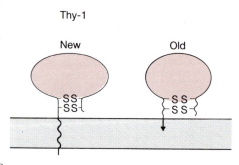

with normal tissues showed that the TL antigen is expressed on thymocytes and on leukaemias in certain mouse strains (e.g. A/J) but is only expressed on leukaemias in other strains (e.g. C57BL/6).

The *Tla* locus codes for at least five alleles and maps close to the H-2 complex on chromosome 17. The finding that leukaemias of TL-negative as well as TL-positive mouse strains can express TL antigens can be explained by postulating that the *Tla* locus contains both a structural gene and a regulator gene. In normal thymocytes of TL-negative strains the regulator gene represses the *Tla* structural locus, but in a proportion of leukaemias of TL-negative mice the regulator gene allows expression of TL antigens. The *Tla* locus may encode more than one structural gene. The product of the *Tla* structural gene is found on the cell surface as the heavy chain glycoprotein of a non-covalent dimer with β_2-microglobulin; this points to a structural homology of the *Tla* product with MHC class I antigens, and this is confirmed by DNA probe analysis (see Chapter 10).

The importance of the TL antigen to the immunologist is as a marker for non-immunocompetent, immature lymphocytes developing in the thymus. TL antigens are absent from bone-marrow cells, expressed on the majority of thymocytes in the cortex but absent from functionally mature peripheral T cells that arise from thymocytes by differentiation.

Lyt alloantigens: markers of mouse T lymphocyte functional subpopulations

The Ly antigen series comprise murine alloantigens with a tissue distribution restricted to lymphocytes. The first three antigens termed Ly A, Ly B and Ly C were detected by Boyse and colleagues (1975) using antisera raised in mice against murine leukaemia cells. Recognition that these three antigens are T cell specific (and a change to a numerical labelling system) led to the nomenclature Lyt-1, Lyt-2 and Lyt-3. *Lyt* congenic mice have been bred for allele-specific antisera production; now monoclonal antibody reagents are available for Lyt antigen typing.

The *Lyt-1* locus maps on chromosome 19, linked to ruby eye. The two alleles are *Lyt-1*[a] and *Lyt-1*[b] and they encode the antigens Lyt-1.1 and Lyt-1.2 respectively. The product of the Lyt-1 gene is found as a cell surface glycoprotein of mol. wt. 67000.

The Lyt-2 and Lyt-3 genes map on chromosome 6 in very close linkage with each other, linked to the lactic dehydrogenase regulatory locus and the micro-ophthalmic locus and associated with the κ light chain gene cluster. No recombinants between Lyt-2 and Lyt-3 have been observed to date. The products of the Lyt-2 and Lyt-3 genes can be separated by sequential precipitation with allele-specific antisera. The specificities are present on distinct cell surface glycoproteins of mol. wt. approximately 35000; the Lyt-2 and Lyt-3 molecules appear to form a complex and may be in that state on the cell surface.

Peripheral T lymphocytes exhibit one of three Lyt phenotypes as defined by cytotoxicity assays using allele-specific alloantisera. The phenotypes are Lyt-1$^+$, 2$^+$, 3$^+$ (Lyt-123), Lyt-1$^+$, 2$^-$, 3$^-$ (Lyt-1) and Lyt-1$^-$, 2$^+$, 3$^+$ (Lyt-23). (Using the more sensitive technique of fluorescence activated cell sorting the cells designated Lyt-23 can be shown to express a low level of Lyt-1.) The importance of these phenotypes lies in the correlation of T cell effector functions to subsets of T cells having the Lyt-1 or Lyt-23 phenotypes (Table 12.1).

Of thymocytes, 95% are Lyt-123 cells and these cells give rise to the Lyt-123 virgin T cells in the peripheral population. Exposure to antigen is involved in the differentiation of Lyt-123 cells to effector or long-lived memory cells, and these mature T cells have either the Lyt-1 or Lyt-23 phenotype. The correlation of the Lyt-1 phenotype with helper T cells and the Lyt-23 phenotype with suppressor cytotoxic T cells is based on studies of cell populations by positive or negative selection methods (Figure 12.11). Lyt-1 cells prepared by negative selection (i.e. by treatment of T lymphocytes with anti-Lyt-23 antiserum plus complement) show helper activity in appropriate assays (e.g. antibody production by B cells). Negatively selected Lyt-23 cells show no helper activity but can suppress antibody responses and effect cell-mediated cytotoxity. Positive selection of cells according to Lyt phenotype can be effected by coating cells with the appropriate murine anti-Lyt antibodies and then passing the cells over an immunoabsorbent column consisting of a matrix bound covalently to rabbit anti-murine Fab antibodies. The results of positive selection agree with the findings from

Table 12.1 Phenotypes of murine T cell subsets

Lyt phenotype	Representation in the periphery	Function
Lyt-1	30–35%	Helper T cells (for both B and T cell responses)
Lyt-23	5–10%	Suppressor T cells Cytotoxic T cells
Lyt-123	50–55%	Precursors of Lyt-1 and Lyt-23 cells

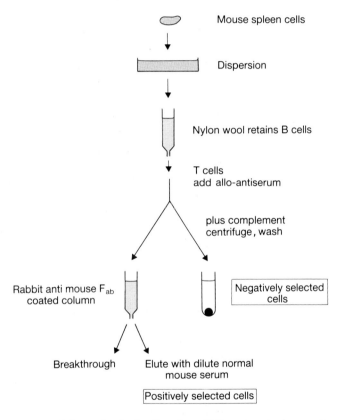

Fig. 12.11 Scheme for the fractionation of T lymphocytes by negative or positive selection according to the presence or absence of alloantigens. Treatment of T cells with alloantiserum plus complement lyses those cells expressing the alloantigen and a differentiation antigen leaving a negatively selected population. T cells carrying the given alloantigen are positively selected by coating cells with the murine alloantibody and then passing them over a column carrying rabbit anti-mouse Fab antibody. T cells bearing the alloantigen are retained by the column and can be eluted with normal mouse serum. Methods due to Cantor *et al.* (1976).

301 DIFFERENTIATION ANTIGENS

negative selection. These methods proved valuable in the elucidation of the patterns of cell–cell interactions in immune responses. The association of Lyt phenotype with the function of T cell subsets, found in population studies, is also evidenced by typing clones of functional T cells grown in culture with the help of interleukin-2 (IL-2, T cell growth factor). Most clones of helper cells display the Lyt-1 phenotype and most clones of cytotoxic T cells display the Lyt-23 phenotype. However, at the clonal level it is found that T cells lacking the Lyt-1 antigen can effect helper function and T cells lacking the Lyt-23 phenotype can kill target cells. On this evidence it seems unlikely that the products of Lyt-1 or Lyt-23 loci are involved in effector functions of mature T cells.

Surface immunoglobulin and Ia antigens are differentiation markers of B lymphocytes

The presence of surface immunoglobulin clearly defines the B lymphocyte of any species; thus, immunoglobulins are exclusive differentiation antigens for B cells. Fluorescent anti-immunoglobulin reagents are used to identify and to separate (using a fluorescence-activated cell sorter, FACS; Figure 12.12) B cells. Clearly the surface immunoglobulin of the B cell is a functional entity, serving as the receptor for a specific antigen. The production and function of surface immunoglobulin is discussed elsewhere (Chapter 8).

Isotypic and allotypic antigens can be used to define B cell subsets. Idiotypic antigens define clonal B cell populations independent of the stage of differentiation.

The diversity of immunoglobulin isotype expression on the surface of B cells serves to identify the stages in B cell differentiation. The initial expression of surface IgM is followed by coexpression of IgM and IgD on B cells (Figure 12.12b). Further switches in the class of immunoglobulin expressed result in B cells bearing any of the heavy chain isotypes either singly or in combinations with each other. The underlying molecular basis of class switching (see Chapter 8) explains why a B cell can actively synthesize μ and δ chains simultaneously; but subsequent class switching by a deletion mechanism allows active synthesis of only a single isotype per B cell. The finding of two or more isotypes on a single B cell may be either an artefact of the detection

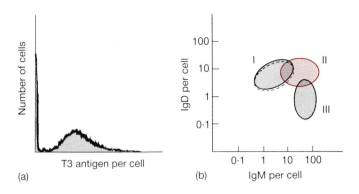

Fig. 12.12 Fluorescence-activated cell sorter (FACS) analysis of lymphocyte populations. (a) Separation of human peripheral blood T lymphocytes by staining with a fluorescein-labelled anti-T3 monoclonal antibody (see p. 307). The green fluorescence of T3-positive cells is recorded for each cell as it passes in a stream (10^4 cells/second) through a laser beam and is plotted on the abscissa. The number of cells counted is plotted as the ordinate. The FACS was devised by L. Herzenberg. (b) Two-colour fluorescence analysis of murine B cell subpopulations. Murine spleen cells were stained with fluorescein-labelled monoclonal anti-IgM (green) and Texas-red-labelled monoclonal anti-IgD. Analysis of the labelled cells using a dual-laser FACS yields a scatter plot representing the amounts of IgM and IgD per cell. The plot shown here is idealized to illustrate the three subpopulations of B cells found in normal mice and in CBA/N (xid) mice. In normal mice the major subpopulations of B cells are type I, whereas in CBA/N mice type I B cells are missing (Hardy *et al.*, 1982).

method or the result of the slow turnover of molecules synthesized prior to class switching events. The observation of B cells actively expressing IgM and IgG or IgM and IgE on their surface necessitates a novel molecular genetic explanation (see Chapter 8). The biological significance of these double producers is unclear.

Expression of surface immunoglobulin isotypes is regulated by the X-linked locus, *xid*. In CBA/N mice this X-linked immune deficiency gene is responsible for an increased proportion, in lymphoid organs, of immature B cells, characterized by a high ratio of IgM/IgD on their surface (Figure 12.12b). The subset of mature B cells that is absent in the CBA/N mouse has been further characterized by the presence of the Lyb-5 antigen (see below).

The Ia antigens are also functional differentiation antigens of B lymphocytes. The products of I-A and I-E subregions are expressed on the majority of B cells. The function of Ia antigens of B cells is probably related to the interaction with T helper cells that are Ia restricted (see Chapter 13). Ia antigens are also expressed on macrophages and other antigen-presenting cells and are involved in the interaction of T cells with presented antigen (see Chapter 13). Epidermal cells and spermatozoa also express Ia antigens. Some T cells have been reported to express I-A region products and suppressor T cells have been defined as expressing I-J region antigens; however, the failure to detect the I-J genes in the predicted region of the H-2 gene complex (see Chapter 9) has raised doubts as to the significance of I-J markers. Only I-A and I-E, α and β structural genes have been detected (see Chapter 10) so I-J antigens may be products of structural genes located outside the MHC or they may be generated by interaction or modification of I-A or I-E products.

Lyb antigens: B-cell differentiation markers

A series of Lyb antigens have been identified as markers of B cells or subsets of B cells. The two most useful are Lyb-3 and Lyb-5, both of which define the mature B cell subset that is absent in CBA/N mice (see above).

Antisera to Lyb-3 are raised in male (CBA/N x BALB/C)F$_1$ mice by immunization with BALB/C spleen cells. The F$_1$ males born to CBA/N mothers exhibit the X-linked defect in B cell maturation. Anti-Lyb-3 antiserum raised in this way detects mature BALB/C B cells by immunofluorescence but not by cytotoxic assay. Note that the Lyb-3 marker has been found in all mouse strains tested, so it is of an isotypic rather than an allotypic nature.

Anti-Lyb-3 antiserum can mimic T cell help in both primary and secondary responses. It is possible that the Lyb-3 molecule (a cell surface protein of mol. wt. 68 000) plays a part in the triggering of B cells by antigen and/or the recognition of T helper effects.

Antisera to Lyb-5 are produced by immunization of C57/B6 mice with DBA/2 lymphocytes and made specific by absorption with spleen cells from male (CBA/N x DBA/2)F$_1$ mice. This alloantiserum detects the Lyb-5.1 specificity. Anti-Lyb-5.2 is raised by the reverse strain immunization.

Unlike anti-Lyb-3 antisera, anti-Lyb-5.1 antisera exhibit complement-dependent cytotoxicity. The use of this cytotoxic reagent converts a normal B lymphocyte population to one resembling the B cell subset found in the CBA/N mouse.

The Lyb-3 and Lyb-5 antigens have been defined

by making use of the CBA/N genetic defect in two distinct ways (immunizing defective mice or using lymphocytes from defective mice for absorbing alloantisera). The Lyb-3 and Lyb-5 loci are not linked to the xid gene of the CBA/N mouse.

Monoclonal antibodies specific for differentiation antigens

The advent of monoclonal antibody production (described in Chapter 6) has greatly extended the detection of differentiation antigens and their use in somatic cell genetic studies of differentiation processes. A monoclonal antibody has a single amino acid sequence and a discrete, characteristic specificity profile. Most importantly, a monoclonal antibody can be continuously produced as required, thus affording reproducible reagents and repeatable experiments. The latter points contrast with the problems of obtaining and using conventional antisera; since they are heterogeneous mixtures of antibody molecules, conventional antisera vary from one animal to another and even from time to time in the same animal.

Another major advantage of the monoclonal antibody production methodology is that antibodies can be readily prepared with specificity for antigens that have not been purified for immunization purposes. It is only necessary that suitable assays are available to identify the specificity of each monoclonal antibody.

Monoclonal antibody reagents specific for each differentiation alloantigen of the mouse are now standard. Even the past previous alloantisera generated by congenic strain immunizations are eclipsed by monoclonal antibody reagents. Given the use of defined conditions experiments should now be reproducible. Monoclonal antibody can be easily produced in quantities sufficient for experiments in vivo that were previously possible only in vitro. Thus, a given subset of lymphocytes can be functionally ablated by administration to mice of a sufficient quantity of monoclonal antibody specific for a defined Ly differentiation antigen.

Human-lymphocyte differentiation antigens

The availability of human alloantisera has depended upon fortuitous alloimmunizations, e.g in multiparous women (see Chapter 5). Attempts to raise xenoantisera specific for human differentiation antigens by immunizing mice with human lymphocytes generate an extremely heterogeneous population of antibodies directed against a large number of different cell surface antigens present on cells of similar or distinct types. Such complex xenoantisera are of limited use even after extensive absorption.

Applying hybridoma technology to mice immunized with human lymphocytes allows the production of many monoclonal antibodies each specific for a single cell surface antigen. The antigen T3 has been defined by murine monoclonal antibodies as being present on all human peripheral T cells (see Figure 12.12a). A set of mouse monoclonal antibodies has been generated that defines human T lymphocyte subsets (see Table 12.2). The monoclonals specific for the T4 antigen define human helper T cells; monoclonals specific for the T8 antigen define suppressor T cells and class I MHC antigen directed cytotoxic T cells. The ratio of T4 to T8 bearing T cells is now measured by clinical immunologists and the value of the ratio is correlated with the immunological status of the patient.

The significance of the definition of functionally distinct T cell subsets by the presence of the T4 and T8 antigens is emphasized by comparison with differentiation antigens found on T cells of the mouse and rat (see Table 12.3). Human and rat T cells express a differentiation antigen analogous to Lyt-1 found on all mouse T cells. The T4 antigen present on human helper T cells has a possibly homologous counterpart (W3/25) on rat helper T cells; the homologous antigen L3T4 has been defined on mouse helper T cells. Suppressor and cytotoxic T cells of human, rat and mouse express proteins homologous to the Lyt-2 antigenic molecule. The evolutionary retention of homologous differentiation antigens in association with specific cell types raises the question of the role of such antigens in cellular function.

It has been proposed that T4 and T8 antigens play a part in the recognition of class II and class I MHC antigens by helper cells and suppressor/cytotoxic cells respectively (see Chapter 13, p. 340). The T4 molecule has been shown to have a pathological function, constituting the receptor (or an important part thereof) for the human immunodeficiency virus (HIV), the causative agent of acquired immune deficiency syndrome (AIDS).

Table 12.2 Human T lymphocyte surface antigens defined by monoclonal antibodies

Antigen nomenclature Old	New	Mol. wt. ($\times 10^3$)	Cell population defined Peripheral T cells	Thymocytes
T1	CD5	69 ⎫		Medullary
T3	CD3	20 ⎬	All mature T cells	thymocytes
T12	CD6	120 ⎭		
T4	CD4	62	50–65% of T cells: all helper T cells and class II MHC specific cytotoxic T cells	75% of thymocytes
T8 (T5)	CD8	30 + 32	25–35% of T cells: all suppressor T cells and class I MHC specific cytotoxic T cells	90% of thymocytes

Table 12.3 Comparison of T lymphocyte differentiation antigens of mouse, rat and human

Species	Antigen	Mol. wt. ($\times 10^3$)	Cellular expression Peripheral T cells	Thymocytes	Function of antigen-bearing T cells
Mouse	Lyt-1	67 ⎫			
Rat	OX-19	69 ⎬	All peripheral T cells	All thymocytes	All functions
Human	T1	69 ⎭			
Mouse	L3T4	52	60%	80%	Helpers,
Rat	W3/25	48–53	85%	70%	class II MHC
Human	T4	55	60%	75%	restriction
Mouse	Lyt 2, 3	30 + 35	40%	90%	Suppressor
Rat	Ox-8	35 + 39	30%	90%	killers, class I
Human	T8(T5)	30 + 32	20%	70–95%	MHC restriction

The future

Differentiation antigens will be defined on a wide variety of cell types, not just those associated with the immune system. The necessity to find alloantigens is obsolete now that monoclonal antibodies can so easily be generated. Where alloantigens are to be defined and used in studies of inheritance, monoclonal antibodies will be the reagents of choice for phenotype determination. Increasingly, genotype will be directly determined by the use of DNA sequence probes. Antigen identification progresses to protein sequencing, then to gene cloning (via complementary DNA made from mRNA, as described in Chapter 9) and gene sequencing. Gene probes can be used to define restriction enzyme site polymorphisms so that inheritance studies can be performed in the absence of alloantigens at the phenotypic level.

Functional analysis of differentiation antigens can now be carried out by the introduction of the structural gene, in expressable form, into any desired cell type. Genes can be introduced by transfection, electroporation or microinjection. The function of a gene product in the whole organism can be investigated by the production of transgenic animals. Cloned genes can be injected into fertilized mouse oocytes which are then re-implanted into foster mothers (pseudopregnant females) that subsequently give birth to the transgenic progeny. A proportion of these neonates have the new gene incorporated into one of their chromosomes so that the gene can be transmitted sexually and a new coisogenic strain derived.

Genetics is now moving into studies that have been termed 'reverse' genetics. Such studies result from extensive sequencing of DNA that reveals open reading frames that could encode proteins. The putative product may be expressed by the introduction of the gene into an appropriate cell type. From the DNA sequence, peptide sequences can be predicted and chemically synthesized. Antibodies (monoclonal or polyclonal) raised against these synthetic peptides are then used to identify the gene product expressed in the transfected cell.

References

Cantor H., Shen F.W. & Boyse E.A. (1976) Separation of helper T cells from suppressor T cells expressing different Ly components. *J. exp. Med.* **143**, 1391.

Hardy R.R., Hayakawa K., Haaijman J. & Herzenberg L.A. (1982) B-cell subpopulations identified by two-colour fluorescence analysis. *Nature* **297**, 589.

Raff M.C. (1969) Theta isoantigen as a marker of thymus-derived lymphocytes in mice. *Nature* **224**, 378.

Reif A.E. & Allen J.M.V. (1964) The AKR thymic antigen and its distribution in leukemias and nervous tissues. *J. exp. Med.* **120**, 413.

Trentin J., Wolf N., Cheng V., Fahlberg W., Weiss D. & Bonhag R. (1967) Antibody production by mice repopulated with limited numbers of clones of lymphoid cell precursors. *J. Immunol.* **98**, 1326.

Further reading

Ahmed A. & Smith A.H. (1982, 1984) Surface markers, antigens and receptors on murine T and B cells: Parts 1, 2 and 3. *CRC Crit. Rev. Immunol.* **3**, 331 and **4**, 19, 95.

Benorst C., Gerlinger P., LeMeur M. & Mathis D. (1986) Transgenic mice: 'new wave' immunogenetics. *Immunol. Today* **7**, 138.

Herzenberg L.A., Sweet R.G. & Herzenberg L.A. (1976) Fluorescence-activated cell sorting. *Sci. Am.* **234**, 108.

Inglis J.R. (Ed.) (1983) *T Lymphocytes Today*. Elsevier, Amsterdam.

Laurence J. (1985) The immune system in AIDS. *Sci. Am.* **253**, 70.

Moller G. (Ed.) (1982) B cell differentiation antigens. *Immunol. Rev.* **69**.

Wong-Staal F. & Gallo R.C. (1985) Human T-lymphotropic retroviruses. *Nature* **317**, 395.

Chapter 13
Genetics of cell—cell interaction

Introduction

The emergence of cellular immunology from a plethora of phenomenology is being guided in a most direct manner by immunogenetics. Genetic studies afford the clearest and most useful description of cellular immunology.

All immune responses require participation of a variety of interacting cell types. For an understanding of an immune response, each cell type involved should be characterized and the pattern and nature of cell—cell interactions should be established. The intensive study of the genetics of the mouse with the development of inbred, congenic and recombinant inbred lines make the mouse the animal of first choice for the basic study of cellular immunology. Extending the hypothesis to man requires different methods.

The involvement of separate cell types in a synergistic response was first recognized in experiments using an adoptive transfer system. Transfer of lymphoid cells with successful expression of their immunological capacity in an adoptive host requires that syngeneic animals are used. Adoptive transfer provides a system for resolving the types of cell required for an immune response. Varying the genetic make-up of donor and host animals allows the genes controlling cell—cell interactions to be mapped.

Genetic methods also allow identification of differentiation antigens characteristic of particular populations of cells involved in the immune system. Production of antisera or monoclonal antibodies specific for a differentiation antigen is described in the previous chapter.

Radiation chimeras constructed by bone-marrow reconstitution of lethally irradiated mice and allophenic (tetra-parental) mice constructed by fusion of blastocysts, have been widely used to investigate cellular immune phenomena. Construction of chimeric mice allows: (1) the cells involved in an immune response to be followed during an ongoing response; and (2) the genetic requirements for cell—cell interaction to be determined.

Cell cooperation in antibody formation

The phenomena of cooperation between thymus-derived T lymphocytes and B lymphocytes was recognized initially by the effect of thymectomy on antibody production. Subsequently the cell transfer system was used by Claman to show that thymocytes and bone-marrow cells together reconstitute an antibody immune response that neither cell population will support separately (Figure 13.1).

The first genetic markers used to study cell co-operation were: (1) murine major histocompatibility differences; and (2) a chromosomal marker. Both marker were used to show that B cells, and not T cells, are the precursors of antibody secreting cells.

1 Neonatally thymectomized CBA mice were reconstituted with the (CBA x C57BL)F_1 T cells, and later antibody response to SRBC was conveniently measured by counting plaque forming cells (PFC). The origin of antibody forming cells was tested by treating the spleen cells with antiserum to either CBA cells or C57BL cells prior to counting PFC. It was found that only anti-CBA

TREATMENT

RESPONSE

(IgM PFC spleen day 4)

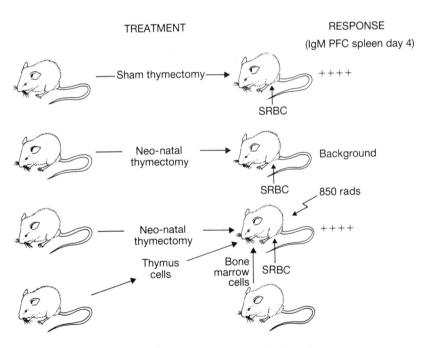

Fig. 13.1 Reconstitution of a lethally-irradiated mouse with either thymocytes or bone-marrow cells separately does not allow an antibody response to sheep red blood cells (SRBC), but injected together these two cell populations cooperate to allow a full antibody response in the host mouse. This experiment was first performed by Claman *et al.* (1966).

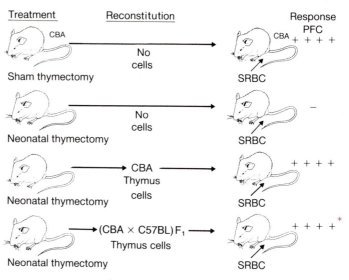

Treatment	Reconstitution	Response

Sham thymectomy — No cells → SRBC — CBA PFC + + + +

Neonatal thymectomy — No cells → SRBC — −

Neonatal thymectomy — CBA Thymus cells → SRBC — + + + +

Neonatal thymectomy — (CBA × C57BL)F₁ Thymus cells → SRBC — + + + + *

*Only CBA antigens on PFC

Fig. 13.2 Experimental demonstration of the role of thymocytes in antibody formation. Neonatal thymectomy abolishes the antibody response in the young mouse. Reconstitution with adult thymocytes restores the antibody response. The use of marked (alloantigen) thymocytes shows that thymus-derived cells are not precursors of antibody-forming cells. Antibody formation is detected by counting the number of plaque forming cells (PFC), i.e. cells causing a lytic plaque in a monolayer of SRBC when complement is supplied. This experiment was performed by Mitchell & Miller (1968).

serum plus complement is lytic for PFC. This shows that the thymectomized host supplies all of the precursors of PFC (Figure 13.2).

2 A chromosomal marker is generated by sub-lethal irradiation of mice, to effect chromosomal damage and/ or rearrangements; selective breeding is then used to obtain a homozygous strain carrying the altered marker chromosome. The strain carrying the marker chromosome is coisogenic (or even syngeneic if only rearrangement has occurred) with the parental strain. CBAT6/T6 is the chromosomally marked strain most used in immunogenetic studies (Figure 13.3).

Pioneering experiments by Davies used irradiation chimeric mice constructed by lethal irradiation followed by reconstitution with grafts of bone-marrow from parental CBA mice together with thymus fragments from CBAT6/T6 mice. Subsequently these chimeric mice were challenged with antigen and the karyotype of

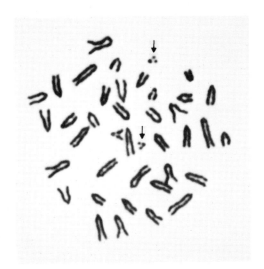

Fig. 13.3 A mitotic cell chromosome spread (karyotype) from a mouse homozygous for the T6 marker chromosome (arrows). The marker chromosome is the product of a radiation induced unequal reciprocal translocation (Ford, 1966). Photograph courtesy of E.P. Evans.

proliferating cells was determined. Both bone-marrow-derived (B) and thymus-derived (T) cells respond to antigen by proliferating. When this method is used with B lymphocytes carrying the T6/T6 marker chromosomes all antibody-secreting plasma cells exhibit the T6 marker. The conclusion is that B lymphocytes (and not T lymphocytes) can proliferate and differentiate into plasma cells secreting antibody. Similar types of experiment were later performed using differentiation antigens as markers of cell types.

Hapten—carrier-protein systems

The individual antigen specificity of lymphocytes can be used as somatic genetic markers. Initially the hapten—carrier phenomenon was seen as a possible challenge to clonal section. When an anti-hapten antibody primary response is evoked to a hapten presented on one carrier-protein then that hapten must be presented on the identical carrier-protein for a successful secondary response; presentation of the hapten on a different carrier-protein fails to evoke a secondary antibody response to the hapten (Figure 13.4). One explanation offered for carrier specificity was that the antibody selected by the hapten—carrier-protein complex was specific for a novel epitope consisting of the hapten and its local environment on the carrier-protein mol-

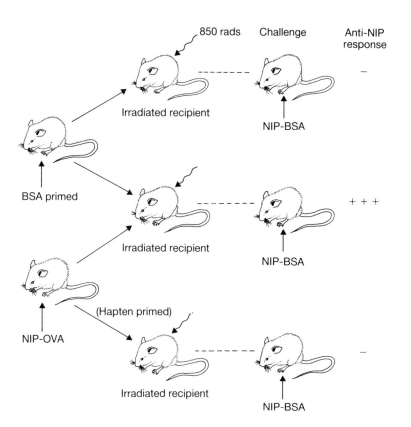

Fig. 13.4 The hapten−carrier-protein system was used by Mitchison (1971) in conjunction with spleen cell transfer to irradiated hosts to show that carrier-protein specificity in the secondary response to hapten is due to a second carrier-specific population of lymphoid cells, T helper cells. In the experiment shown here splenic B cells from a mouse primed with the hapten NIP (see Fig. 6.10) on ovalbumin will respond to NIP-BSA only when supplied with splenic T cells from a mouse primed with BSA.

ecule. The competing hypothesis of carrier-protein-specific T cells cooperating with hapten-specific B cells is now accepted on the strength of considerable evidence. (These alternative hypotheses, i.e. novel antigenic determinant and dual recognition, recurred as competing explanations of MHC restriction of T cell specificity for antigen; see Figure 13.18.)

The original Mitchison version of the antigen-bridging hypothesis invoked T and B lymphocytes specifically recognizing two different epitopes of the same antigen molecule simultaneously. The demonstration of the involvement of non-antigen-specific antigen presenting cells (APC) led to a bridging hypothesis in which the surface of the APC carries antigen molecules that T and B cells specifically recognize (Figure 13.5).

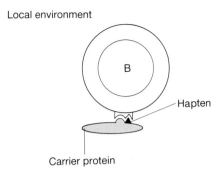

Local environment

B

Hapten

Carrier protein

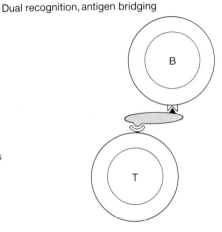

Dual recognition, antigen bridging

B

T

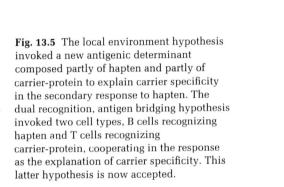

Fig. 13.5 The local environment hypothesis invoked a new antigenic determinant composed partly of hapten and partly of carrier-protein to explain carrier specificity in the secondary response to hapten. The dual recognition, antigen bridging hypothesis invoked two cell types, B cells recognizing hapten and T cells recognizing carrier-protein, cooperating in the response as the explanation of carrier specificity. This latter hypothesis is now accepted.

Immune response (Ir) genes map onto cellular interactions in immune responses

Ir gene expression is manifested via cellular interactions in immune responses. This finding has provided one of the most effective genetic approaches to the elucidation of the function and mechanism of such interactions.

The first link observed between cell—cell interactions and Ir gene function was in the response to poly-L-lysine (PLL) in guinea pigs. The failure of strain 13 guinea pigs to respond to PLL, either with antibody production or by DTH, extended to a failure to produce anti-DNP antibody when challenged with DNP-PLL. These results are consistent with a defect in carrier (PLL)-specific T cells. Presentation of DNP-PLL on bovine serum albumin (BSA) results in production of anti-DNP antibody but there is still no direct T-cell mediated response to PLL as measured by delayed-type hypersensitivity; the BSA acts as the carrier molecule, in place of PLL, allowing helper T cells specific for BSA to be recruited and to function (Figure 13.6).

Genetic understanding of cellular interactions in immune responses stems mainly from studies with inbred mice, initially using synthetic polypeptides as antigens. A series of branched chain polypeptides is based on a poly-L-lysine backbone with poly-D,L-alanine side chains terminating in oligomers of glutamic acid with one of three amino acids—tyrosine, histidine or phenylalanine (Figure 13.7). Perversely, immunologists call these polymers (TG)-AL, (HG)-AL and (PheG)-AL, ignoring the one-letter code for amino acids employed by biochemists.

Mouse strains have been classified as high-responder and low-responder strains to each of these synthetic antigens on the basis of high or low levels of antibody production. The difference is not only quantitative. Low responders make IgM antibody, but not IgG antibody, against the specific challenge. After thymectomy, mice of high-responder strains lose the ability to switch from IgM to IgG and so behave like low responders. The defect in low-responder mice is not a total absence of T cells, only an apparent functional deficit of T cells

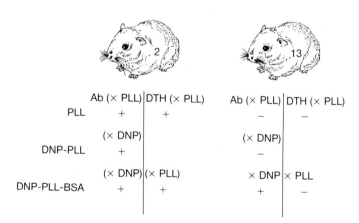

	Ab (× PLL)	DTH (× PLL)	Ab (× PLL)	DTH (× PLL)
PLL	+	+	−	−
	(× DNP)		(× DNP)	
DNP-PLL	+		−	
	(× DNP)	(× PLL)	× DNP	× PLL
DNP-PLL-BSA	+	+	+	−

Fig. 13.6 Strain 2 and strain 13 guinea pigs differ in their immune response to poly-L-lysine (PLL). Strain 2 guinea pigs are high responders making antibody (Ab) to PLL and mounting a delayed type hypersensitivity (DTH) response to PLL. Strain 13 guinea pigs are low responders making little or no anti-PLL Ab and showing no DTH response to PLL. PLL acts as a carrier for an anti-DNP Ab response in strain 2 but not in strain 13. A complex of DNP-PLL with bovine serum albumin (BSA) elicits an anti-DNP Ab response in either strain. Experiments reviewed by Benacerraf & McDevitt (1972).

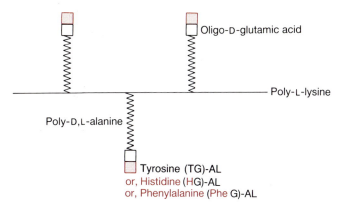

Fig. 13.7 Diagrammatic representation of the structure of the series of synthetic polypeptide antigens based on the poly-L-lysine backbone with poly-D,L-alanine branches ending in oligo-D-glutamic acid and optionally tyrosine, histidine or phenylalanine residues. These synthetic antigens were devised by Sela and his colleagues and have proven very useful in studies of immune response control.

specific for the challenge antigen. This deficit is attributable to the expression of genes mapping in the H-2 complex and, more specifically, at the I region, as shown by recombinant inbred strains (Table 13.1).

Tetra-parental mice: evidence that the defect in low-responder mice is not at the B cell level

Tetra-parental or allophenic mice result from fusion of blastocysts of different genetic origin and subsequent transfer to pseudopregnant females for development to parturition. The neonates contain, in varying proportions, cells of both conceptuses (i.e. genes from four parents) in a mosaic phenotype (Figure 13.8).

High- and low-responder (with respect to (TG)-AL) mice, differing also in their immunoglobulin allotype, have been fused to yield tetra-parentals. Upon challenge with (TG)-AL these high/low mosaic mice produce anti(TG)-AL antibody bearing the allotypes of both high- and low-responder strains in similar proportions to those seen in the bulk serum immunoglobulin. The conclusion was drawn that low-responder B cells are not defective. The hypothesis that the defect lies at the T cell level is not disproved by the tetra-parental mouse experiment. However, more recent experiments (see below) point to the defect in responsiveness being at the level of B cells and APC, with an associated genetic control over T cell maturation in the thymus.

Table 13.1 Level of immune response to synthetic polypeptides by inbred mice

Immunogen	MHC haplotype Ab response	
	High	Low
(TG)-AL	b	a,l,q,s
(HG)-AL	a,k	b,q,s
(PheG)-AL	a,b,k,q	s

NB   The *a* haplotype is a recombinant between *k* and *d* haplotypes and is I^k (see Chapter 9).

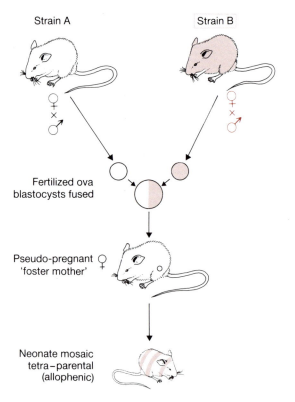

Strain A   Strain B

Fertilized ova
blastocysts fused

Pseudo-pregnant
'foster mother'

Neonate mosaic
tetra–parental
(allophenic)

Fig. 13.8 Tetra-parental (allophenic) mice are produced by fusing together, at the blastocyst stage, ova (fertilized *in vitro*) taken from genetically differing matings. The fused blastocyst is implanted in a pseudopregnant foster mother. The neonatal tetra-parental mice are mosaics of cells of strain A and strain B. If these strains differ in a coat colour marker then the tetra-parental mice will have a patchy or even a striped appearance. The proportion of cells of A and B genotypes will vary from mouse to mouse and between cell differentiation types within one mouse. This method of producing tetra-parental mice is due to Mintz (1967).

Genetic requirements for cellular interactions in immune responses: T−B cooperation

Kindred and Shreffler reported in 1972 that successful cooperation between T and B cells in the production of an antibody response in mice to sheep erythrocytes requires that the T and B cells are identical at the H-2 locus. This phenomenon was extensively explored in both hapten−carrier systems and synthetic polypeptide systems by Benacerraf, Katz and colleagues (Figure 13.9). Helper T cell function in anti-hapten responses was studied by cell transfer to F_1 recipient mice (Table 13.2). In this system effective help was seen only when T and B cells were syngeneic or semi-syngeneic at the H-2 locus. These experiments demonstrated the involvement of a gene(s), located in the murine major histocompatibility complex (MHC), in the control of antigen-specific T−B cooperation. Study of antibody responses to synthetic polypeptides under H-2 linked Ir gene control showed that F_1 T cells can cooperate to effect antibody production by B cells from responder mice, but not from non-responder mice. Thus, genetic identity at one of the two H-2 alleles is sufficient for successful T and B cell cooperation, but not in specific responses for which the single shared H-2 haplotype carries a low-responder Ir gene.

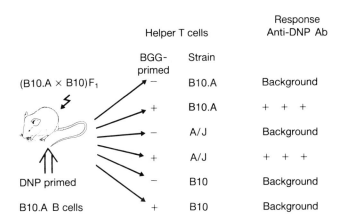

	Helper T cells		Response Anti-DNP Ab
	BGG-primed	Strain	
(B10.A × B10)F_1	−	B10.A	Background
	+	B10.A	+ + +
	−	A/J	Background
	+	A/J	+ + +
DNP primed	−	B10	Background
B10.A B cells	+	B10	Background

Fig. 13.9 Physiological cooperation between T and B lymphocytes requires that the cells are histocompatible at the MHC. In the experiment shown DNP-primed B10.A B cells will accept help from specifically carrier primed B10.A or A/J T cells but not from unprimed cells or from primed B10 cells. Experiments of this type were devised by Katz et al. (1973) to show that identity of the MHC was essential and sufficient for physiological cooperation between correctly primed B and T cells.

Table 13.2 Antibody production requires syngeneic T helper cells

Irradiated (A × B10)F$_1$ host		Anti-hapten Ab
T Carrier primed	B Hapten primed	
B10.A	B10.A	+
B10.A	B10	−
B10.A	A/J	+
(A × B10)F$_1$	B10.A	+
(A × B10)F$_1$	B10	+

Experiments aimed at exploring the genetic requirements for cell−cell interaction in immune responses
are fraught by the problem of allogeneic effects, either
positive or negative, caused by allo-reactive T cells that
seem to dominate the specificity repertoire of antigen-
specific T cell receptors. The positive allogeneic effect
has been demonstrated in many different experiments;
one clear example is the use of a positive allogeneic
effect to overcome an Ir-gene controlled response defect (Figure 13.10). Mice of the F$_1$ generation of crosses
H-2^k by H-2^q are low responders to (TG)-AL, as are both
parental strains. Injection of (H-2^k × H-2^q)F$_1$ mice with
H-2$^{k/k}$ spleen cells non-specifically stimulates production
of a specific antibody response to (TG)-AL at responder-
strain levels. The interpretation is that H-2^q-specific
allo-reactive H-2^k T cells are substituting for the deficiency of antigen-specific syngeneic helper T cells in
supplying a stimulus to F$_1$ B cells.

Evidence that the defect in H-2-linked low-responder
mice is linked to I-A expression on B cells and APC
comes very clearly from experiments of Kappler and
Marrack (1977−78). They showed in cell cultures that
(TG)-AL primed helper T cells derived from (high- ×
low-responder)F$_1$ mice induce anti-(TG)-AL antibody
production when co-cultured with B cells and APC,
both of the high-responder parental type, but not when
either B cells or APC were of low-responder type.

The importance of the genotype of the host in which
helper T cells are primed can be demonstrated by
making parent-into-F$_1$ and F$_1$-into-parent chimeric mice
(Figure 13.11). T cells of low-responder genotype
(B10.A) first exposed to (TG)-AL in the F$_1$ host can
specifically cooperate with B cells and APC of high-
responder genotype (B10), but not with syngeneic B10

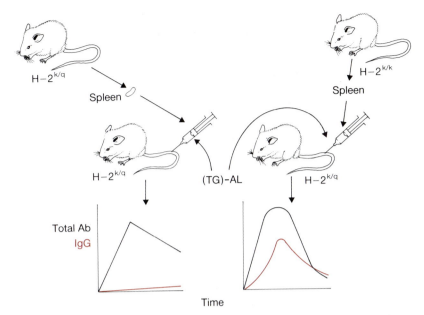

Fig. 13.10 A positive allogeneic effect can substitute for physiological cooperation. In this experiment, devised by Ordal & Grumet (1972) low-responder (H-2^k × H-2^q)F$_1$ mice are induced to respond fully to (TG)-AL by injection of semi-allogeneic H-2$^{k/k}$ cells.

B cells and APC, to generate an anti-hapten plaque-forming cell (PFC) response. It can be concluded that the defect seen in helper T cells from low-responder mice results from the regulation of T-cell differentiation by expression of I region genes on non-T cells. (Thymic epithelial cells are crucial in governing T-cell differentiation, as described below.)

Genetic requirements for cellular interactions in immune responses: helper T cell interaction with APCs

Proliferation of antigen-specific T cells, generating a population of helper T cells, is triggered physiologically by antigen presented in 'an appropriate manner' on the surface of various types of APC (see Chapter 12). Rosenthal and Shevach (1973) showed that in the guinea pig allogeneic macrophages are much less efficient than syngeneic macrophages in presenting antigen to T cells, as measured by the ensuing increase in DNA synthesis indicative of T-cell proliferation. This antigen presentation assay is conveniently performed *in vitro*, thus facilitating experimental intervention. The proliferative

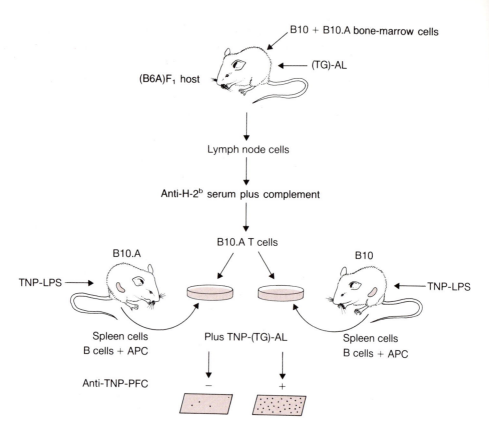

Fig. 13.11 A lethally-irradiated (B6.A)F$_1$ host is repopulated with mixed parental bone-marrow cells. The resultant chimeric mouse is immunized with (TG)-AL. Primed B10.A T cells are recovered from chimeric lymph node cells by negative selection and tested in culture for their ability to act as carrier-primed helper T cells when presented with TNP-(TG)-AL and TNP-primed B cells together with antigen presenting cells (APC) from either B10.A or B10 mice. Low-responder B10.A T cells primed in a chimera can only cooperate with B cells and APC of high-responder (B10) genotype; the response is measured by a plaque-forming cell assay using TNP-coated SRBC. (B6.A)F$_1$ mice are heterozygotes H-2$^{b/a}$; B10.A mice are H-2^a; and B10 mice are H-2^b.

response of T cells to antigen presented by macrophages requires histocompatibility at the I region between the two cell types. In the mouse the required identity maps at I-A. It was first shown with guinea pig cells that antibodies against I region products can specifically block the proliferative response of T cells to antigen presented by macrophages. By using antigens that are under Ir gene control it can be shown that the Ir gene defect is functionally expressed during antigen presentation. Strains 2 and 13 guinea pigs differ at the I region of the MHC. Assayed either *in vivo* or by the antigen presentation, T-cell proliferation assay *in vitro*, strain 2 responds to the hapten—polypeptide conjugate DNP-GL but not to peptide GT. Strain 13 responds to poly GT but not to DNP-GL. Strain-specific alloantiserum (containing antibodies against I region products) blocks the proliferative response of T cells to antigen presented on high responder APCs. In the $(2 \times 13)F_1$ system, T-cell proliferation stimulated by DNP-GL on APCs is blocked by anti-strain 2 sera, but not by anti-strain 13 sera; conversely, the response to GT-APCs is blocked by anti-strain 13 sera and not by anti-strain 2 sera.

That the presence of identical Ia antigens on T cells and APCs is not sufficient for functional antigen presentation is shown by the specificity of secondary responses. Elicitation of a secondary proliferative response of F_1 T cells requires that the identical primary antigen be presented on APC syngeneic with those used for the primary stimulus, i.e. specifically either of the parental haplotypes (Table 13.3).

The conclusion is that helper T cells are specific for both the foreign antigen and the I-A phenotype of the APC on which antigen is first presented. Taking this conclusion together with the finding that it is the genotype of the host environment in which T cells differentiate, and not the genotype of the T cells, that determines their ability to cooperate successfully with appropriate B cells, places the APC, together with the B cell, in a crucial role in determining the Ir-gene-controlled responder status of the animal.

Klein (1984) has written an obituary for the term Ir gene: 'Now, however, when it is known that the Mhc-linked Ir genes *are* the Mhc genes, perpetuation of the former term is good only for confusing immunology students. (This comment applies also to perpetuation of all terms derived from "Ir genes", such as "I-region",

Table 13.3 Linked specificity of primary and secondary antigen presentation

T cell	APC		Secondary response
	Primary	Secondary	T-cell proliferation
F_1	P_1	P_1	+
		P_2	−
F_1	P_2	P_1	−
		P_2	+

"Ia antigens", "*I-A* subregion", etc. All of them are obsolete and misleading.)' In this, and in other aspects of the subject, the warning should be taken that terminology must not obscure understanding.

'Soluble factors' involved in cell—cell regulation of immune responses

Elucidation of the genetics of cellular interactions in the immune system ignores the nature of non-genetically specific mechanisms that are involved in intercellular regulation. However, the emergence of a clear picture of the genetics of cellular interactions has been delayed by a plethora of conflicting results, some of which may be attributed to the mechanics of cell—cell communication via soluble factors.

From the extensive genetic evidence it is clear that cellular cooperation in immune responses involves direct cell—cell contact. Where soluble factors are part of the cooperative mechanism they apparently pass between cells in close proximity so that the action of such factors is limited temporarily and topographically. When these physiological limitations are circumvented (as in certain cell culture experiments *in vitro*) exceptions to the genetic requirements for cellular cooperation may be seen.

Three classes of soluble factor have been described: (1) non-antigen specific factors; (2) antigen-specific helper and suppressor factors; and (3) antigen-specific, MHC-specific helper and suppressor factors. The nature of the various types of antigen-specific factor is controversial. The most reasonable hypothesis is that antigen-specific factors derived from T cells represent some soluble form of the antigen-specific receptor found on the T cell surface. As the nature of antigen-specific receptors on T cells is revealed, so the nature and physiological relevance of antigen-specific T cell factors may be clarified.

The nature and role in immune responses of some non-antigen specific factors is emerging and these are briefly described here. An overview of the position of these factors in immune responses is shown in Figure 13.12.

Interleukin-1 (IL-1)

IL-1 was originally described as lymphocyte activating factor (LAF). IL-1 is a product of macrophages, and

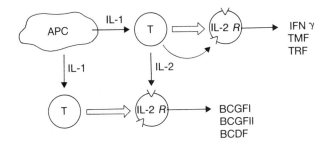

Fig. 13.12 The origins and targets of some of the non-antigen-specific factors involved in regulating immune responses.

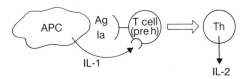

Fig. 13.13 Interleukin-1 (IL-1) is produced by the APC and acts on T cells together with antigen associatively presented with class II MHC molecules on the surface of the APC. The result is the activation of the T cell with induction of IL-2 production and expression of receptors for IL-2 on T cells.

probably of all APC (Figure 13.13). The best defined role of IL-1 in immune responses is as a second signal in the induction of interleukin-2 (IL-2) production by helper T cells; the primary signal in IL-2 induction is antigen associatively presented with class II MHC products on APCs.

The immunological function of murine IL-1 is associated with a partially purified polypeptide of mol. wt. ~15 000. Several other biological properties (e.g. growth factor activities and pyrogenic activity) are found in these IL-1 preparations (Figure 13.14). Whether or not a single polypeptide mediates all of these functions remains to be proved. Recently (1985) two IL-1 genes have been cloned. The encoded polypeptides IL-1α and IL-1β are homologous; different properties are yet to be defined.

Interleukin-2 (IL-2)

IL-2 was originally described as T cell growth factor (TCGF) and this useful functional description is still often used. Helper T cells are the main producers of

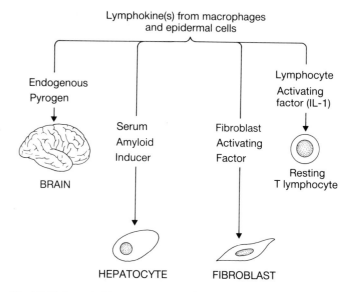

Fig. 13.14 Lymphokine preparations derived from macrophages and epidermal cells show a multiplicity of actions including endogenous pyrogen, serum amyloid induction, fibroblast activation and T-lymphocyte activation; the latter activity is defined as being due to IL-1 but whether the other activities are due to IL-1 remains to be defined.

IL-2, though other subclasses of T cells can release IL-2.

IL-2 acts via a specific IL-2 receptor expressed on the surface of T cells and delivers a proliferative stimulus. Any T-cell-bearing IL-2 receptors will proliferate if IL-2 is supplied. Antigen and genetic specificity are preserved because the induction of IL-2 receptor expression is dependent on the interaction of T cells with an antigen – APC complex.

IL-2 is one of a family of lymphokines, polypeptides produced by T cells and acting as soluble factors in immune regulation. Since IL-2-driven proliferation is involved in the release of other lymphokines from T cells, the properties of some of these factors may on occasion have been ascribed incorrectly to IL-2. It may be that IL-2 does not function solely as T cell growth factor but this remains to be proved.

The amino acid sequence of human IL-2 has been derived from a cloned cDNA encoding IL-2. The IL-2 gene has also been cloned (Figure 13.15). IL-2 cDNA encodes a 153 amino acid chain of which the N-terminal 20 residues are the signal peptide. The mature 133 residue IL-2 is O-glycosylated at position 3. The IL-2 receptor has been

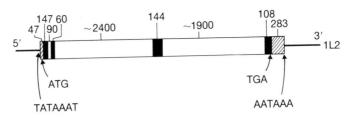

Fig. 13.15 The exons and introns mapped in the human IL-2 gene.
■ coding exons; ▨ non-translated regions; ☐ introns. (From
Degrave *et al.*, 1983.)

partially characterized using monoclonal antibodies that
block IL-2 binding and the response to IL-2. The mode of
action of IL-2 will be the subject of intense investigation
in the near future. Already IL-2 is being tested in man as a
treatment for acquired immune deficiency syndrome
(AIDS) and as an adjuvant to cancer therapy.

In the laboratory IL-2 is extensively used to grow
T cell lines and clones. Such T cell clones, in some
cases converted to hybridomas by fusion with T cell
tumour lines, provide a source of other lymphokines.
The sequences of cloned cDNA molecules encoding in-
terleukin-3 and γ-interferon have been derived.

The finding that γ-interferon induces increased transcription, translation and surface expression of MHC
molecules may be relevant to the requirement for such
products in immune recognition by T cells. Ectopic
production of class II MHC molecules by endothelial
cells may be caused by exposure to virally-induced
γ-interferon. The possibility that these cells expressing
Ia may then present self-antigens to T cells has been
proposed as an explanation for the induction of autoimmunity.

Two B cell growth factors (BCGF1 and 2) and a B cell
differentiation factor (BCDF) have been defined by
biological assays and are subject to further characterization. Recently (1986) two B cell growth factor genes
have been cloned and expressed in bacterial cells. The
properties of the pure factors (probably to be designated
IL-4 and IL-5) thus derived can now be studied.

Elucidation of the physiological roles of each of the
interleukins will be facilitated when antagonists are
found. Antagonists, or agonists of interleukins should
prove useful for therapeutic intervention in certain
immune disorders (see Part 4).

MHC-restricted recognition of antigen: cytotoxic T cells see self-MHC products plus foreign antigen

In 1974, while the genetic requirements for interactions between helper T cells and APC or between helper T cells and B cells were being defined, Zinkernagel and Doherty published a paper on the genetic restriction of virus-specific cytotoxic T cells. This paper marked a turning point in the rationalization of the phenomenon of genetically restricted recognition of foreign antigens.

It was found that cytotoxic T cells, generated *in vivo* by infecting mice systemically with vaccinia virus or with lymphocytic choriomeningitis virus (LCMV) and assayed *in vitro*, are capable of lysing only those target cells that are acutely infected with the original virus type *and* have the MHC antigens in common with the infected mouse (Figure 13.16). Antigen specificity is expected in immune responses but the requirement for the correct MHC antigens to allow the cytotoxic effector cells to function was seen as a new phenomenon; it was termed MHC restriction or, in the mouse, H-2 restriction. Clearly, a similar phenomenon was contemporaneously described in studies on interactions of helper T cells with either APC or B cells.

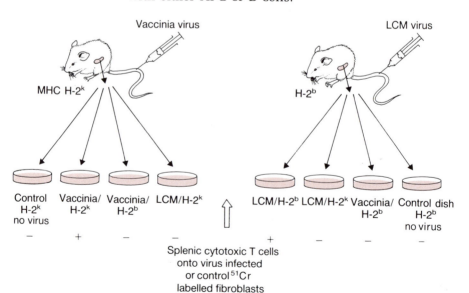

Fig. 13.16 MHC restriction. Cytotoxic T cells specific for virus-infected cells also show specificity for the MHC antigens of the infected cell. Mice infected with a given virus type yield cytotoxic T cells that will kill, in culture, only cells infected by the original virus and also having MHC antigens in common with the infected mouse.

MHC-restricted specificity for cytotoxic T cells has been demonstrated for minor histocompatibility antigens and for hapten-modified target cells. Thus, the origin of the target antigen (viral gene expression, chromosomal gene expression or external chemical modification of self antigen) does not dictate the need for recognition of MHC antigens to allow lysis by cytotoxic T cells.

The original experiments revealed a simple set of rules.

1 Class I MHC antigens are the restricting elements for cytotoxic T lymphocytes (CTL).

2 In systems involving helper T cell recognition of antigens class II MHC antigens are the restricting elements.

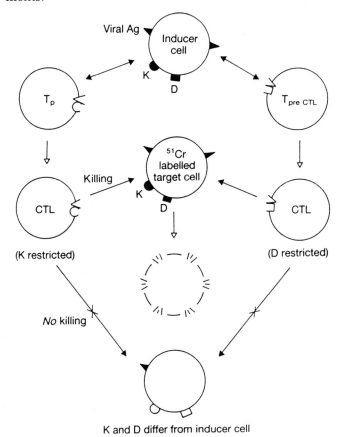

Fig. 13.17 Each clone of cytotoxic T lymphocytes (CTL) is restricted by a single class I antigen specificity. Killing is assayed *in vitro* by loading virus-infected target cells with ⁵¹Cr-chromate and measuring lysis of target cells by counting the radioactivity released into the medium.

329 GENETICS OF CELL–CELL INTERACTION

3 Like antigen specificity, MHC restriction is a clonal phenomenon (Figure 3.17). A given clone of CTL is specific for a single foreign antigen and for a particular class I MHC antigen, e.g. either K, D or L in the mouse. Subsequent experiments in other systems have confirmed clonality but have also shown that class II MHC antigens can be restricting elements for CTL.

Two types of hypothesis were advanced to accommodate MHC-restricted recognition of antigens (Figure 3.18).

1 Dual recognition hypothesis. T cells express on their surface two separately encoded receptors, one recognizing MHC antigens and the other recognizing foreign antigens.

2 Single recognition or 'altered self' hypothesis. T cells express a single type of receptor that is capable of recognizing only new antigenic determinants formed at the presenting cell surface by the association of foreign antigens with self-MHC antigens, i.e. 'altered self' antigens.

An acceptable hypothesis should apply to both class I and class II restricted recognition.

Allo-reactivity and the special nature of MHC antigens

The apparently high frequency of allo-MHC-reactive T cells must be accommodated by any hypothesis for MHC restriction. There is both a qualitative and quantitative difference in the perception by T cells of MHC antigens relative to all other antigens, even other (minor) histocompatibility antigens. The quantitative difference was first observed by Simonsen (1968) in a

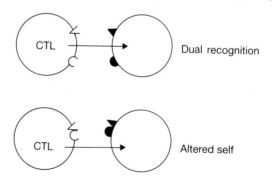

Fig. 13.18 Diagram of the extreme hypothesis of dual recognition and altered self.

study of allo-reactive cells in chickens. In many systems allo-MHC-reactive cell frequencies in excess of 1% have been found. Given the extensive polymorphism of the H-2 locus, with of the order of 100 alleles known, the frequency of allo-reactive T cells poses a problem for clonal selection. The special nature of MHC antigens was recognized in the original allograft experiments. The differentiation of major from minor transplantation antigens can now be seen to be fundamental since *minor* histocompatibility antigens are recognized only in the context of *major* histocompatibility antigens; class I MHC antigens are recognized directly by allo-reactive T cells without the requirement for a self-MHC-restricting element.

Restriction specificity of T cells is acquired, not genetically predetermined

Initial demonstrations of class I MHC-restricted recognition of antigens were explicable by invoking the requirement that immunizing and target cells presenting a foreign antigen share MHC antigens with each other and with the cytotoxic T cell. A similar explanation was taken to apply to class II MHC-restricted interactions.

That identity of MHC antigens between presenting cells, target cells and effector T cells is not necessary for MHC-restricted antigen recognition was revealed by experiments with bone-marrow (lymphohaemopoietic) chimeric mice. Helper T cells (H-2^k) were shown to cooperate with B cells (H-2^d) across H-2 barriers when the T cells had 'adaptively differentiated' in a chimeric host constructed by reconstituting an irradiated (H-2^k × H-2^d)F$_1$ mouse with H-2^k bone-marrow cells (P F$_1$ chimera).

Bevan constructed chimeric mice by injecting (A × B)F$_1$ bone-marrow (where A and B are different MHC genotypes) into an irradiated parent A; in this F$_1$ P chimera, cytotoxic T cells generated against minor histocompatibility antigens recognize the minor antigens preferentially in association with the MHC haplotype of parent A as opposed to that of parent B. In a similar system, Zinkernagel demonstrated exclusive restriction for the MHC haplotype of parent A in cytolysis of either virally infected or hapten-modified target cells. T cells from the appropriate chimera will see as self-MHC antigens of a totally foreign MHC haplotype and will not recognize as self antigens that are genotypically a self haplotype.

T cell antigen receptor

The story of the elucidation of the nature of the antigen receptor on T cells (TCR) is one of many surprising twists and turns. It is a story in which the classical genetic methods play little part, and that part was probably confusing. The recent clarification of the nature of TCR is due to the incisive use of somatic cell genetics and molecular genetics.

The simplest hypothesis to explain the recognition of antigen by T cells is that, like B cells, T cells use antibody genes to encode their receptor. Support for this idea came from studies showing that anti-idiotype antibodies raised against allo-MHC-specific T cells cross-react against similarly allo-MHC-specific antibodies. Alone, this is a weak argument since antibodies to insulin can have idiotypic determinants cross-specific with the insulin receptor, a protein unrelated genetically to immunoglobulins. The link between TCR genes and immunoglobulins appeared to have been confirmed by genetic studies showing a formal linkage between the gene encoding TCR idiotype and the *Igh* locus in mice. The correct explanation of the latter findings is unclear but recent evidence shows that TCR genes are not on the same chromosome as the *Igh* locus.

With the various demonstrations of MHC restricted recognition of antigen by T cells came the realization that an antibody molecule could not serve as a single TCR. *Either* T cells have two types of receptor, one (that could use Ig genes) to recognize antigen and a second novel receptor to recognize class I or class II MHC molecules; *or* there is a unique TCR encoded by non-immunoglobulin genes and capable of recognizing antigen and an appropriate MHC molecule simultaneously. Present evidence strongly favours the latter hypothesis.

The most compelling evidence for a single TCR molecule comes from an ingenious somatic-cell genetic experiment performed by the Kappler and Marrack group (Figure 13.19). Antigen-responsive T cell hybridomas were constructed that could be induced to secrete IL-2 when presented with the specific antigen on APC bearing the same class II MHC antigen as on the original presenting cell. One such hybridoma specific for the

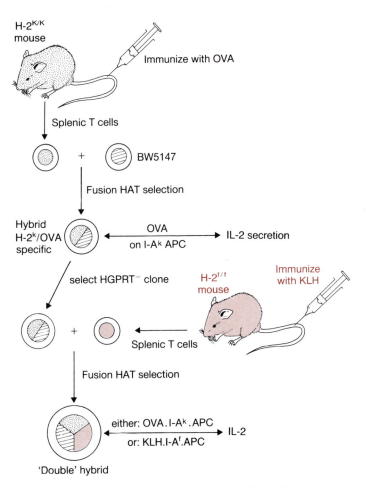

Fig. 13.19 The 'double hybridoma' experiment of Kappler and Marrack. See text for explanation. (From Kappler et al., 1981.)

antigen chicken ovalbumin (OVA) responded to this protein only when the APC carries the Ia antigen specified by the I-A^k allele. These hybrid cells were used to generate a set of 'double' hybrid cells by fusion of the first hybridoma with splenic T cells from B10.M (H-2^f) mice immunized against keyhole limpet haemocyanin (KLH). Upon initial screening a high proportion of double hybrids secrete IL-2 in response to either OVA presented with I-A^k or KLH presented with I-A^f; no response was seen to either OVA with I-A^f or KLH with I-A^k. It proved possible to clone out a double hybridoma cell line that retained responsiveness to both OVA and KLH but only in combination with the original presenting Ia antigen. These data are best explained by a single receptor molecule seeing the foreign antigen together with the Ia antigen. If separate receptors were used to recognize the foreign and self antigens, then cross-recognition would have been expected in the double hybridoma. The full pattern of responsiveness, shown in Table 13.4, reveals that this double hybridoma cross-reacts with an H-2^b antigen encoded at either K or I-A loci. Such cross-reactions are not uncommon with monoclonal T cell lines or hybridomas.

Clonotypic, or idiotypic, antibodies have been raised against T cell clones and hybridomas and selected as monoclonal antibodies. This approach has proved more definitive than the earlier use of anti-idiotype antisera raised against polyclonal T cells. Monoclonal, clonotypic antibodies have been used to isolate TCR molecules from human and murine T cells. In both species the TCR is a hetero-dimeric glycoprotein of mol. wt. ~ 85 × 10^3; reduction of the inter-chain disulphide bond and dissociaton yields the α and β chains (Table 13.4). In each case the α chains are acidic and the β chains neutral or slightly basic.

Peptide maps of separated α and β chains can be interpreted as showing that the two types of chain come from different gene families, both can vary in amino acid sequence and there are variable and constant region sequences within each chain.

Table 13.4 The polypeptide chains of the T cell receptor

	Approximate mol. wt. ($\times 10^3$)	
	α	β
Human	46	40
Mouse	43	43

Molecular genetics of the T cell receptor

The nature of the T cell receptor has been defined by nucleotide sequences of cloned genes. The starting point was the cloning of cDNA encoding the β chain of TCR. This was accomplished for murine and human β chains by the strategy of subtractive selection of cDNA populations as illustrated in Figure 13.20. Similar approaches have yielded cloned cDNA molecules encoding the α chain of TCR and a third type of T cell

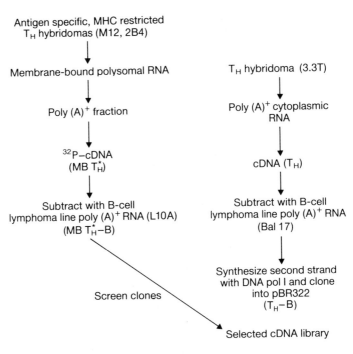

Fig. 13.20 Strategy adopted by Davis and colleagues to select a T cell-specific cDNA clone library. cDNA was synthesized from mRNA of two different T helper hybridoma cell lines and in each case hybridized with mRNA from a (different) B cell line to subtract cross-hybridizing cDNA molecules. One selected cDNA population (from 3.3T cells) was cloned in the plasmid vector pBR322. The other selected cDNA population (from M12.2B4 cells) had been labelled with ^{32}P and was used as a probe to screen the cloned cDNA library from 3.3T cells and select clones containing sequences common to the two T_H hybridoma lines. One selected clone was identified as encoding the β chain of the T cell receptor. (From Hendrick *et al.*, 1984.)

specific sequence (the γ chain) having properties (e.g. V and C regions, rearrangement of genes in going from germ-line to T cell) expected of a TCR chain DNA.

The α and β cDNA sequences are unambiguously identified by comparison with the N-terminal amino acid sequences determined for purified α and β chains of the TCR proteins. Inspection of the nucleotide and derived amino acid sequences of α and β chains reveals putative V, D, J and C regions. Cloning of the germ-line genes and of rearranged genes from T cell hybridomas confirms the pattern of distinct V, D, J and C genetic elements, with the exception that no D elements for α and γ chains have been found.

The arrangements of known or predicted genes in the

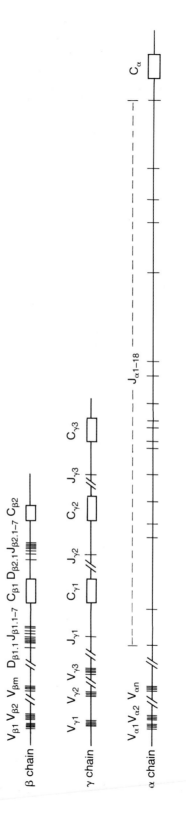

Fig. 13.21 Genomic organization of the murine T cell receptor α, β and γ genes. C genes are shown (☐) without indication of exons and introns. Indeterminate distances are indicated by —//—. Data collected from the laboratories of Davis, Hood, Tak Mak and Tonegawa.

α, β and γ families is as shown in Figure 13.21. There is a great similarity between the arrangement of the TCR gene families and the immunoglobulin gene families. Rearrangement of TCR genes apparently occurs by a mechanism closely similar to the immunoglobulin rearrangement mechanism. Each gene segment is flanked by a conserved heptamer separated by a non-conserved spacer of either one DNA turn (twelve nucleotides) or two DNA turns (twenty-three nucleotides) in length from an A/T-rich nonomer, the combination of these three units being similar to the joining recognition sequences found adjacent to immunoglobulin gene segments. As with joining of immunoglobulin genes, TCR recognition elements with a one-turn spacer always join with elements having a two-turn spacer. Also, in both systems the joining process is imprecise, allowing a junction to be formed at different nucleotides around the V, D and J gene boundaries. A significant difference between th D_H and D_β recognition elements is that D_H is flanked on both sides by one-turn elements, whereas D_β has a one-turn element upstream, for joining to V_β, and a two-turn element downstream, for joining to J_β (Figure 13.22). In theory, this latter arrangement would allow D_β elements to join with each other.

Diversity of T cell receptor genes

The multiplicity of V, D and J elements forms the basis of TCR diversity. The numbers of V_β and D_β elements appear to be small. On the evidence of the first nineteen murine β chain sequences, that include only eleven different V_β sequences (with two V_β sequences seen three times and three V_β sequences seen twice), the total number of V_β genes is predicted to be twenty-nine (95% confidence). Only two D_β genes are required to explain the sequences of the nineteen β chains. Of course there may be many V_β genes that are rarely expressed. Using V_β gene probes in Southern blotting experiments reveals the presence of very small V_β gene

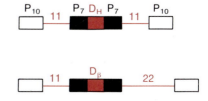

Fig. 13.22 The assymmetrical arrangement of recognition elements flanking D_β exons as compared with the symmetrical arrangement around D_H exons (cf. Fig. 7.25).

sets, each comprising one to four genes (this contrasts with V_H sets of four to fifty genes).

From more limited data on murine α chains the diversity of V_α may be greater than that of V_β with a repertoire of more than 100 V_α genes being predicted. Only three V_γ genes have been identified and each is associated with its own J_γ-C_γ cluster. The expression and utility of V_γ genes is now being investigated.

Combinational diversity and junctional diversity play an important part in creating novel TCR diversity. The presence of twelve functional J_β genes enhances the possibilities for both combinatorial and junctional diversity. Four β chain D elements certainly contribute to diversity. The evidence points to only a few germ-line D genes but in addition to junctional diversity, of the sort seen in antibody V_H–D_H–J_H joining, it appears that D_β genes can be joined to V_β and J_β genes so as to be read in any of the three possible reading frames (Figure 13.23). Each of the three reading frames is used randomly in both D_β genes. By contrast D_H genes are infrequently joined and read in more than one reading frame.

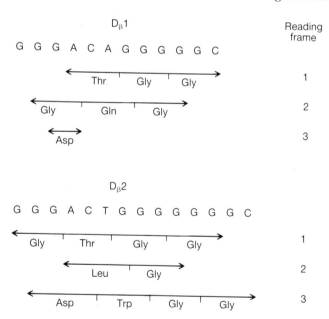

Fig. 13.23 Two D_β exons are read in each of the three reading frames when used in rearranged β chain genes. The germ-line sequences of $D_\beta 1$ and $D_\beta 2$ are shown above the sets of β chain D region amino acid sequences deriving therefrom (Barth *et al.*, 1985). Examples of the use of each of the three reading frames, numbered on the right, is given for each D exon. Note that not all of the D exon sequence is included in the rearranged gene.

Sequences flanking D regions in β chains are not readily attributable to germ-line elements; they are N regions analogous to those described for antibody H chains. There is extensive N region diversity in rearranged β chain genes, and N region sequences have also been found at the Vγ–Jγ junction but no N region insertions have been found in functional α gene chains.

Somatic hypermutation, as occurring in expressed antibody V genes, does not appear to be an important feature of V_α, V_β or V_γ diversity. It is an intriguing possibility that the mechanisms for somatic hypermutation of antibody V genes evolved after the divergence of TCR and Ig gene families.

Expression of T cell receptor genes

On the surface of mature T cells the αβ dimer (often referred to as the Ti molecule) exists as a complex with a number of invariant polypeptide chains, in particular a set of three chains that in man are termed T3 (CD3). The three chains of the T3 molecule are 28K, 21K and 16K in size. There is evidence for an involvement of T3 in signal transduction with phosphorylation of at least one of the chains during T-cell activation. The Ti/T3 complex can be immunoprecipitated by antibodies specific for either the Ti molecule or the T3 molecule. Chemical cross-linking agents can join the β chain of Ti to the 28K chain of T3 indicating the nature of the physical association.

The γ chain gene is not expressed in the large majority of T cells. Rearrangement of the γ locus occurs in T cells of various phenotypes (helper, suppressor and cytotoxic) but the protein encoded by the rearranged γ gene proved elusive until it was found in a population of T cells that do not express either α or β chains. A suitable population of cells was obtained by fluorescent-activated cell sorting of allo-reactive T cell lines (grown with repeated IL-2 stimulation) derived from immuno-deficient paients (Brenner et al., 1986). These cells express T3 and, after chemical-cross-linking, immunoprecipitation with anti-T3 antibody reveals two polypeptide chains (55K and 40K M_R) associated with T3. The higher M_R chain is precipitable by two antisera each raised against a different synthetic peptide predicted from the γ chain gene sequence. The functional significance of γ chains, and the putative fourth type of

TCR chain associated with γ chains and T3, remain to be discovered.

During development of T cells in the thymus the order of gene rearrangement is that γ chain genes show the earliest detected joining events, followed by β chain gene rearrangement, and with joining events at the α chain locus being last. The existence of a definite developmental order of joining at the various TCR loci correlates with the fact that joining at the immunoglobulin H chain locus precedes L chain rearrangement. It is noteworthy that N region diversity is only seen in γ and β chains of the TCR and H chains of immunoglobulin, and this may reflect the activity of the enzyme polynucleotide terminal transferase (postulated to be the enzyme that forms N regions, see Chapter 8, p. 212) during the early developmental stages of both B and T cells. Available evidence for the functional rearrangement and expression of TCR α and β loci show that the principle of allelic exclusion is followed (see Chapter 8, p. 201 for discussion of allelic exclusion of immunoglobulin gene expression).

What determines the MHC specificity of T cell subsets?

This remains a major unanswered question. There is sufficient knowledge of V_α and V_β sequences, correlated with antigen/MHC binding specificity and with T cell function, to conclude that both helper and suppressor/cytotoxic T cell subsets use the same sets of V_α and V_β genes. This leaves open the question as to how the recognition of distinct MHC specificities by each subset of T cells is determined.

One hypothesis is that the subset-specific antigens, CD8 and CD4 in man, are involved. This proposal would involve CD8 in the recognition of class I MHC antigens and CD4 in the recognition of class II MHC antigens. Even if there is a role for CD8 and CD4, the $V_\alpha-V_\beta$ binding site must also play a part in recognition of MHC antigens, either alone in alloantigenic responses or as complexes in presenting peptides derived by processing of protein antigens. To explain how distinct class I and class II recognition is partitioned between T cell subsets when a common pool of recognition structures is used, requires that a powerful selection process operates (probably in the thymus) during the maturation of T cells.

A definitive explanation is still required for the high frequency of allo-reactive T cells in the periphery (see p. 330). Bevan (1984) has suggested that the high density of MHC antigens on lymphocyte surfaces allows stimulation of T cells expressing very low affinities (much lower than the affinities needed for response to non-MHC antigens). Response to non-MHC antigens requires presentation by a class I or class II MHC molecule; the weak association between antigen fragment and MHC molecule will ensure that the density of self MHC-plus foreign antigen will be very much less than the density of MHC antigens.

References

Barth R.K., Kim B.S., Lan N.C. et al. (1985) The murine T-cell receptor employs a limited repertoire of expressed V_β gene segments. Nature 316, 517.

Benacerraf B. & McDevitt H.O. (1972) Histocompatibility-linked immune response genes. Science 175, 273.

Bevan M.J. (1984) High determinant density may explain the phenomenon of alloreactivity. Immunol. Today 5, 128.

Brenner M.B., McLean J., Dialynas D.P. et al. (1986) Identification of a putative second T-cell receptor. Nature 322, 145.

Claman H.N., Chaperon E.A. & Triplett R.F. (1966) Thymus—marrow cell combinations — synergism in antibody production. Proc. Soc. Exp. Biol. Med. 122, 1167.

Degrave W., Tavernier J., Duerinck. F., Plaetinck G., Devos R. & Fiers W. (1983) Cloning and structure of the human interleukin 2 chromosomal gene. EMBO J. 2, 2349.

Ford C.E. (1966) The use of chromosome markers (Appendix). In Tissue Grafting and Radiation by Micklem H.S. & Loutit J.F. Academic Press, New York.

Hendrick S.M., Cohen D.I., Nielsen E.A. & Davis M.M. (1984) Isolation of cDNA clones encoding T cell-specific membrane-associated proteins. Nature 308, 149.

Kappler J.W., Skidmore B., White J. & Marrack P. (1981) Antigen-inducible, H-2-restricted, interleukin-2-producing T cell hybridomas. Lack of independent antigen and H-2 recognition. J. exp. Med. 153, 1198.

Katz D., Hamaoka T. & Benacerraf B. (1973) Cell interactions between histoincompatible T and B lymphocytes. II Failure of physiologic cooperative interactions between T and B lymphocytes from allogeneic donor strains in humoral responses to hapten-protein conjugates. J. exp. Med. 137, 1405.

Kindred B. & Shreffler D.C. (1972) H-2 dependence of cooperation between T and B cells in vivo. J. Immunol. 109, 940.

Klein J. (1984) What causes immunological nonresponsiveness? Immunol. Rev. 81, 177.

Mintz B. (1967) Gene control of mammalian pigmentary differentiation.

I. Clonal origin of melanocytes. *Proc. natn. Acad. Sci. USA* **58**, 344.

Mitchell G.F. & Miller J.F.A.P. (1968) Cell to cell interaction in the immune response. II The source of hemolysin-forming cells in irradiated mice given bone marrow and thymus or thoracic duct lymphocytes. *J. exp. Med.* **128**, 821.

Mitchison, N.A. (1971) The carrier effect in the secondary response to hapten-protein conjugates. II Cellular cooperation. *Eur. J. Immunol.* **1**, 18.

Ordal J.C. & Grumet F.C. (1972) Genetic control of the immune response. The effect of graft-*versus*-host reaction on the antibody response to poly-L(Tyr,Glu)-poly-D,L-Ala−poly-L-lys in non-responder mice. *J. exp. Med.* **136**, 1195.

Rosenthal A.S. & Shevach E.M. (1973) Function of macrophages in antigen recognition by guinea pig T lymphocytes. I Requirement for histocompatible macrophages and lymphocytes. *J. exp. Med.* **138**, 1194.

Simonsen M. (1968) The clonal selection hypothesis evaluated by grafted cells reacting against their hosts. *Cold Spring Harb. Symp. quant. Biol.* **32**, 517.

Zinkernagel R.M. & Doherty P.C. (1974) Activity of sensitized thymus-derived lymphocytes in lymphocytic choriomeningitis reflects immunological surveillance against altered self components. *Nature* **251**, 547.

Further reading

Arden B., Klotz J.L., Siu G. & Hood L.E. (1985) Mouse T-cell antigen receptor genes: diversity and structure of genes of the α family. *Nature* **316**, 783.

Dalton R.W. & Swain S.L. (1982) Regulation of the immune response: T cell interactions. *CRC. Crit. Rev. Immunol.* **3**, 209.

Gowans J.L. & Uho J.W. (1966) The carriage of immunological memory by small lymphocytes in the rat. *J. exp. Med.* **124**, 1017.

Kronenberg M., Siu G., Hood L.E. & Shastri N. (1986) The molecular genetics of the T-cell antigen receptor and T-cell antigen recognition. *Ann. Rev. Immunol.* **4**, 529.

Lanzavecchia A. (1985) Antigen-specific interaction between T and B cells. *Nature* **314**, 537.

Zinkernagel R.M. (1978) Thymus and lymphohemopoietic cells: their role in T cell maturation in selection of T cells' H-2-restriction-specificity and in H-2 linked Ir gene control. *Immunol. Rev.* **42**, 224.

Zinkernagel R.M. & Doherty P.C. (1979) MHC-restricted cytotoxic T cells: studies on the biological role of polymorphic major transplantation antigens determining T-cell restriction — specificity, function and responsiveness. *Adv. Immunol.* **27**, 51.

Part 4
Genetic Aspects of
Immunological Diseases

The following three chapters will focus on three areas of immunological disease and will discuss the genetic aspects of each. These areas are chosen because immunogenetics is beginning to illuminate certain elements of the underlying pathological processes.

Immunodeficiency diseases are considered first and in most detail (Chapter 14). Here studies of single complement component deficiencies have helped to resolve aspects of the normal physiological mechanisms of complement activation. The paraimmunoglobulinopathies discussed in Chapter 15 are of particular interest because they provided some of the first evidence of multi-gene control of immunoglobulin structure. Finally, one of the major areas of expanding activity in clinical immunology throughout the last decade has been the dissection of the MHC in relation to both natural immunity and disease susceptibility. Chapter 16 discusses various autoimmune disorders, many of which show HLA-linkage.

Although the aetiology of these diseases is still often shrouded in mystery, we are now beginning to identify categories of prognosis and genetic risk for such patients and their relatives.

Chapter 14
Genetics of the immunodeficiency disorders

Introduction

Most immunodeficiency is probably secondary to some other process or event and is frequently transient and self limiting. The primary immunodeficiencies have been the subject of numerous attempts at classification and the most recent version, compiled by the WHO Scientific Group on Immunodeficiency, is presented in Table 14.1.

Many immunodeficiencies are clearly familial although for most of the late onset variable deficiencies the mode of inheritance is still unknown. Six of the immunodeficiency diseases listed in Table 14.1 are known to be X-linked disorders. This suggests that the X-chromosome has an important role in the control of immunological function, but what this is remains unclear. In other immunodeficiency diseases, particularly those associated with deficiencies of adenosine deaminase, purine nucleoside phosphorylase and transcobalamin 2, there is evidence of autosomal recessive inheritance. Autosomal dominant inheritance appears to be more rare but has been associated with some types of common variable immunodeficiency.

Selected examples from the classification in Table 14.1 will now be considered in more detail, but for a comprehensive coverage of the field the reader should consult Wedgwood et al. (1983).

Defects of specific immunity

Predominantly antibody defects

X-linked agammaglobulinaemia (classification (a)1 in Table 14.1)

In 1952 an American army paediatrician, Colonel Bruton, described a boy aged 4 years who was experiencing frequent severe infections, including pneumonia and meningitis. Using the recently developed technique of

Table 14.1 A classification of primary immunodeficiencies

Designation	Serum immunoglobulin	Cell-mediated immunity	Presumed defect	Inheritance
(a) Predominantly antibody defects				
1 X-linked agammaglobulinaemia	All isotypes decreased	Normal	Intrinsic defect in differentiation from pre-B to B cell	X-linked
2 X-linked hypogammaglobulinaemia with growth hormone deficiency	All isotypes decreased	Normal	Not known	X-linked
3 Autosomal recessive agammaglobulinaemia	All isotypes decreased	Normal	Not known	Autosomal recessive ?
4 Ig deficiency with increased IgM and IgD	(A) Increased IgM and IgD (B) Decreased IgG and IgA	Normal Normal	Isotype switching defect	Autosomal recessive in some
5 IgA deficiency	(A) Decreased IgA1 and IgA2 (B) Decreased IgA1, IgA2 and IgG2 (± IgG4 deficiency) (C) Decreased IgA1 or IgA2	Usually normal Normal Normal	Defective maturation of IgA B cells (± IgG subclass)	Generally unknown but autosomal
6 Selective deficiency of other Ig isotypes	Decrease in IgM or IgG1, 2, 3, 4	Normal	Differentiation defect	Unknown
7 κ-chain deficiency	Ig(κ) decreased	Normal	Unknown	Unknown
8 Antibody deficiency with normal or hypergammaglobulinaemia	Normal	Variable	Differentiation defect; defective T cell help	Unknown
9 Immunodeficiency with thymoma	All isotypes decreased	Variably decreased	Unknown	None
10 Transient hypogammaglobulin-aemia of infancy	IgG and IgA decreased	Variable	Unknown	Unknown

(b) Common variable immunodeficiency (CVID)

11 CVID with predominant B cell defect				
(A) Normal B cell numbers (primarily $\mu^+ \delta^+$, without $\mu^+ \delta^-$, γ^+ or α^+ cells)	Decreased	Variable	Intrinsic differentiation defect of immature to mature B cells	Unknown, autosomal recessive or autosomal dominant
(B) Very low B cell number	Decreased	Variable	Intrinsic differentiation defect of pre-B to B cell	Unknown, autosomal recessive or autosomal dominant
(C) $\mu^+ \gamma^+$ or γ^+ non-secretory B cells with plasma cells	Decreased	Normal	Intrinsic defect in B-cell maturation at plasma cell level	Unknown
(D) Normal or increased B cell number with $\mu^+ \delta^+ \gamma^+$, $\mu^+ \delta^+$ α^+, γ^+ and α^+ B cells	Decreased	Variable	Intrinsic defect in maturation of B cells to plasma cells	Unknown
12 CVID with predominant immuno-regulatory T cell disorder				
(A) Deficiency of T helper cells	Decreased	Variable	Immunoregulatory T cell disorder: defect in differentiation of thymocyte to T helper cell	Unknown
(B) Presence of activated T suppressor cells	Decreased	Variable	Immunoregulatory T cell disorder	Unknown
13 CVID with autoantibodies to B cells or T cells	Decreased	Variable	Unknown	Unknown

Table 14.1 Continued

Designation	Serum immunoglobulin	Cell-mediated immunity	Presumed defect	Inheritance
(c) Predominantly cell-mediated immunity (CMI) defects				
14 Combined immunodeficiency with predominant T cell defect	Near normal	Decreased	Unknown	Unknown, autosomal recessive
15 Purine nucleoside phosphorylase (PNP) deficiency	Normal	Progressive decrease	T cell defect from toxic metabolites due to enzyme deficiency	Autosomal recessive
16 Severe combined immunodeficiency (SCID) with adenosine deaminase (ADA) deficiency	Decreased	Decreased	T and B cell defects from toxic metabolites due to enzyme deficiency	Autosomal recessive
17 SCID				
(A) Reticular dysgenesis	Decreased	Decreased	Defective differentiation of T and B cells, lymphomyeloid maturation defect	Autosomal recessive
(B) Low T and B cell numbers	Decreased	Decreased	Lymphoid maturation defect of both T and B cells	Autosomal recessive or X-linked
(C) Low T, normal B cell ('Swiss type')	Decreased	Decreased	Lymphoid maturation defect of both T and B cells	Autosomal recessive or X-linked
(D) 'Bare lymphocyte syndrome'	Decreased	Decreased	Differentiation defect with lack of HLA determinants on T and B cells	Autosomal recessive
18 Immunodeficiency with unusual response to EBV	Decreased after EBV infection in some	Normal by conventional tests	Unknown	X-linked or autosomal recessive

(d) Immunodeficiency associated with other major defects

19 Transcobalamin 2 deficiency	All isotypes decreased	Normal	Defect in B12 transport resulting in defective cell proliferation; B cell to plasma-cell proliferation	Autosomal recessive
20 Wiskott – Aldrich syndrome	Increased IgA and IgE; decreased IgM	Progressive decrease	Cell surface glycoprotein defect affecting all haematopoietic stem cell derivatives	X-linked
21 Ataxia telangiectasia	IgA, IgE and IgG often decreased; increased IgM (monomeric)	Decreased	Unknown; defective T cell maturation	Autosomal recessive
22 3rd and 4th pouch/arch syndrome (Di George)	Normal (?)	Decreased	Embryopathy; abnormal thymus with resultant T cell defects	None

paper electrophoresis Bruton was able to demonstrate the absence of gammaglobulin in the boy's serum. Other cases were soon described and, after some initial confusion with the more profound defect of severe combined immunodeficiency, it became possible to define the syndrome. Patients usually present with recurrent pyogenic infections between 3 and 24 months of age. The disease is characterized by low levels of IgG and, usually, very low levels of IgM and IgA. B cells are absent from the peripheral blood and antibody responses are usually undetectable. Cell-mediated immunity is normal.

The X-linked nature of the disease suggests a single gene defect but the nature of the gene product is obscure. It is now clear that the immunoglobulin structural genes are not involved since these have now been mapped elsewhere. It is more likely that the gene involved codes for a protein (probably an enzyme) which is centrally important to B cell development. Recent studies have localized the gene to the region near band q 21.22 on the long arm of the X chromosome.

The inheritance of this disorder follows that of all classical X-linked diseases and a carrier female has one normal and one abnormal X chromosome. Consequently, 50% of her male offspring are likely to have the disease and 50% of the female offspring will be carriers. In view of the high mortality in the absence of treatment it seems likely that the disease has persisted through spontaneous germ-line mutations, but this is not clearly established. There is, at present, no satisfactory method for the detection of the carrier state. In a family with one affected boy, fetal sexing by amniocentesis and termination of male fetuses may be offered.

Treatment of hypogammaglobulinaemia is replacement injection of immunoglobulin. This is usually given by intramuscular injection at a dosage of 25 mg/kg body weight per week and effectively prevents the serious bacterial infections that characterize the disorder. Prompt antibiotic therapy during the early stages of any infection and regular medical care throughout life are essential for these patients.

IgA deficiency (classification (a)5 in Table 14.1)

This is the most common primary immunodeficiency affecting approximately 1 in 700 of Caucasian populations. As the name suggests there is deficiency of IgA

(<10 iu/ml) with essentially normal IgG and IgM levels. At least three variants of IgA deficiency are now recognized. Both subclasses may be absent, one subclass only may be involved or both IgA subclasses and one or more IgG subclasses may be deficient. It was claimed by Oxelius et al. (1981) that approximately 20% of IgA deficient patients had an associated IgG2 deficiency and that this subgroup was far more likely to suffer problems with recurrent infections than those with normal IgG2 but low IgA levels.

Many of the individuals with IgA deficiency are apparently asymptomatic, but it is believed by many investigators that they all eventually express disease. The form that such expression takes ranges from autoimmune disease such as rheumatoid arthritis, thyroiditis, lupus and pernicious anaemia to gastrointestinal diseases (especially coeliac disease) and respiratory tract infections.

There is much evidence that the defect is familial (see Figure 14.1) but no clear mode of inheritance has emerged to date. In many the data are consistent with an autosomal recessive model but in some rare cases autosomal dominant inheritance with incomplete penetrance has been proposed.

SELECTIVE IgA DEFICIENCY

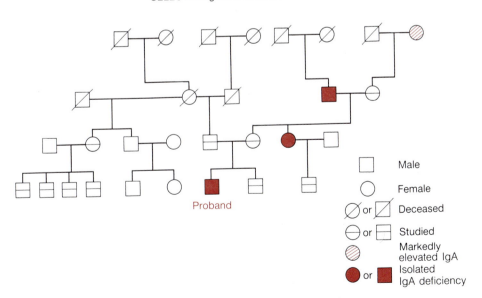

Fig. 14.1 Pedigree of a boy with selective IgA deficiency whose maternal aunt and maternal grandfather also had absent IgA. One maternal great-grandmother had a markedly elevated serum IgA concentration of 3.2 g/100 ml. (Adapted with permission from Buckley, 1975a.)

The application of DNA hybridization technology to the immunodeficiency diseases is beginning to illuminate more precisely the molecular basis of a number of disorders including IgA deficiency. It has been known for some years that IgG2 and IgG4 deficiencies are more commonly associated with IgA2 deficiency than with lack of IgA1 and are not infrequently linked with IgE deficiency. Such observations are compatible with defective switching in the heavy chain locus where the human gene sequence is $5'-\gamma_3-\gamma_1-(\varepsilon)_{pseudo}-\alpha_1-\gamma_2-\gamma_4-\varepsilon-\alpha_2-3'$. Using a human IgA heavy chain constant region probe, deletions of the gene coding for IgA1 have been described in three individuals with selective IgA1 deficiency (Lefranc *et al.*, 1982) suggesting a structural gene defect. In contrast, a recent study of 66 individuals with selective IgA1 deficiency and one individual with selective IgA2 deficiency showed that both α chain constant region genes were present in the genome (Hammarström *et al.*, 1985). This latter study included four familial cases of IgA deficiency and the authors concluded that the defects probably arose from mutations interfering with the processing and translation of mRNA.

In the management of IgA deficiency there is as yet no satisfactory counterpart to the IgG given to patients with X-linked hypogammaglobulinaemia. A minority of IgA deficient patients develop antibodies to IgA when given blood products and subsequently manifest anaphylactic reactions on further exposure.

Immunodeficiency with thymoma (classification (a)9 in Table 14.1)

This is an example of late onset immunodeficiency. Most cases present between the ages of 40 and 70 with recurrent respiratory tract infections. A thymoma may be diagnosed by chest X-ray several years before symptoms of immunodeficiency are manifest. Most often this immunodeficiency takes the form of hypogammaglobulinaemia but neutropenia, lymphopenia and candidiasis have also been observed. The disease is slightly more common in women than men: no clear pattern of inheritance has yet been reported.

Transient hypogammaglobulinaemia of infancy (classification (a)10 in Table 14.1)

The concept of transient hypogammaglobulinaemia is somewhat controversial. Many of the 3- to 6-month-old infants presenting with repeated pyogenic infections and having low serum immunoglobulins subsequently achieve normal levels. There is probably no single aetiological cause and a proportion probably represent the low end of the normal range.

Common variable immunodeficiency (CVID)

CVID with predominant B cell defect (classification (b)11 in Table 14.1)

The majority of patients with hypogammaglobulinaemia fall into this group. Males and females are equally represented and disease may present at any age. As with the X-linked form there is an increased frequency of infections of both the upper and lower respiratory tracts. In addition many patients have gastrointestinal disease which may be severe and cause malabsorption.

Most of the patients in this group have immunoglobulins on the surface of their lymphocytes, clearly differentiating them from the patients with the X-linked form of the disease. The abnormality expressed by the serum immunoglobulins may range from low levels of all classes (panhypogammaglobulinaemia) to selective deficiencies of only one or two classes or subclasses. For several years it has been considered likely that many of these defects involve proteins controlling the maturation of B cells to plasma cells, rather than reflecting gene deletion. This view is supported by a recent study in which a human IgA heavy chain constant region probe was used to search for gene deletions in seven patients with variable immunodeficiency (Smith & Hammarström, 1984). Although serum IgA was undetectable, both the IgA1 and IgA2 genes were present in all seven individuals.

These patients are treated in an essentially similar fashion as the X-linked patient group and with similar success. There is, however, an increased tendency for such patients to develop polyarthritis and autoimmune diseases.

Predominantly cell-mediated immunity (CMI) deficiency syndromes

Purine nucleoside phosphorylase (PNP) deficiency (classification (c)15 in Table 14.1)

Patients with this deficiency have a primary T lymphocyte deficit but essentially normal serum immunoglobulins. Clinically all have suffered from repeated infections of both the upper and lower respiratory tracts. A general inability to cope with virus infections is a common feature.

The disease appears to have a biochemical basis somewhat similar to ADA deficiency (see below). PNP is an enzyme present in lymphocytes, erythrocytes and fibroblasts which catalyses the conversion of inosine to hypoxanthine. Healthy individuals are believed to be homozygous for the NP^1 structural gene allele (known to be on chromosome 14). In contrast, children with PNP deficiency have no detectable enzyme activity and are believed to be homozygous for a 'silent' gene, NP^0. This suggests that the inheritance is autosomal recessive. In affected children the plasma levels of the PNP substrates inosine, deoxyinosine, guanosine and deoxyguanosine are raised. Toxic levels of deoxyguanosine triphosphate (dGTP) build up in the lymphocytes and, as with ADA deficiency, these impair DNA synthesis by inhibiting ribonucleotide reductase.

As in ADA deficiency regular red cell transfusions and gammaglobulin therapy have been attempted in some of these children, but bone-marrow transplantation may offer the best hope of long-term success.

It is now possible to measure the PNP enzyme either in cultured amniotic fibroblasts or directly in fetal red cells obtained at fetoscopy (Simmonds *et al.*, 1983) so that prenatal diagnosis can be offered for subsequent 'at risk' children. In addition the increased levels of guanosine and inosine and their deoxy forms in the plasma and urine of affected individuals makes the measurement of purines in amniotic fluid a valuable alternative approach to prenatal diagnosis. In contrast, prenatal leucocyte phenotyping is not thought likely to be of diagnostic value in this condition.

In the mouse, purine nucleoside phosphorylase is inhibited by high doses of allopurinol riboside. Such animals appear to develop diminished cellular immunity whilst maintaining normal humoral immune mechanisms, but whether this is a valid experimental model of the disease remains to be explored.

Severe combined immunodeficiency (SCID) with adenosine deaminase (ADA) deficiency (classification (c)16 in Table 14.1)

Severe combined immunodeficiency syndromes are characterized by defects of both humoral and cell-mediated mechanisms. Clinically these patients are normal at birth but most present in the first 3 months of life with recurrent infections, diarrhoea and weight loss. Nevertheless, the classical course is variable. The most severely affected rarely survive beyond 2 years of age without a bone-marrow graft — preferably from an HLA-matched sibling.

A significant (~20%) proportion of SCID patients with an autosomal pattern of inheritance have been shown to lack adenosine deaminase (ADA) activity in their red cell lysates. This enzyme is an aminohydrolase which catalyses the conversion of adenosine or deoxyadenosine to inosine or deoxyinosine. It is probable that ADA is the product of a single gene locus and exists in three common phenotypic forms: ADA1, ADA2, ADA1−2. Tests on relatives of SCID patients have revealed a silent gene (ADA°) which has no detectable gene product. It is highly likely that SCID patients with profound ADA deficiency are homozygous for this gene (i.e. ADA°/ADA°). The ADA gene has been mapped to chromosome 20 (20q13-qter) — see McKusick (1984).

Although ADA is a widely distributed enzyme, it is solely the lymphocyte which is affected in patients with ADA deficiency and it was suggested that this might reflect selective trapping of a metabolite produced in ADA deficiency tissue. In the absence of ADA, deoxyadenosine is phosphorylated by a purine kinase which is found predominantly in lymphocytes. The product of phosphorylation — deoxydenosine triphosphate (dATP) — is probably the major toxic metabolite responsible for the disease manifestations. It inhibits the enzyme ribonucleotide reductase, which seems to play an important role in the control of DNA synthesis. Figure 14.2 shows some of the metabolic pathways which become involved in this disorder. Treatment

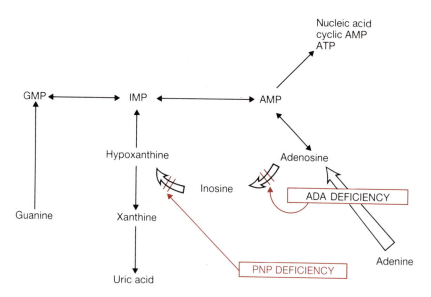

GMP ← → IMP ← → AMP

Nucleic acid
cyclic AMP
ATP

Guanine

Hypoxanthine

Xanthine

Uric acid

Inosine

Adenosine

ADA DEFICIENCY

Adenine

PNP DEFICIENCY

Fig. 14.2 Simplified scheme showing metabolic pathways of purine nucleotides and sites of action of adenosine deaminase (ADA) and purine nucleoside phosphorylase (PNP). Deficiencies of these two enzymes lead to increased levels of the toxic metabolites dATP and dGTP respectively.

with regular red cell transfusions has been attempted with variable success, although a bone-marrow transplant from an HLA-identical sibling remains the treatment of choice. It is probable that the first attempt at gene replacement therapy for immunodeficiency will be made in an ADA deficient individual. ADA deficiency has been diagnosed prenatally by assaying the ADA levels of cultured amniotic fibroblasts but this is time consuming and of uncertain outcome. Recently, rapid prenatal diagnosis of ADA deficiency has been achieved by direct measurement of fetal red cell ADA and dATP levels.

It is possible to produce adenosine deaminase deficiency in mice by daily injections of the ADA inhibitor deoxycoformycin (0.2 – 1.0 mg/kg). This results in greatly increased levels of dATP in thymus and erythrocytes but not in liver, spleen, intestinal mucosa, bone-marrow or peripheral lymphocytes. After three days of such treatment there is usually a marked lymphopaenia.

Severe combined immunodeficiency (other than ADA deficiency (classification (c)17 in Table 14.1)

Eighty per cent of patients characterized by severe defects of both humoral and cell-mediated mechanisms have normal ADA levels. Normal numbers of granulocytes are usually present (except in reticular dysgenesis) and it is believed that the syndromes reflect maturation defects involving all lymphoid cells.

The leucocyte abnormalities in SCID are demonstrable during the second trimester of pregnancy. In studies reviewed by Linch and Levinsky (1983) fetal blood samples of $200-500$ µl obtained at this stage of gestation were analysed by a microstaining technique using ten monoclonal antibodies to leucocyte antigens. By such investigations it was possible to exclude SCID in one case and make a positive diagnosis in three others. There were less than 100 T cells/mm^3 of fetal blood in the three affected cases compared to 2500/mm^3 in fourteen immunologically normal fetuses. Prenatal diagnosis of SCID has previously only been possible in patients with a defined defect such as ADA deficiency (see above), but these recent studies suggest that prenatal diagnosis and the option of termination may be offered for most at-risk pregnancies.

Reticular dysgenesis (classification (c)17(A) in Table 14.1)

This particularly rare form of SCID presents early and proceeds rapidly to a fatal outcome unless a successful engraftment is achieved. These patients have agranulocytosis as well as the usual SCID pattern and this points to an early stem cell defect. Even within the small number of patients recorded in the literature (~10) there are three pairs of twins, but there is no clearly discernible inheritance pattern.

Low T, normal B cell ('Swiss type') (classification (c)17(C) in Table 14.1)

The description of severe combined immunodeficiency was for several years used synonymously with the alternative nomenclature Swiss-type agammaglobulinaemia and, if restricted to the autosomal recessive form associated with normal ADA, this is probably the most common form of SCID. The population incidence of this particular group is about 1 in 100 000 in the UK, but higher in certain selected populations. It remains a heterogeneous entity. Serum immunoglobulin and antibody levels are decreased despite normal levels of circulating B cells. This may reflect an abnormality at an early stage of T-cell differentiation, possibly due to faulty induction by thymic epithelium.

Immunodeficiency associated with other major defects

Wiskott—Aldrich syndrome (classification (d)20 in Table 14.1)

The Wiskott—Aldrich syndrome is an X-linked recessive disorder (frequency 1 in 250 000 live births) characterized by thrombocytopenia, eczema and immunodeficiency. A low platelet count is a major diagnostic feature and the immunological abnormalities are much more variable. At first the patient may have normal levels of IgG and either normal or raised levels of IgA, IgM and IgE. However, after 1 year of age the usual pattern is one of high IgE and IgA levels but low IgM and normal or low IgG. Antibody responses to polysaccharide antigens have been reported as poor in contrast to responses to protein antigens which appear normal. There is evidence of abnormal T cell function and it has been suggested that this may underly the development of eczema and high levels of IgE.

At present there is no available method for the detection of female heterozygotes and most current interest centres on attempting to identify the biochemical abnormality responsible for the dysfunction expressed by both the platelets and lymphocytes. Parkman *et al.* (1981) reported preliminary studies which suggested that a glycoprotein of mol. wt. 115 000 (GPL-115) was either absent or reduced on the surface of lymphocytes and platelets of patients with the Wiskott—Aldrich syndrome and that this was the primary defect. The absence of GPL-115 was not due to the effect of the spleen since the protein was absent from lymphocyte membranes both before and after splenectomy. The gene defect is believed to be located on the long arm of the X chromosome, close to the centromere.

It should be possible to produce specific antisera able to differentiate between normal lymphocytes and those of Wiskott—Aldrich patients. This should eventually lead to prenatal diagnosis by leucocyte phenotyping. In a sex-linked disease such as this it would only be necessary to perform fetoscopy after amniocentesis had confirmed the presence of a male fetus.

Allogeneic bone-marrow transplantation has been used to treat successfully two patients with the Wiskott—Aldrich syndrome. The clinical improvement and associated restoration of both haematologic and immunologic functions in these two patients further supports the view that this syndrome represents a primary T lymphocyte defect.

Ataxia telangiectasia (classification (d)21 in Table 14.1)

Ataxia telangiectasia is a progressive multisystem disease characterized by neurological degeneration, oculocutaneous telangiectasia and increased susceptibility to infection. Most patients first present in childhood with ataxia, and this progresses to a range of other neurological problems. Later, conjunctival telangiectasia is a common finding and many other abnormalities may arise. The most common clinical manifestation of immunodeficiency is frequent chest infection. Low levels of serum and secretory IgA are common, as is a low serum IgE. Many patients also produce a monomeric form of IgM. Lymphocyte responses may be impaired in older children and young adults.

The disease has been shown to be transmitted by an autosomal recessive mode of inheritance (see Figure 14.3). The population frequency is approximately 2—3 per 100 000 live births. The basic defect appears to be a failure in the DNA repairing mechanisms, and obligate heterozygote parents also show some evidence of this together with an increased incidence of chromosomal abnormalities.

The gene coding for susceptibility to ataxia telangiectasia is apparently not closely linked to the Gm locus of the immunoglobulin heavy chain complex on chromosome 14 (band q 32), a region where translocations have been reported in patients with the disease.

The molecular basis of the common IgA deficiency is unclear. DNA preparations from five patients with ataxia telangiectasia and no detectable serum IgA were investigated using a human IgA heavy chain constant region probe and both the IgA1 and IgA2 structural genes were present (Waldmann et al., 1983). Thus, as in selective IgA deficiency itself, it appears likely that the immunoglobulin deficiency arises from faulty translation or processing of the relevant mRNA.

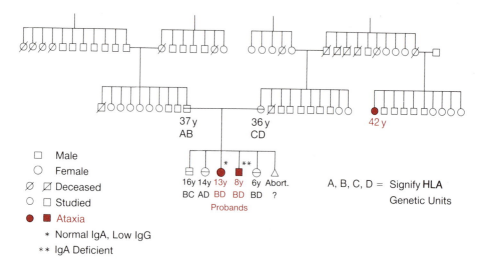

Male
Female
Deceased
Studied
Ataxia
* Normal IgA, Low IgG
** IgA Deficient

37 y
AB

36 y
CD

42 y

16y 14y 13y 8y 6y Abort.
BC AD BD BD BD ?
Probands

A, B, C, D = Signify **HLA**
Genetic Units

Fig. 14.3 Pedigree of a pair of sibs with ataxia telangiectasia. The probands were HLA and MLR compatible with each other and with a normal younger sister. One affected sib had selective IgA deficiency, whereas the other had a normal serum IgA concentration but low IgG and elevated IgM concentrations. A 42-year-old maternal second cousin had had ataxia throughout her life but no telangiectasia or susceptibility to infection. Serum Ig concentrations were normal in all first-degree relatives of the probands. (Adapted with permission from Buckley, 1975b.)

These patients have an increased susceptibility to cancer and it is hoped that a better general understanding of the factors predisposing to oncogenesis may follow from successful mapping and cloning of the AT susceptibility gene.

Treatment is mainly palliative since the basic defect is beyond correction. Most patients die in the second decade, but a few have survived into their twenties.

3rd and 4th pouch/arch syndrome (Di George) (classification (d)22 in Table 14.1)

These patients present in the first few days of life with wide ranging developmental abnormalities. The thymus is absent or fragmented and many patients have an associated transposition of the aorta, absent parathyroid glands and other features. Although in general the syndrome appears to arise following an environmental insult to the fetus between 6 and 8 weeks of gestation, there is a reported association in two families with partial deletions of chromosome 22.

Predictably these patients have impaired cell-mediated function but essentially normal humoral immunity. Successful treatment with grafts of fetal thymus and thymic epithelial cells have been reported.

Defects of non-specific immunity

Phagocytic defects

Primary defects of phagocytosis

An effective mobilization of neutrophils is frequently the major mechanism for the removal of much potentially pathogenic material and antigen—antibody complexes. Primary neutrophil defects do occasionally occur and some patterns of inheritance have been established. Table 14.2 summarizes some of the defects.

Neutropenia

Severe congenital neutropenia is of two main types. An infantile genetic agranulocytosis has been described in Sweden. Infants presented with skin infections, otitis media and pneumonia, and the disease was usually fatal within the first year of life. An autosomal recessive inheritance was established for this variant. In contrast, a more benign form of chronic neutropenia was described by Hitzig in which the inheritance was clearly autosomal dominant. Neutropenia may also be cyclical,

Table 14.2 Primary neutrophil defects

Neutropenia
Morphological abnormalities of phagocytes
   Chediak—Higashi syndrome
   Other primary morphological abnormalities
Defective neutrophil mobility
   Lazy leucocyte syndrome
   Job's syndrome
   Actin dysfunction syndrome
   Schwachman's syndrome
   Defective mobility associated with delayed cord separation
Defective bacterial killing
   Chronic granulomatous disease (CGD)
   Glucose 6-phosphate dehydrogenase (G6PD) deficiency
   Neutrophil pyruvate kinase deficiency
   Myeloperoxidase deficiency

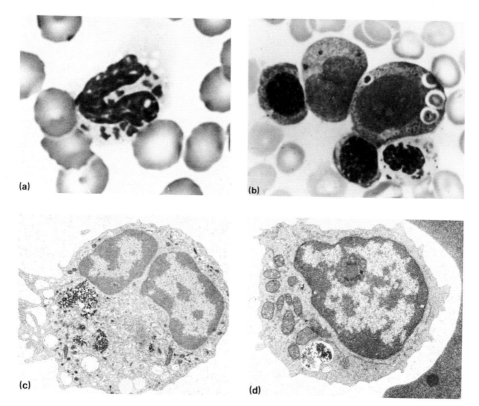

Fig. 14.4 Various cells from a patient with Chediak–Higashi syndrome. Note the giant cytoplasmic granules. (a) Light microscope preparation of peripheral blood neutrophil (Mag. × 1700). (b) Light microscope preparation of bone-marrow eosinophil (Mag. × 1120). (c) Electron microscopy preparation of peripheral blood neutrophils (Mag. × 6000). (d) Electron microscopy preparation of peripheral blood lymphocyte (Mag. × 7200). Photographs kindly supplied by Prof. B.D. Lake; © Prof. B.D. Lake.

with the neutropenic phase lasting for a few days every 15–35 days. Skin infections are common at the time of neutropenia but recovery is usually rapid and spontaneous. Autosomal dominant inheritance has been demonstrated in a number of these patients.

Morphological abnormalities of phagocytes

Chediak–Higashi syndrome

This rare autosomal recessive disorder is characterized by the presence of giant granules in the neutrophils and certain tissue sites such as the gastric mucosa and pancreas (see Figure 14.4). The neutrophil granules contain both myeloperoxidase and lactoferin suggesting that they are derived from a fusion of primary and secondary granules. Several abnormalities of neutrophil function have been described in Chediak—Higashi patients. The cells respond poorly to chemotactic stimuli, and bacterial killing is slow.

Clinically the patients have frequent severe pyogenic infections, particularly of the upper and lower respiratory tract and of the skin.

Although initially the infections respond to antibiotic therapy the disease becomes increasingly difficult to manage and few patients reach adult life. There may be raised cyclic AMP levels in the polymorphs of these patients and treatment with high doses of ascorbic acid has been shown both to correct this abnormality and also to improve mobility and bacterial killing function in the short term. Recent studies suggest that these patients have very poor natural killer (NK) cell activity and defective interferon production.

There are well characterized models of the Chediak—Higashi syndrome in a number of animals including the Aleutian mink, Hereford cattle and mice. In each case giant cytoplasmic granules in the neutrophils of the species confirms the diagnosis but other features of the human disease such as photophobia and oculocutaneous albinism are also commonly observed (see Figure 14.5).

A world literature of 78 cases of Chediak—Higashi syndrome was reviewed by Klebanoff and Clark in 1978, and this provided clear evidence of an autosomal recessive inheritance with consanguinity in approximately 50% of cases. Breeding experiments in the animal models of the disease also support a pattern of autosomal recessive inheritance with similar frequencies of affected males and females.

Although heterozygotes are apparently healthy and lack the oculocutaneous features of the homozygotes there are, nevertheless, reports of abnormal granules and cytoplasmic inclusions in the circulating lymphocytes of such individuals. As these observations are controversial they do not as yet constitute a satisfactory basis for the diagnosis of heterozygosity.

Other primary morphological abnormalities

Other rare abnormalities of neutrophil morphology

have been described but seem to be of much less clinical significance than the Chediak–Higashi syndrome. The May–Hegglin anomaly is a rare autosomal dominant disorder characterized by the presence of large, basophilic inclusions in the neutrophil cytoplasm. In the Pelger–Hüet anomaly (also autosomal dominant inheritance) there is incomplete segmentation of the nucleus.

Defective neutrophil mobility

The measurement of neutrophil mobility is not standardized and there exist a wide range of function tests — both *in vivo* and *in vitro*. Furthermore, this function is particularly influenced by factors such as malnutrition and infection so that secondary defects are probably rather common. Progress in the delineation of the primary defects has been slow but may now accelerate as more sophisticated biochemical studies are undertaken. Any formal classification of defects would be unsatisfactory at this stage and the small selected group

(a)

Fig. 14.5 (above and facing page) Chediak-Higashi syndrome (CHS) in various animals. (a) Group of cattle showing a CHS calf with the background genes for the Hereford colour. It has a light grey coat colour in a grade Hereford distribution. In the background are two Hereford cattle and on the left is a portion of a black Angus calf. (b) Neutrophil of Aleutian mink with CHS. (c) Neutrophil of cat with CHS. (d) Monocyte of cat with CHS. (Courtesy of David J. Prieur, Washington State University, USA.)

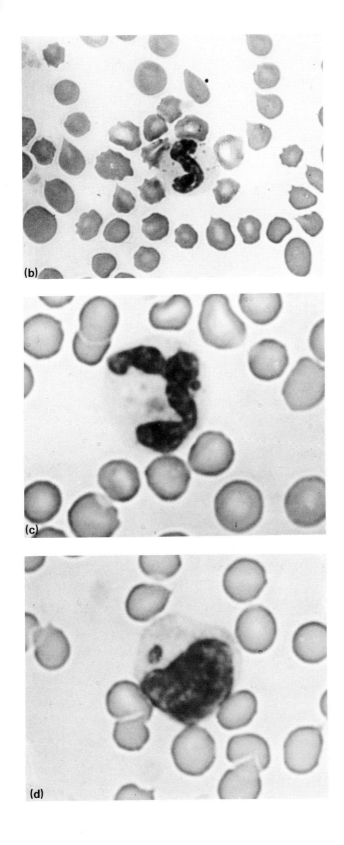

(b)

(c)

(d)

of syndromes that follows is, therefore, probably rather unrepresentative.

Lazy leucocyte syndrome

The first report of primary defective neutrophil mobility was in two infants with frequent infections of mouth and ears. The association of severe neutropenia with some, but not all, patients raises doubts about the diagnostic criteria for the syndrome.

Job's syndrome and hyperimmunoglobulinaemia E

This group of syndromes has also proved difficult to define. The first patients to be described were red-haired girls with eczema and recurrent cold subcutaneous staphylococcal abscesses. Subsequently, boys were also claimed to have the syndrome and the range of clinical and laboratory features was extended. For example, many of these patients are reported to have very high IgE levels. The study by Blum et al. (1977) provides some support for the view that there may be a genetic basis for some of these patients. An infant and his father were both found to have severe staphylococcal infections, eczema and elevated IgE levels. The infant had impaired neutrophil chemotactic responses but the father appeared to be normal in this respect.

Actin dysfunction syndrome

The neutrophils of a single male patient with repeated staphylococcal skin infections were shown to be defective in both phagocytosis and mobility assays. Evidence was obtained to suggest a primary defect of the actomysin system. More recently, the neutrophils of a similar patient with defective neutrophil adhesion, chemotaxis and phagocytosis were found to lack a plasma membrane glycoprotein with a mol. wt. of 110 000.

Shwachman's syndrome

This is a multisystem disease characterized by pancreatic insufficiency, neutropenia, growth retardation and frequent infections. Aggett et al. (1979) studied neutrophil mobility in 14 patients with this syndrome and 13 of their parents. As shown in Figure 14.6 both

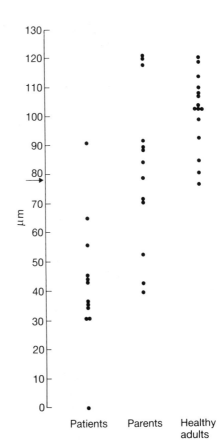

Fig. 14.6 Neutrophil mobility is frequently defective in patients with Shwachman's syndrome. The neutrophils of the parents of such patients also show a less marked mobility defect consistent with a state of heterozygosity. Migration distances in a Boyden chamber type of assay are given for cells stimulated with chemotactic factor. (Reproduced with permission from Aggett et al., 1979.)

groups gave values significantly lower than appropriate controls but with the healthy heterozygotes half-way between the patients and the controls. This is strong presumptive evidence that the defect is primary and suggests a pattern of autosomal recessive inheritance. There is some evidence for a cytoskeletal defect in the neutrophils of such patients.

Defective neutrophil mobility with delayed umbilical cord separation

An association has been described between frequent skin infections, septicaemia, delayed cord separation (>4 weeks) and defective neutrophil mobility. The defect has been corrected *in vitro* by the addition of ascorbic acid, and remission of symptoms and defect correction have also been reported with *in vivo* therapy. Recently it has been shown that the CR3 surface glycoprotein is absent from the neutrophil membrane of such patients (Thompson *et al.*, 1984). This protein (see Chapter 4, p. 92) is the receptor for fixed C3bi and therefore of central importance in phagocytosis. Other patients with recurrent bacterial infections have since been shown to be deficient in this and/or other cell surface molecules such as LFA-1 and p150, 95. The techniques of modern molecular biochemistry are beginning to open up exciting possibilities for meaningful investigations in the heterogeneous group of syndromes discussed in this section.

Secondary defects of neutrophil mobility

Defective neutrophil mobility frequently occurs secondary to some other agent or event and often poses problems in the diagnosis of the primary defects. Some of the factors responsible are listed in Table 14.3.

Table 14.3 Secondary defects of phagocyte function: contributory environmental factors

Defective mobility	Defective killing
Immaturity	Immaturity
Malnutrition	Malnutrition
Infection	Infection
Burns	Burns
Anaesthetics	Irradiation
Diabetes	

Defects of bacterial killing

Chronic granulomatous disease (CGD)

This disease is characterized by repeated pyogenic infections and recurrent discharging abscesses in the groin and neck. Hepatic abscesses and osteomyelitis are also common. Some of these clinical features are summarized in Figure 14.7.

The neutrophils of the CGD patient phagocytose organisms normally but are unable to kill catalase-positive strains, which consequently accumulate and survive in the cell. Eventually these infected neutrophils are themselves phagocytosed by the tissue-fixed mononuclear cells — hence the granuloma formation in macrophage-rich regions such as the lung, liver, lymph nodes and bone marrow. In CGD, catalase-negative organisms are killed normally, probably because they

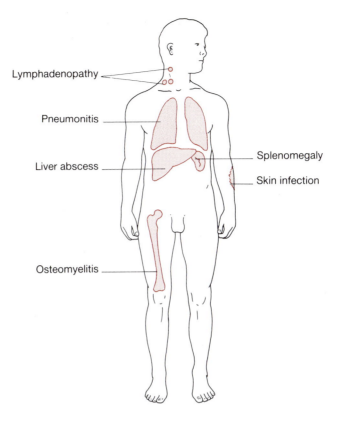

Fig. 14.7 Chronic granulomatous disease affects many organ systems. Some of the major sites likely to be involved are shown here. (Modified with permission from Hosking *et al.*, 1977.)

are able to provide hydrogen peroxide. The latter seems to be an essential prerequisite for effective bacterial killing. A defective oxidase system in the CGD neutrophil results in almost no activation of the hexose monophosphate shunt and no 'respiratory burst' following ingestion. This constitutes the basis of the standard NBT test for CGD since the neutrophils of these patients fail to reduce the dye nitroblue tetrazolium to blue formazan following phagocytosis.

Most CGD patients are boys and in general there is a familial incidence suggesting X-linked inheritance. However, there are some girls with CGD and some boys in whom the inheritance appears to be autosomal recessive. The genetics of CGD are thus clearly heterogeneous and much further work is required in this area. The presumed heterozygotic mothers of boys with the X-linked form of the disease are frequently found to be intermediate in the NBT and bactericidal killing tests, but there is a wide scatter and exclusion of heterozygosity can be difficult. These individuals may not always be in full health — some appear to suffer from an excess of infections and others have lupus erythematosus — both discoid and systemic.

Progress in the understanding of CGD genetics appears to have been further confused by observations linking the disorder with certain antigens of the Kell blood group system which are *not* inherited on the X chromosome. Thus, it was noted following blood transfusion reactions that boys with the X-linked form of CGD may lack the K_x antigen, whereas patients with the autosomal recessive form do not. A possible explanation of such a discrepancy is that a gene controlling production of a precursor of the blood group substance is inherited on the X chromosome but this remains totally speculative.

As previously mentioned, the basic defect in CGD is failure to manifest the respiratory burst which accompanies phagocytosis and normally precedes killing. Many enzyme systems have been implicated by a succession of investigators and the field remains controversial. One of the more recent suggestions by Segal and Jones (1978, 1980) is that an electron transporting system containing a *b* type of cytochrome is absent from the neutrophil plasma membrane of CGD patients. In contrast, patients with the autosomal recessive form of CGD appear to have normal amounts of this cytochrome but, unlike that of normal individuals, this is not reduced following stimulation of the respiratory activity of the patients' neutrophils with the agent phorbol myristate acetate. Segal and Jones (1980) have suggested that in this group of patients there is either a defect in the process of activation or absence of a proximal component in the electron transport chain.

The progression of this disease is variable. Although some patients die in infancy, many survive into their teens and a number of adult patients are now on record. Septrin (cotrimoxazole) is now widely used prophylactically in these patients and has markedly reduced the incidence of infections.

Prenatal diagnosis of CGD by NBT testing is possible using samples of fetal blood obtained at fetoscopy. Using phorbol myristate acetate as stimulator in the NBT test, clear distinctions between healthy carriers and CGD neutrophils are readily achieved (Linch & Levinsky, 1983).

Glucose 6-phosphate dehydrogenase (G6PD) deficiency

Although G6PD deficiency in red cells is well recognized, the analogous leucocyte defect is much more rare. The enzyme acts in the early stage of the HMP shunt and so the biochemical effect of deficiency is similar to that seen in classical CGD and clinically the two diseases are indistinguishable. All affected male siblings in one family had totally absent G6PD in their leucocytes and their neutrophils failed to reduce NBT. Neutrophils from their mother were intermediate in leucocyte function tests suggesting an X-linked transmission of the deficiency.

Neutrophil pyruvate kinase deficiency

A female patient giving a CGD-like history of recurrent abscesses in multiple sites was found to be normal in classical NBT testing. However, an unstable abnormal PK isoenzyme was detected in her neutrophils. Enzyme activity in other tissues was apparently normal.

Myeloperoxidase deficiency

Large numbers of asymptomatic individuals with virtual deficiency of neutrophil myeloperoxidase have now been described. However, inability to kill Candida albicans, Staphylococcus aureus, S. marcescens and E. coli may also be associated with the defect, probably reflecting the importance of the myeloperoxidase—halide system for effective intracellular killing.

Deficiency of myeloperoxidase has been observed in both male and female siblings, and parents of patients generally have half-normal levels, suggesting an autosomal recessive mode of inheritance.

Patients with myeloperoxidase deficiency are usually managed in the same way as CGD patients, i.e. prophylactic antibiotic therapy.

Secondary defects of neutrophil killing

Table 14.3 lists the environmental factors which may give rise to impaired bactericidal killing by neutrophils.

As with the secondary mobility defects, these may sometimes make diagnosis of primary defects difficult.

Deficiencies of complement components

General introduction

Although primary deficiencies of the complement system are rare they have evoked a widespread interest, particularly the genetic aspects. Genes controlling C4, C2 and Factor B production are known to be located in the HLA 'super gene' complex and the significance of this linkage is being actively investigated. In addition to the genetic aspects, deficiencies of specific complement components have also helped to clarify the biological roles of various components.

The clinical problems associated with deficiencies of complement proteins fall into a small number of major categories, as summarized schematically in Figure 14.8.

Defects of early acting classical pathway components frequently give rise to a lupus-like illness and/or glomerulonephritis. This suggests that the clearance of antigen − antibody complexes is an important effector role for the classical pathway. Deficiencies of C3 and the control protein Factor I are both associated with repeated pyogenic infections, probably reflecting the importance of C3b in opsonization. Deficiencies of the late components may also be grouped together since an increased frequency of Neisserial infections is a commonly observed association. This suggests that lytic events are an important part of the immune response to these organisms.

Table 14.4 lists the known deficiencies, their frequencies and general clinical presentation. The true frequencies of these diseases cannot be ascertained from the number of known pedigrees. Many cases of complement deficiency must go undetected even in countries with advanced medical facilities. Nevertheless, as a group these remain rare diseases compared to the immunoglobulin deficiency diseases. Some well-established animal models of complement deficiencies are recognized, and these are listed in Table 14.5. These animals appear to be strikingly healthy.

Primary deficiencies of all the classical pathway components are known to occur in man. In the case of

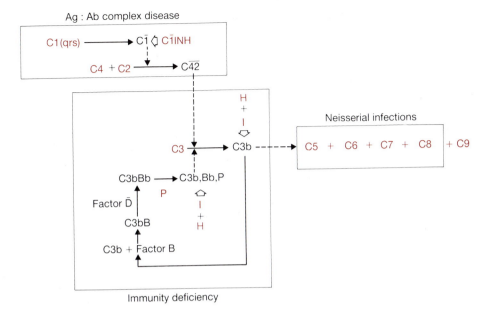

Fig. 14.8 Schematic diagram of complement activation pathways illustrating the predominant disease manifestations associated with different primary deficiencies of individual components. Known deficiencies are shown in red. No primary deficiencies of Factor B or Factor D components are yet recognized.

the alternative pathway only properdin deficiency has been described to date, although yeast opsonization (a process requiring an intact alternative pathway) is defective relatively frequently. Brief clinical and genetic details of each deficiency are given in the following sections.

C1q deficiency

The complete absence of C1q and a biologically inactive variant of the molecule have both been described. In three patients the associated clinical feature was SLE. Low levels have also been observed both in patients with Bruton's type sex-linked agammaglobulinaemia and in patients with severe combined immunodeficiency.

C1r deficiency

A primary deficiency of C1r was first described in a 13-year-old child with chronic glomerulonephritis. Subsequently two families with C1r deficient siblings were described. SLE was the presenting symptom and C1r was undetectable in all. C1q was normal and C1s was

Table 14.4 Complement deficiency in man

Complement component	Number of pedigrees	Clinical presentation
C1q	15	Immune
C1r,s	8	complex disease
C1-INH	>500	Hereditary angioedema
C4	16	Immune complex
C2	>60	disease
C3	11	Frequent
Factor I	5	pyogenic
Factor H	2	infections
C5	12	Predominantly
C6	17	Neisserial
C7	14	infections —
C8	14	some immune complex disease
C9	Many within the Japanese population	Apparently healthy

From P.J. Lachmann (personal communication).

Table 14.5. Complement deficiencies in laboratory animals.

Complement component	Species	Number of strains
C4	Guinea pig	1
C5	Mouse	Many
C6	Rabbit	3
	Hamster	1

present at approximately half-normal levels. Total haemolytic complement activity was, of course, undetectable in each case. The mode of inheritance is unclear.

C1s deficiency

C1s deficiency has been described in a single 6-year-old child who had gradually developed SLE. Serum from the patient lacked both haemolytic complement and haemolytic C1 activity. Addition of purified C1s restored full haemolytic activity.

C1 esterase inhibitor deficiency

An inherited deficiency of the C1 esterase inhibitor protein gives rise to one of the most common complement deficiency diseases, namely hereditary angioedema (HAE). The clinical characteristics of the disease are recurrent attacks of acute circumscribed subepithelial oedema of the skin, gastrointestinal tract or respiratory tract. Patients often present before 10 years of age and there is usally a marked exacerbation of the condition during adolescence. The oedema, which may last as long as 2 days, is sometimes associated with urticaria and usually involves the face or an extremity. Some patients present with nausea and vomiting or abdominal pain and signs of obstruction. Acute laryngeal oedema has led to death by asphyxiation in a number of patients (see Donaldson & Rosen, 1966 for further details). Differential diagnosis of HAE from benign, non-familial angioedema is clearly most important and fortunately there is now available both a functional assay for the inhibitory protein which these patients lack and an effective form of treatment.

The C1 esterase inhibitor protein is the only known inhibitor of activated C1s. It does, however, also have inhibitory activity against a number of other serum enzymes such as Hageman factor, kallikrein, plasmin and Factor XI.

Hereditary angioedema is transmitted as an autosomal dominant deficiency disease and affected patients are heterozygotes. The defect has been demonstrated in three generations in several kindreds (see Figure 14.9). The gene locus for the inhibitor protein appears not to be linked to the HLA region.

Rosen and colleagues (1971) found that individuals with HAE had mean serum concentrations of the inhibitor which were somewhat less than half-normal levels (in the range 5−31%). During attacks serum levels are usually very low and it is assumed that unrestrained activation of C1 occurs periodically at extravascular sites and this results in unchecked cleavage of C4 and C2. One of the products of C2 cleavage (C2b) is further split by plasmin to generate a vasoactive subfragment which produces the effects of angioedema. It is probable that the activation of C1 itself is due to the unrestrained action of plasmin, suggesting that C1 esterase inhibitor is a major control protein for this plasma enzyme.

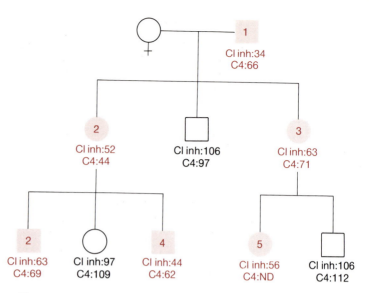

Fig. 14.9 Autosomal dominant inheritance of C1 esterase inhibitor deficiency. Many family members over three generations have exhibited symptoms of hereditary angioedema (shown in pink).
Clinical notes
1   Removal of suspicious lump from colon (7th decade). Histology showed only oedematous tissue.
2   Cutaneous and oral angioedema.
3   Recurrent cutaneous and visceral angioedema. Multiple hospital admissions for acute intestinal obstruction.
4   Recurrent mild intestinal symptoms. Oedema of face and tongue.
5   Recurrent cutaneous and visceral angioedema. Admissions to hospital for 'salpingitis', 'ectopic pregnancy' and 'appendicitis'.
   Functional levels of C1 esterase inhibitor protein and immunochemical levels of C4 are given for pretreatment samples obtained between clinical attacks. (Values are expressed as %; lower limit of normal functional C1 esterase inhibitor activity is 70% of normal adult value and the normal range for C4 is 85–130% of median adult value; ND = not determined.) (Data kindly provided by Professor J.F. Mowbray, St Mary's Hospital Medical School, London.)

Patients with hereditary angioedema have an increased susceptibility to SLE and glomerulonephritis, probably as a result of the secondary deficiencies of C2 and C4 which occur in such patients. This further supports the view that these immune complex diseases are more likely to occur when there is failure to eliminate antigen by classical pathway activation mechanisms.

In most patients with HAE there is defective biosynthesis of the inhibitor but in approximately 15% of cases there is evidence of a non-functional protein present at normal levels. This non-functional allotypic variant is immunochemically indistinguishable from the normal protein and makes the use of a functional assay essential if a correct diagnosis is to be reached. The clinical presentation appears to be similar in both types of patient and the disease is dominantly transmitted in both groups.

For several years inhibitors of serine proteases, such as epsilon amino-caproic acid, have been used to treat patients with HAE. More recently, patients in the UK have usually been treated with the androgen preparation danazol at a dosage of 800 mg/day. The drug induces synthesis of the inhibitor and normal levels are usually reached within a few weeks. Surprisingly, in those patients with the non-functioning allotypic variant the drug suppresses synthesis of the variant and 'switches on' synthesis of the normal gene product. There is no place for plasma therapy in these patients, but replacement therapy with purified inhibitor protein is becoming available.

C4 deficiency

Several cases of homozygous C4 deficiency have now been described. SLE appears to be the most common presenting syndrome, but one patient manifested multiple immunologic defects including an abnormal antibody response and also developed severe membrano-proliferative glomerulonephritis. Characteristically, total haemolytic complement activity and C4 are undetectable in these patients. Parents of patients appear to have half-normal C4 levels.

C2 deficiency

Isolated C2 deficiency was the first true complement deficiency disease to be described and more than sixty cases are now recorded. Approximately half of the C2 deficient individuals appear to be in good health whilst the remainder have presented with SLE, dermatomyositis, membrano-proliferative glomerulonephritis, anaphylactoid purpura and chronic vasculitis.

Patients with homozygous C2 deficiency have no detectable total haemolytic complement and no haemolytic C2 activity. Functional C2 levels in heterozygotes are approximately 30−70% of normal, suggesting autosomal co-dominant inheritance (see Figure 14.10). There is an increased frequency of the HLA antigens A10 and

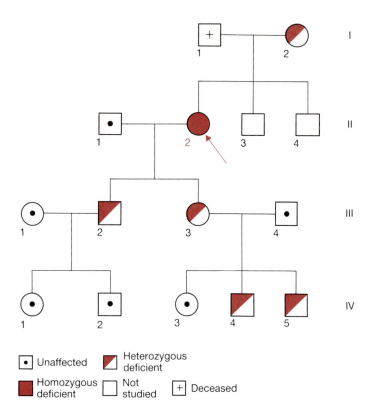

Unaffected •

Heterozygous deficient ◨

Homozygous deficient ■

Not studied ☐

Deceased +

Fig. 14.10 Pedigree of a family with C2 deficiency. The propositus had symptoms of SLE, but none of the heterozygous C2 deficient members of the kindred showed any clinical abnormalities. (Adapted from *J. Immunol.* (1972) **108**, 838, with permission of the copyright holder, Williams & Wilkins Co.)

W18 in patients with homozygous C2 deficiency. Gene mapping studies have failed to show any recombinants between C2 and the B and C loci of the HLA complex, indicating a close linkage between the C2 gene and these histocompatibility loci. The mechanisms underlying the associations between C2 deficiency, immune complex disease and HLA antigens have yet to be established.

C3 deficiency

Following earlier descriptions of a number of healthy individuals with half-normal levels of C3 (presumably heterozygotes) a small number of cases of homozygous C3 deficiency have now been reported. In each case the clinical presentation was one of recurrent infections, particularly otitis media and acute lobar pneumonia due to *Streptococcus pneumoniae, Klebsiella pneumoniae* and *Staphylococcus aureus*. Serum from each patient had almost zero total haemolytic complement and C3 was undetectable by both immunochemical and functional assays. Characteristically the parents and siblings of these patients had half-normal levels of C3 and were in good health, suggesting an autosomal co-dominant inheritance.

Factor I (C3b inactivator) deficiency

Hypercatabolism of C3 has been described in two patients with a primary deficiency of the control protein Factor I (C3b inactivator). Predictably their clinical picture was very similar to that of primary C3 deficiency, both having histories of frequent infections, particularly recurrent meningitis, otitis media and pneumonitis.

Deficiency of the control protein I disturbs the normal homeostatic balance between C3 convertase production and degradation so that C3 levels become depleted. The defect may be transiently corrected by plasma infusions and the inheritance is probably of an autosomal co-dominant type (see Figure 14.11).

Factor H deficiency

Two Asian brothers (aged 8 months and 3 years) with very low levels of Factor H (less than 10% of a reference standard serum) were described by Thompson and

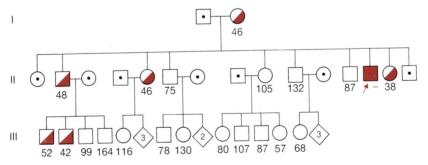

Fig. 14.11 Pedigree of a family with Factor I (C3bINA) deficiency. The propositus with no detectable Factor I is indicated by the arrow. Heterozygous individuals are indicated by half-red symbols. Untested individuals are shown by symbols with dots in their centres, and a diamond with a number within it represents untested sibs. The concentration of Factor I as a percentage of that in normal pooled serum is given below the symbol for each tested person. (Reproduced with permission from Alper & Rosen, 1975.)

Winterborn (1981). Both boys had a low haemolytic complement, a very low C3 level and a normal C4 level. The younger boy had developed a haemolytic uraemic syndrome following infection but the elder brother was healthy. Both parents, who were first cousins, had half-normal levels of Factor H and low levels were detected in other family members, suggesting that the defect was inherited. As with Factor I deficiency there is increased alternative pathway activation in such individuals and a secondary C3 deficiency is the most significant effect.

C5 deficiency

Twelve individuals with C5 deficiency have been described, mostly belonging to four kindred (one Asian and three Negro). One individual had SLE, diffuse membrano-proliferative glomerulonephritis and frequent infections from the age of eleven. Nine of the twelve cases had histories of repeated systemic Neisserial infections (both *N. meningitidis* and *N. gonorrhoeae*). Characteristically these patients have no detectable haemolytic activity and are unable to generate chemotactic activity. Family studies suggest that C5 deficiency is inherited as an autosomal co-dominant disorder.

C6 deficiency

Seventeen individuals with C6 deficiency have been described and although immune complex disease (SLE and Sjögren's syndrome) was present in two individuals the predominant clinical feature was recurrent systemic

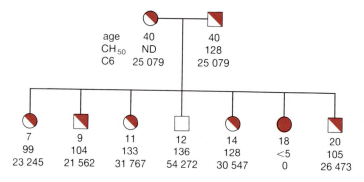

Normal CH$_{50}$ 80–160
  C6 47 587 + 1798 (S.E.)

Fig. 14.12 Pedigree of a family with C6 deficiency. (Adapted with permission from Frank *et al.*, 1975.)

Neisserial infections. One of the patients was an 18-year-old girl with disseminated gonococcal infection, an absent haemolytic complement activity and no detectable C6. Another one was a 6-year-old boy with relapsing meningococcal meningitis associated with similar laboratory findings. The parents and siblings of both these patients had approximately half-normal C6 levels suggesting an autosomal co-dominant disorder (see Figure 14.12).

C7 deficiency

C7 deficiency has been described in fourteen individuals covering all ages and diverse clinical features, ranging from apparently normal health to Raynaud's disease with scleroderma. Nevertheless frequent Neisserial infections were again a feature in six individuals. The major laboratory findings in these patients were a low total haemolytic complement in association with no detectable functional C7. This deficiency also appears to be transmitted as an autosomal co-dominant trait in which heterozygotes have approximately half-normal serum levels.

C8 deficiency

Fourteen individuals have been described with C8 deficiency and, as with C7, there appears to be a range of associated clinical disorders. In eight patients a history of repeated Neisserial infections was prominant but SLE and xeroderma pigmentosa have also been described.

There is evidence that in some of these individuals the C8 molecule is a defective variant lacking the β-subunit.

C9 deficiency

Deficiency of the ninth component of complement has recently been reported to occur relatively frequently in the Japanese population, although those individuals with the defect are apparently healthy.

Properdin deficiency

The only example of an alternative pathway component deficiency has recently been reported by Sjoholm et al.

(1982). These workers described a large Swedish family in whom three males were found to be totally properdin deficient. One of these individuals died from a fulminant infection with *Neisseria meningitidis* and subsequent investigation of the family history revealed that there had been three previous deaths following short meningitis-like illnesses. However, two other family members aged 25 and 40 years were essentially healthy and free of repeated bacterial infections at the time of the report. Since only males have been affected the authors suggest an X-linked mode of inheritance. Several females in the family appeared to have half-normal levels of properdin (see Figure 14.13) which would support the X-linkage hypothesis but unfortunately parents of the defective males appeared to have normal levels. The assumption that this is an X-linked defect may, therefore, be premature and is at variance with other recent studies of partial properdin deficiency in which autosomal recessive inheritance was suggested.

Defective opsonization

The clinical importance of antibody independent

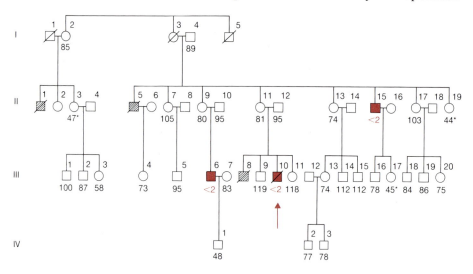

Fig. 14.13 Pedigree of a family with properdin deficiency. Generations are indicated by Roman numerals, individuals in each generation by numbers. Males are represented by squares and females by circles. Deceased members of the family are indicated by an oblique line across the symbol. Closed symbols denote individuals with a complete deficiency in properdin; the index patient is marked with an arrow. Hatched symbols are used for individuals previously dying of fulminant infections. The properdin values in percentage of normal are given below the symbols. Values lower than 2 s.d. below the normal mean (adults) are marked with asterisks. (Reproduced with permission from Sjoholm et al., 1982.)

mechanisms of opsonization was first suggested by the work of Miller et al. (1968), who described two patients with a plasma-associated phagocytic defect which was revealed using baker's yeast. Patients with this defect characteristically present in early life with recurrent infections, failure to thrive, persistent diarrhoea and severe dermatitis. Although the precise nature of the defect remains to be established it is known that the levels of C3b bound to yeast surfaces are much less using the sera of such individuals than are those observed using normal sera. Standard haemolytic tests of complement activity (both clinical and alternative pathways) are apparently intact. Although the defect is familial the exact pattern of inheritance is not established.

References

Aggett P.J., Harries J.T., Harvey B.A.M. & Soothill J.F. (1979) An inherited defect of neutrophil mobility in Shwachman's syndrome. J. Paediatr. **94**, 391.

Alper C.A. & Rosen F.S. (1975) Increased susceptibility to infection in patients with defects affecting C3. In Immunodeficiency in Man and Animals. Birth Defects: Original Article Series, Vol. XI, No. 1, (ed. Bergsma D.), p. 301. Sinauer Associates, USA.

Blum R., Geller G. & Fish L.A. (1977) Recurrent severe staphylococcal infections, eczematoid rash, extreme elevations of IgE, eosinophilia and divergent chemotactic responses in two generations. J. Paediatr. **90**, 607.

Bruton O.C. (1952) Agammaglobulinaemia. Pediatrics **9**, 722.

Buckley R.H. (1975a) Clinical and immunologic features of selective IgA deficiency. In Immunodeficiency in Man and Animals. Birth Defects: Original Article Series, Vol. XI, No. 1, (ed. Bergsma D.), p. 134. Sinauer Associates, USA.

Buckley R.H. (1975b) Bone marrow and thymus transplantation in ataxia-telangiectasia. In Immunodeficiency in Man and Animals. Birth Defects: Original Article Series, Vol. XI, No. 1, (ed. Bergsma D.), p. 421. Sinauer Associates, USA.

Donaldson V.H. & Rosen F.S. (1966) Hereditary angio-neurotic oedema: a clinical survey. Pediatrics **37**, 1017.

Frank M.M., Leddy J.P., Gaither T., Heusinkveld R.S., Breckenbridge R.T. & Klemperer M.R. (1975) Hereditary C6 deficiency in man. In Imunodeficiency in Man and Animals. Birth Defects: Original Article Series, Vol. XI, No. 1, (ed. Bergsma D.), p. 318. Sinauer Associates, USA.

Hammarström L., Carlsson B., Smith C.I.E., Wallin J. & Wieslander L. (1985) Detection of IgA heavy chain constant region genes in IgA deficient donors: evidence against gene deletions. Clin. exp. Immunol. **60**, 661.

Hosking C.S., Fitzgerald M.G. & Shelton M.J. (1977) The immunological investigation of children with recurrent infections. Aust. Paediatr. J. **13** (suppl.), 54.

Lefranc M.P., Lefranc G. & Rabbits T.H. (1982) Inherited deletion of immunoglobulin heavy chain constant region genes in normal human individuals. *Nature* **300**, 760.

Linch D.C. & Levinsky R.J. (1983) Prenatal diagnosis of fetal immunological disorders. *Brit. Med. Bull.* **39**, 399.

McKusick V.A. (1984) The human gene map. *Clin. Genet.* **27**, 207.

Miller M.E, Seals J., Haye R. & Levitsky L.C. (1968) A familial plasma associated defect of phagocytosis. *Lancet* **ii**, 60.

Oxelius V.-A., Laurell A.B., Lindquist B., Golebiowska H., Axelson U., Bjorkander J., Hanson L.A. (1981) IgG subclasses in selective IgA deficiency. *New Engl. J. Med.* **304**, 1476.

Parkman R., Remold-O'Donnell E., Kenny D., Perrine S. & Rosen F.S. (1981) Surface protein abnormalities in lymphocytes and platelets from patients with Wiskott–Aldrich syndrome. *Lancet* **ii**, 1387.

Rosen F.S., Alper C.A., Pensky J., Klemperer M.R. & Donaldson V.H. (1971) Genetically determined heterogeneity of the C1 esterase inhibitor in patients with hereditary angioneurotic oedema. *J. clin. Invest.* **50**, 2143.

Segal A.W. & Jones O.T.G. (1978) A novel cytochrome b system in phagocytic vacuoles from human granulocytes. *Nature* **276**, 515.

Segal A.W. & Jones O.T.G. (1980) Absence of cytochrome b reduction in stimulated neutrophils from both female and male patients with chronic granulomatous disease. *FEBS Lett.* **110**, 111.

Simmonds H.A., Fairbanks L.D., Webster D.R., Rodeck C.H., Linch D.C. & Levinsky R.J. (1983) Rapid prenatal diagnosis of adenosine deaminase deficiency and other purine disorders using foetal blood. *Biosci. Rep.* **3**, 31.

Sjoholm A.G., Braconier J.H. & Soderstrom C. (1982) Properdin deficiency in a family with fulminant meningococcal infections. *Clin. exp. Immunol.* **50**, 291.

Smith C.I.E. & Hammarström L. (1984) Detection of an α1 and α2 heavy-chain constant region genes in common variable hypogammaglobulinaemia patients with undetectable IgA. *Scan. J. Immunol.* **20**, 361.

Thompson R.A., Candy D.C.A. & McNeish A.S. (1984) Familial defect of polymorph neutrophil phagocytosis associated with absence of a surface glycoprotein antigen (OKM1). *Clin. exp. Immunol.* **58**, 229.

Thompson R.A. & Winterborn M.H. (1981) Hypocomplementaemia due to a genetic deficiency of β1H globulin. *Clin. exp. Immunol.* **46**, 110.

Waldmann T.A, Brider S., Goldman C.K., Frost K., Korsmeyer S.J. & Medici M.A. (1983) Disorders of B cells and helper T cells in the pathogenesis of the immunoglobulin deficiency of patients with ataxia telangiectasia. *J. clin. Invest.* **71**, 282.

Wedgwood R.J., Rosen F.S. & Paul N.W. (Eds.) (1983) *Primary Immunodeficiency Diseases*. Birth Defects: Original Article Series, Vol. XIX, No. 3. Alan R. Liss Inc., New York, for the National Foundation — March of Dimes.

Further reading

Asherson G.L. & Webster A.D.B. (1980) *Diagnosis and Treatment of Immunodeficiency Diseases*. Blackwell Scientific Publications, Oxford.

Klebanoff S.J. & Clark R.A. (1978) *The Neutrophil: Function and Clinical Disorders*. North-Holland Publishing Co., Amsterdam.

(Various chapters on deficiencies of complement components, primary neutrophil abnormalities, etc.)

Rother K. & Rother U. (Eds.) (1986) *Hereditary and Acquired Complement Deficiencies in Animals and Man*. Progress in Allergy Vol. 39. Karger, Basel.

Soothill J.F., Hayward A.R. & Wood C.B.S. (Eds.) (1983) *Paediatric Immunology*. Blackwell Scientific Publications, Oxford.

Wedgwood R.J., Rosen F.S. & Paul N.W. (Eds.) (1983) *Primary Immunodeficiency Diseases*. Birth Defects: Original Article Series, Vol. XIX, No. 3. Alan R. Liss Inc., New York, for the National Foundation — March of Dimes.

Chapter 15
Genetics of paraimmunoglobulinopathies and B lymphocyte neoplasia

Introduction

The paraimmunoglobulinopathies are neoplastic diseases involving cells secreting immunoglobulins. In general, these conditions are characterized by the presence of excessive quantities of homogeneous immunoglobulin (the M component) in the serum and urine. The major forms of paraimmunoglobulinopathy are classified in Table 15.1.

The aetiology of these disorders is obscure. Irradiation, viruses and mineral oils have been cited as possible environmental 'triggers' but firm evidence is lacking. There is suggestive evidence of individual genetic susceptibility from at least two studies. Williams *et al.* (1967) reported several monoclonal M components in healthy members of some families in which the index case had myelomatosis. Similarly, Seligmann *et al.*, in the same year, reported M components in healthy relatives of patients with Waldenström's macroglobulinaemia. Long-term follow-up of some cases of benign M components has revealed the late development of myelomatosis.

In order to evaluate the incidence of paraimmunoglobulinopathies in a general population, Axelsson *et al.* (1966) studied 6995 persons living in Varmland, a county of Sweden. This represented 60% of the population over the age of 25. Of this number, a total of 64 (7 of whom were over 80) showed an M component on serum electrophoresis. Further investigations indicated that the M components belonged to the following immunoglobulin classes: IgG (39), IgA (17) and IgM (5). In two individuals there was evidence to suggest biclonal M components. The latter are particularly interesting and have occasionally been shown to share the same idiotype, suggesting that they originated from the same clone. It has been suggested that a defective class switch mechanism may be the underlying mechanism involved. Deletion of intervening H chain exons may be prevented by mutations in either (1) the class switching

Table 15.1 Classification of the paraimmunoglobulinopathies

Paraimmunoglobulinopathy	Associated serum protein
Multiple myelomatosis	IgG, IgA, IgD, IgE, Bence Jones protein, deleted molecules
Waldenström's macroglobulinaemia	IgM, Bence Jones protein
Heavy chain diseases: γ-HCD	Deleted γ chains
α-HCD	Deleted α chains
μ-HCD	Deleted μ chains
δ-HCD	Free δ chains
Light chain disease	Bence Jones protein

DNA recombinases or (2) their intronic DNA recognition sites. Subsequent transcription of this DNA would carry the H chain gene for two classes and the RNA splicing of these with a single V region would lead ultimately to the simultaneous production of two different H chain classes.

Clinical and laboratory features of main syndromes

Multiple myelomatosis

This is a malignant neoplastic disease of plasma cells manifested by invasive destruction of the skeleton, particularly the 'punched out' osteolytic lesions seen on X-ray. Serum electrophoresis usually reveals an M component which further immunochemical studies with specific antisera will identify with one of the immunoglobulin classes. In many patients light chain synthesis is excessive and free light chains (Bence Jones proteins) are secreted by the proliferating plasma cells in addition to the complete immunoglobulin. In such cases the light chain of the myeloma protein is structurally identical to the Bence Jones protein. In general, the frequencies of the various classes of paraprotein are proportional to the serum concentrations of non-pathological immunoglobulins. Males and females are apparently equally vulnerable.

Waldenström's macroglobulinaemia

Waldenström was the first to describe this neoplastic condition, which is characterized by a clinical picture of lymphocytosis, enlarged lymph nodes and spleen, bone-marrow infiltration and the presence of abnormal levels of IgM in blood. Bone pain is infrequent and X-rays of bones rarely reveal the punched-out lesions characteristic of myelomatosis.

Macroglobulinaemia is associated with an older group of patients (60−80 years) than myelomatosis (50−65 years) and carries a strikingly better prognosis (median survival range 23−152 months, compared to 9−35 months for multiple myelomatosis). There is also a sex difference: males with macroglobulinaemia are twice as common as females.

The heavy chain diseases

γ-heavy chain disease

Independent studies in 1964 by Franklin et al. and by Osserman and Takatsuki led to the recognition of this condition, which has now been described in more than sixty individuals. Clinically, patients present with evidence of malignant lymphoma, the main symptoms being a generalized painful lymphadenopathy associated with hepatosplenomegaly and pyrexia. The diagnosis is made on the basis of careful laboratory investigations of serum and urinary proteins. The characteristic finding in the serum is an abnormal γ-heavy chain disease (γ-HCD) fragment, which usually sediments at 3.5−4.0S and has a mol. wt. of 45 000−55 000. These fragments are structurally similar to the Fc fragment produced by the action of papain, and proteins antigenically related to all of the four IgG subclasses have now been reported. Structural studies have indicated that none of the γ-HCD proteins is merely an Fc fragment. All have additional peptide material derived from the N-terminal regions of the IgG molecule and the nature of this portion may be used as the basis of a classification of the whole group. Two such types are illustrated in Figure 15.1 and will be briefly described.

V_H and C_H1 region deletions

These proteins consist of γ chains in which the C_H3, C_H2 and hinge regions are all present but the C_H1 region and parts of the V_H region are absent. The first γ-HCD protein to be studied in depth (protein Cra; Franklin et al., 1964) was subsequently shown to be a $γ_1$ V_HIII

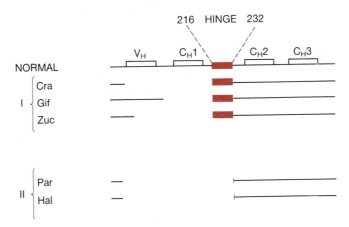

Fig. 15.1 Two types of γ-HCD protein compared to normal γ chain. In Type I proteins the C_H1 region is deleted and part of the V_H region may also be absent; a normal amino acid sequence resumes at position 216. In Type II proteins both the hinge and C_H1 regions are absent, together with much of the V_H region.

protein with only 10–11 V_HIII amino acid residues linked directly to residue 216 (glutamic acid). The protein was shown to differ from normal IgG1 in another important respect: it had three rather than two inter-heavy chain S–S bridges (see Figure 15.2) since the cysteines which are normally involved in γ-L chain S–S bridge formation were bonded to each other. Two other examples are also shown in Figure 15.2. Protein Gif is a $γ_2$ protein lacking approximately 100 residues from the C_H1 region and protein Zuc is a $γ_3$ protein with a deletion of about 200 residues starting some 17–18 residues after the N-terminus.

V_H, C_H1 and hinge region deletions

Three proteins of this type have been described in detail. Proteins Par and Baz are both antigenically $γ_1$ chain proteins and protein Hal is a $γ_4$ protein. The proteins have mol. wts. close to 55 000 in neutral solvents and chemical investigations have shown that most of the Fd region and all of the hinge region (with the inter-heavy chain S–S bridges) are absent. In each case two portions of γ chain exist as a dimer stabilized by non-covalent bonds.

Some heavy chain disease proteins are not readily classified as above and may have arisen by enzymatic cleavage of a larger chain. Some of these proteins have

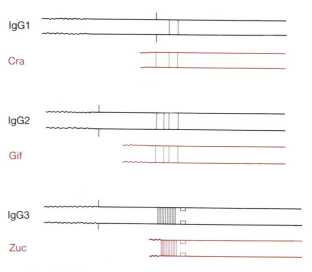

Fig. 15.2 Details of the structure of three Type I γ-HCD proteins showing the peptide chains and inter-chain S−S bonds. Normal immunoglobulins of the same subclass are included for comparison. Intra-chain S−S bridges are omitted for the sake of clarity. Note the absence of the H−L bond on all γ-HCD proteins. In the case of protein Cra the cysteines normally forming such bonds have joined together to give an additional inter-heavy chain bond.

N-terminal residues in the hinge region and others in the inter-domain region between the V_H and C_H1 regions. Normally there is insufficient data to establish whether such proteins represent undegraded gene products or are the result of post-synthetic enzymic degradation.

Lam and Stevenson (1973) reasoned that the proteins secreted by many lymphoid system tumours have subsequently been identified at low levels in normal individuals. Bence Jones protein (= light chain), IgD and IgE are classic examples of this. These authors therefore sought to demonstrate γ-HCD in normal plasma using the following criteria for identification: (1) a mol. wt. appreciably lower than that of IgG; (2) antigenic identity with Fcγ but not Fabγ; and (3) the presence in the N-terminal peptide of PCA (pyrrolidone carboxylic acid), which is characteristic of many human γ chains. The authors isolated 2 mg of Fcγ-like protein from 12 litres of human plasma by a combination of gel filtration, ion-exchange chromatography and immunoadsorption. The isolated protein sedimented at 3.9S, reacted with anti-Fcγ but not anti-Fabγ or anti-light chain and was shown to contain N-terminal PCA. Thus the protein showed the same structural properties as γ-HCD proteins. The authors concluded that the pathological protein deletions were not attributable to some characteristic of the neoplastic process but more probably to the result of a genetic event which could occur in either normal or neoplastic lymphocytes. The frequency of such events must be very low and the products only recognized when neoplastic cell clones are involved.

Mouse IgG structural mutants

Mutant mouse myeloma cell lines have been obtained by selecting single cells from MOPC21 myeloma cells in continuous culture. Four such cell lines producing immunoglobulin mutants have been characterized by a combination of amino acid sequencing and messenger RNA studies. Two of these proteins (IF1 and IF3) were found to lack the entire C_H3 region and another (IF2) resembled some of the human heavy chain disease proteins in that there was an internal deletion of the C_H1 region (see Figure 15.3). Milstein *et al.* (1977) have discussed the possibility that these mutants are the result of defects in the heavy chain cistron and that the IF1 and 3 mutants may be due to early termination and a frame shift.

Genetic aspects

The human γ-HCD proteins were initially regarded as a mere pathological curiosity, probably arising from enzymatic degradation of intact IgG. The structural revelation that they were proteins with deletions (Prahl, 1967) therefore caused a minor immunological shockwave. Subsequently, Frangione and Franklin (1973) suggested various possible genetic mechanisms to account for such deletions. In particular, the frequent observation of a deletion ending at position 216 led them to suggest that a codon specifying glutamic acid at residue 216 could represent the start of another gene and implied separate genetic control for the Fc region and possibly for each homology region. The description of a myeloma protein (Mcg) with a deleted hinge region was also cited at this time as evidence for a separate gene controlling this portion of the molecule. With the application of recombinant DNA technology to the study of immunoglobulin genes it became clear that

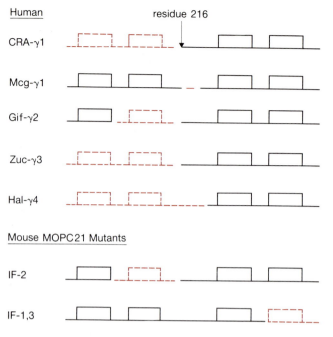

Fig. 15.3 Comparisons of deletions observed in human heavy chains and in mouse mutants. (Adapted with permission from Milstein *et al.*, 1977.)

these predictions were probably largely correct. All of the most frequently observed γ-HCD mutants are similar in that the deleted region is identical to a separately coded domain exon or terminates at the end of an exon. It seems probable that these mutations result from errors of exon recombination during RNA splicing. Thus a mutation in the intron adjacent to a heavy chain constant region domain may eliminate the recognition site for RNA splicing enzymes and prevent the incorporation of that domain into H chain messenger RNA. The hinge region deletions may also result from mutations within the introns adjacent to the hinge and domain exons.

α-heavy chain disease

α-heavy chain disease (α-HCD) is the most common of the heavy chain diseases (in excess of 150 cases reported) and occurs in two major clinical forms. The first patient to be described presented with abdominal lymphoma and diffuse lymphoplasmocytic infiltration of the small intestine. This variant of the disease is the most common and has been known for several years as 'Mediterranean lymphoma' although it is not confined to this geographical area. Patients usually have severe malabsorption and diarrhoea and the disease generally pursues a progressive and fatal course. Patients with the second form of the disease have shown no intestinal involvement but a diffuse lymphocytic plasmocytic reticular cellular infiltrate of the respiratory tract.

The diagnosis of α-HCD is difficult and requires careful laboratory investigation to show the presence of the α chain fragment in the serum without associated light chain. This task is complicated by the tendency of many IgA myeloma proteins to fail to react with anti-light chain antisera. Furthermore, the α-HCD proteins are less commonly reported in urine, a feature of γ-HCD which makes diagnosis considerably easier.

The mol. wts. of α-HCD proteins reportedly range from 36 900 to 56 000 and most of the proteins are highly glycosylated. Light chains are invariably absent and most of the proteins appear to be of the α_1 subclass. Antigenic analyses have indicated the presence of an intact Fc region in most of these proteins although the N-terminus is often heterogeneous. An α-HCD protein (Def) was sequenced by Wolfenstein-Todel et al. (1974) and was found to comprise a short heterogeneous N-terminal stretch of peptide probably corresponding to part of a V region, absence of the rest of the V region and the Cα1 domain and initiation of the α chain sequence at a valine residue just before the hinge region (see Figure 15.4). The authors postulated that the protein was synthesized as an internally deleted α_1-heavy chain and subsequently underwent N-terminal proteolysis. Frangione (1975) has suggested that the valine residue before the α_1 hinge region may be the counterpart of the glutamic acid residue at position 216 of the γ chain, i.e. the translation product marking a switch point from one gene to another.

μ-heavy chain disease

This variety of HCD has been described in only fifteen patients and is still relatively poorly characterized. The general clinical picture is one of long standing chronic lymphocytic leukaemia involving spleen, liver and abdominal lymph nodes but with little peripheral lymphadenopathy. The diagnosis of μ-HCD is suggested by

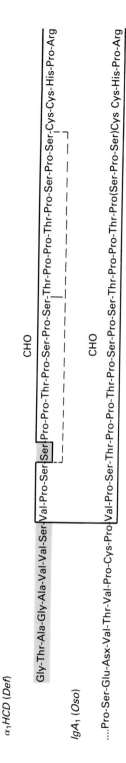

Fig. 15.4 Comparison of amino acid sequences in the hinge regions of α-HCD protein Def and the normal α₁ chain from myeloma OSO. The sequences are essentially identical within the boxed section. The amino acid sequence of α₁-HCD Def which is shown on the N-terminal side of the valine residue (shown in pink) is probably a portion of the V region. The dashed line indicates a duplicated region and CHO indicates a carbohydrate moiety. (Adapted from Wolfenstein-Todel *et al.*, 1974.)

immunoelectrophoretic identification of μ chain anti-genicity in the α_1 position. The abnormal protein is usually present in low concentrations in the serum but generally not in the urine. In contrast to γ-HCD and α-HCD, Bence Jones protein appears to be synthesized and secreted in many of these patients.

Most of the μ-HCD proteins appear to exist in the serum as pentamers resembling the $(Fc\mu)_5$ fragment. Mol. wts. ranging from 180 000 to 300 000 have been reported and J chain has been associated with some but not others. Despite this heterogeneity alanine has been reported to be the N-terminal residue of each protein. A partial amino acid sequence of one of the proteins iden-tified the N-terminus of the molecule as residue 338 (Ou sequence numbering). In this instance the deletion comprised the whole of the V region and both the $C_\mu 1$ and $C_\mu 2$ domains (see Figure 15.5).

The major properties of γ-, α- and μ-HCD proteins are summarized in Table 15.2.

δ-heavy chain disease

A single case of IgD HCD has been reported by Vilpo *et al.* (1980). This 70-year-old male patient presented with a clinical picture resembling multiple myelomatosis; osteolytic lesions were present in the skull and a marked marrow plasmacytosis was recorded. A small M com-ponent detected on serum electrophoresis was shown by immunofixation to react with monospecific anti-IgD but not with antisera to other heavy chain classes or light chains. No Bence Jones protein was detected. Mol. wt. determinations suggested that the δ-heavy chains were present in serum as tetramers and there was no clear evidence of a domain deletion such as is found in all other HCD proteins. Final acceptance of this abnormal protein as a representative HCD entity awaits further structural characterization.

Miscellaneous abnormalities of heavy and light chain structure

Heavy chain abnormalities

In addition to the HCD proteins, which lack light chains, a number of immunoglobulin molecules with

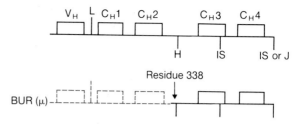

Fig. 15.5 Structure of the μ-HCD protein BUR compared to the homology regions and disulphide bridges of a normal μ chain; H is the inter-heavy chain S−S bridge; IS is the inter-subunit S−S bridge and J is the μ-J chain S−S bridge. The dotted lines indicate deleted regions. (Adapted from Frangione, 1975.)

Table 15.2 Some properties of γ-, α- and μ-heavy chain disease proteins

Property	γ-HCD	α-HCD	μ-HCD
Electrophoretic mobility	β−fast γ	β	$α_1$
Molecular weight	35−55 000	29−34 000	180−300 000
Content of carbohydrate	High	High	High
Associated J chain	No	Often	Some
Associated light chain production	No	No	Yes
Presence of protein in urine	Common	Rare	No

Table 15.3 Some examples of abnormalities of human heavy and light chains

Heavy chain	Light chain
Heavy chain disease proteins	Light chain disease
IgA molecule with deleted $C_α3$ domain	IgG molecule with internal V region deletions of both light and heavy chains
IgG molecule with deleted hinge region	
IgM molecule with deleted $C_μ3$ and $C_μ4$ domains	
Hybrid molecules: IgG1−3, IgG2−4	

structural abnormalities of the heavy chain are recorded (see Table 15.3). Among the most interesting of these is the IgG molecule Mcg, which was shown to lack only 15 amino acids from the hinge region (residues 216 to 232), an IgA myeloma protein lacking the C_H3 domain, and an IgM protein (Klo) which lacked the $C_μ3$ and $C_μ4$

domains and was apparently synthesized and secreted as F(ab')$_2\mu$ fragments with a mol. wt. of 130 000.

However, the most informative group of aberrant proteins are the hybrid molecules. In 1969 Kunkel et al. described a family with one member whose serum was devoid of all the usual Gm genetic antigens. The serum lacked ordinary IgG1 and IgG3 proteins but contained instead hybrid molecules of the type IgG3–IgG1. Evidence was adduced to show that in the hybrid protein the N-terminal portion was antigenically related to IgG3 whereas the C-terminal portion was clearly IgG1 related. Kunkel and his colleagues drew parallels between this hybrid molecule and the $\alpha\beta$ chain hybrids established for Lepore-type haemoglobins. An unequal homologous cross-over involving mispairing of heavy chain cistrons would readily explain the deletion of Gm antigens. (See Chapter 5 for a detailed account of this protein and the probable underlying defect.)

A second hybrid molecule, a myeloma protein, was subsequently identified in a Negro patient. This protein appeared to be a hybrid of IgG4 and IgG2 with a probable cross-over point between the C_H2 and C_H3 domains, i.e. the C_H1 and C_H2 regions were derived from an IgG4 molecule and the C_H3 region from an IgG2 molecule.

Light chain abnormalities

Light chain disease occurs as a neoplastic condition in which Bence Jones protein is detected in both serum and urine but no associated myeloma protein is secreted. Hobbs (1969) found that 19% of 212 cases of myelomatosis fell into this category. The mean age at diagnosis for this group is 56 years and the median survival time 11–28 months. Bence Jones proteinuria is universal and osteolytic lesions very common.

Abnormalities of light chains have also been recorded and some of these are listed in Table 15.3. An IgG myeloma protein with internal deletions of both the light and heavy chain V regions has been studied in detail.

Immunoglobulin gene rearrangements in patients with lymphoid neoplasia

The advent of DNA-hybridization technology is beginning to have a significant influence on the differential diagnosis of lymphoid neoplasia. Investigations of some of these clinical conditions have provided essential basic information on the sequence of immunoglobulin gene rearrangements. For example, in patients with acute lymphoblastic leukaemia of the 'non-B, non-T' form (i.e. B cell precursor leukaemias) rearrangements of the heavy chain genes have been demonstrated in every case (Korsmeyer et al., 1983). These were characterized as complete $V_H-D_H-J_H$ sequences or as intermediate D_H-J_H forms lacking a V_H segment. In a large number of these B cell precursors incomplete D_H-J_H intermediates or aberrant $V_H-D_H-J_H$ rearrangements were identified. In the same study approximately half of the B cell precursor leukaemias showed evidence of light chain rearrangements. The gene patterns observed in cells arrested at particular stages of differentiation are consistent with a sequence of gene rearrangements in which heavy chain precedes light chain and κ precedes λ.

Chromosomal translocations involving immunoglobulin genes

Certain neoplastic disorders are associated with specific chromosomal translocations. One that has been the subject of extensive recent studies is that frequently found in human Burkitt's lymphoma, a tumour that is largely confined to people resident in central Africa. In a number of patients with this neoplastic condition, gene loci originally located on the long arm of chromosome 14 at band q 32 and on the long arm of chromosome 8 at band q 24 become reciprocally translocated. Less frequently, Burkitt lymphoma translocations have been shown to involve sites for the κ light chain gene (chromosome 2, band p 13) and the λ light chain gene (chromosome 22, band q 11). Thus, in all three types of translocation the chromosomal break was flanked by an immunoglobulin locus and, furthermore, the translocated flanking site was uniformly found to be 8 q 24 (see Figure 15.6). These observations suggested that the latter locus might in some way contribute to the malignant phenotype itself. Subsequently a group of cancer related genes, the so-called cellular oncogenes, thought to represent the normal cellular homologues of transforming agents in various animal viruses, were considered plausible candidates for such a role. A discovery

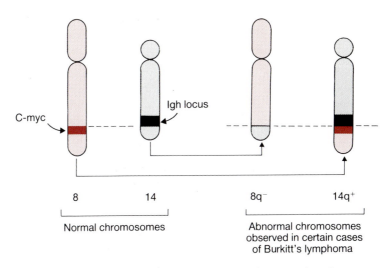

C-myc

Igh locus

8   14   8q⁻   14q⁺

Normal chromosomes   Abnormal chromosomes observed in certain cases of Burkitt's lymphoma

Fig. 15.6 The abnormal chromosomes found in a number of patients with Burkitt's lymphoma arise from a reciprocal recombination involving the long (q) arms of chromosomes 8 and 14. One effect of this is the juxtaposition of the c-myc gene within the immunoglobulin heavy chain locus on chromosome 14 q⁺.

of major importance was then made by Hayward *et al.* (1981), who showed that avian leukosis virus, a slow transforming retrovirus, almost always integrates near the *c-myc* gene, a cellular homologue of the avian myelocytomatosis virus. Subsequently the human c-myc gene was localized by *in situ* hybridization studies to chromosome 8 q 24 and found to be rearranged in a number of patients with Burkitt's lymphoma. Cross-species somatic cell hybrids between Burkitt's lymphoma cells and mouse plasmacytomas have provided valuable data about the nature of the recombination event leading to the Burkitt's tumour. In all cases analysed to date the 14 q⁺ translocated chromosome has possessed an aberrantly rearranged Ig gene together with c-myc. The reciprocally translocated chromosome 8 q⁻ receives portions of the Ig gene locus, most commonly including some of the V_H genes. This translocation affects only one allele, usually the immunoglobulin allele that has been subject to allelic exclusion, i.e. the non-expressed allele. It seems that the introduction of c-myc into the Ig gene locus is an event of some regulatory importance. Gillies *et al.* (1983) have suggested that since this occurs at a time of active transcription of Ig genes in B-cell differentiation, mechanisms augmenting Ig transcription may also influence transcription of c-myc.

Nevertheless, the precise role of c-myc in oncogenesis remains controversial and nothing is known about the normal function of the protein encoded by the gene except that it is found within the nucleus and binds to DNA.

References

Axelsson U., Bachmann R. & Hallen J. (1966) Frequency of pathological proteins (M-components) in 6995 sera from an adult population. *Acta med. scand.* **179**, 235.

Frangione B. (1975) Structure of human immunoglobulins and their variants. In *Immunogenetics and Immunodeficiency*, (ed. Benacerraf B.), p. 1. MTP, Lancaster.

Frangione B. & Franklin E.C. (1973) Heavy chain diseases: clinical features and molecular significance of the disordered immunoglobulin structure. *Semin. Hematol.* **10**, 53.

Franklin E.C., Lowenstein J., Bigelow B. & Meltzer M. (1964) Heavy chain disease — a new disorder of serum γ-globulins. Report of the first case. *Am. J. Med.* **37**, 332.

Gillies S.D., Morrison S.L., Oi V.T. & Tonegawa S. (1983) A tissue specific transcription enhancer element is located in the major intron of a rearranged immunoglobulin heavy chain gene. *Cell* **33**, 717.

Hayward W.S., Nell B.G. & Astrin S.M. (1981) Activation of a cellular onc gene by promoter insertion in ALV-induced lymphoid leukosis. *Nature* **290**, 475.

Hobbs J.R. (1969) Immunochemical classes of myelomatosis: including data from a therapeutic trial conducted by a Medical Research Council working party. *Brit. J. Haematol.* **16**, 599.

Korsmeyer S.J., Arnold A., Bakhshi A., Ravetch J.V., Siebenlist U., Hieter P.A., Sharrow S.O., Le Bien T.W., Kersey J.H., Poplack D.G., Leder P. & Waldmann T.A. (1983) Immunoglobulin gene rearrangement and cell surface antigen expression in acute lymphocytic leukaemias of T-cell and B-cell precursor origins. *J. clin. Invest.* **71**, 301.

Kunkel H.G., Natvig J.B. & Joslin F.G. (1969) A 'Lepore' type of hybrid γ-globulin. *Proc. natn. Acad. Sci. USA* **62**, 144.

Lam C.W.K. & Stevenson G.T. (1973) Detection in normal plasma of immunoglobulin resembling the protein of γ-chain disease. *Nature* **246**, 419.

Milstein C., Adetugbo K., Cowan N.J., Köhler G., Secher D.S. & Wilde C.D. (1977) Somatic cell genetics of antibody-secreting cells: studies of clonal diversification and analysis by cell fusion. *Cold Spring Harb. Symp. quant. Biol.* **41**, 793.

Osserman E.F. & Takatsuki K. (1964) Clinical and immunochemical studies of four cases of heavy (Hγ2) chain disease. *Am. J. Med.* **37**, 351.

Prahl J.W. (1967) N- and C-terminal sequences of a heavy chain disease protein and its genetic implications. *Nature* **215**, 1386.

Seligmann M., Danon F., Mihaesco C. & Fudenberg H.H. (1967) Immunoglobulin abnormalities in families of patients with Waldenström's macroglobulinaemia. *Am. J. Med.* **43**, 66.

Vilpo J.A., Irjala K., Viljanen M.K., Klemi P., Kouvonen I. & Ronnemaa T. (1980) δ-heavy chain disease. A study of a case. *Clin. Immunol. Immunopath.* **17**, 584.

Williams R.C., Erikson J.L., Polesky H.F. & Swain W.R. (1967) Studies of monoclonal immunoglobulins (M-components) in various kindreds. *Ann. Int. Med.* **67**, 309.

Wolfenstein-Todel C., Mihaesco E. & Frangione B. (1974) 'Alpha chain disease' protein Def: internal deletion of a human immunoglobulin A, heavy chain. *Proc. natn. Acad. Sci. USA* **71**, 974.

Further reading

Korsmeyer S.J. & Waldmann T.A. (1984) Immunoglobulin genes: rearrangement and translocation in human lymphoid malignancy. *J. clin. Immunol.* **4**, 1.

Chapter 16
Genetics of autoimmune disease

Introduction

Our understanding of the genetics of the autoimmune disorders has been transformed by the work associating various HLA antigens with these disorders. As described elsewhere it is now believed that the HLA-A, -B, and -C gene products physically associate with viral and other antigens to facilitate cytotoxic T cell killing, whereas DR locus gene products mediate macrophage – helper T cell interactions. Many autoimmune diseases have been shown to occur in individuals having the DR3 and B8 antigens (see Table 16.1). It has been noted that within Caucasian populations individuals with these antigens tend also to show evidence of much higher immune responsiveness compared, for example, to DR2 and B7 individuals. It is possible that antigen handling by the DR3 individual is so efficient that responses to altered-self antigens are also initiated. Equally, since the activities of suppressor T cells are also thought to be influenced by the DR products, it is possible that auto-immune disorders arise from inefficient suppression.

Many investigators have been tempted to speculate on the possible advantages and disadvantages of main-taining the extensive polymorphism of the major histo-compatibility complex (MHC) antigens. It has been suggested that in the past the individuals able to mount a vigorous immune response (e.g. DR3, B8 individuals) were at a considerable advantage in dealing with life-threatening infections. Nevertheless, the genes associa-ted with poor responses have been maintained and it is postulated that the particular advantage enjoyed by in-dividuals having 'low-responder' genes (e.g. DR2, B7) is a decreased susceptibility to many of the autoimmune diseases. Similar pressures may operate on the Gm genes, which have a completely different chromosomal location but which are increasingly seen to show signi-ficant associations with certain autoimmune disorders such as Graves' disease and myasthenia gravis.

Table 16.1 Tissue type associations with suspected autoimmune diseases

Disease	Antigen	Relative risk
Diabetes (Type 1) (Caucasian)	B8	2.6
	B15	2.0
	DR3	5.7
Graves' disease	A1	1.5
Caucasian	B8	2.3
	Dw3	3.6
	DR3	3.5
Japanese	Bw35	3.9
	DHO	4.6
Anti-GRM nephritis (Goodpasture's syndrome)	DR2	13.1
Idiopathic membranous nephropathy	DR3	12.0
Myasthenia gravis		
Females: age of onset <35 years	B8	12.7
Females: age of onset >35 years	B8	2.6
Males: age of onset <35 years	B8	5.1
Males: age of onset >35 years	B8	0.9
Systemic lupus erythematosus	B5	1.7
	B8	2.11
	DR2	3.7
	DR3	3.0
Rheumatoid arthritis	Dw4	4.2
	DR4	5.8
Juvenile rheumatoid arthritis	B27	4.5
Chronic active hepatitis	B8	9.0
	DR3	13.9
Coeliac disease	B7	0.41
	B8	8.3
	Dw3	10.9
	DR3	64.5
Dermatitis herpetiformis	B8	8.7
	Dw3	13.5
	DR3	56.4

Diabetes

Diabetes is a heterogeneous disease. There now appear to be sound reasons for regarding juvenile onset and maturity onset disease as quite distinct and for adopting the classification Type 1 and Type 2. Some of the characteristic features of both types are listed in Table 16.2. Carefully acquired family histories suggest that the two variants breed true and that there must be major but distinct genetic mechanisms involved in each. Twin studies have revealed a concordance rate (for monozygous twins) approaching 100% for Type 2 diabetes but only 50% for Type 1 diabetes. This suggests both an important role for environmental factors in the development of Type 1 disease and also the existence of a large pool of healthy but genetically susceptible individuals.

After initial observations had indicated an association between juvenile onset diabetes (i.e. Type 1) and the tissue type antigens B8 and B15, it was established that the association occurred strongly with the D locus antigen DR3. The frequency of the latter antigen in European patients ranges from 36 to 59% compared with 11−24% in healthy non-diabetic controls (relative risk of 5.7). HLA-DR3 is in linkage disequilibrium with HLA-Dw3, B8, Cw7 and A1 and each of these has been shown to be increased in Type 1 patients. A further complication, however, is that in a proportion of European and all Japanese patients there is an association with DR4 which is in linkage disequilibrium with Dw4, Bw62 (a subcomponent of B15), B40, Cw3 and A2. A typical HLA inheritance pattern for three diabetic siblings is illustrated in Figure 16.1.

Family studies reviewed by Gorsuch and Cudworth (1983) have shown that in patients with Type 1 disease there is a striking linkage with particular HLA haplotypes. If there is a random distribution of haplotypes

Table 16.2 Various features of Type 1 and Type 2 diabetes

	Type 1	Type 2
Age of onset	Usually before 30 years	Usually after 30 years
Sex	Predominantly males	Predominantly females
Family history of:		
Type 1 diabetes	Common	Rare
Type 2 diabetes	Relatively uncommon	Common
HLA linkage	Yes	No
Organ-specific autoimmunity	Common	Relatively uncommon
Islet cell antibodies	Usual at onset	Absent
Other autoantibodies	Increased	Not increased

Adapted with permission from Gorsuch & Cudworth (1983).

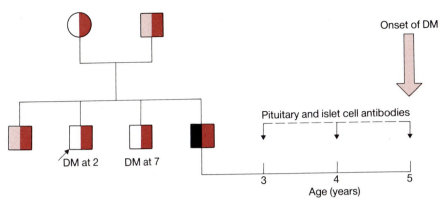

Fig. 16.1 The inheritance of HLA antigens in siblings with Type I diabetes (DM). Haplotypes: □ A3, B14, DR6; ■ A3, B7, DR4; ■ A28, B51, DR4; and ■ A2, B62, C3, DR4. Disease is linked to possession of the A2, B62, C3, DR4 haplotype. The 5-year-old brother had complement-fixing antibodies to the islet cell surface for 2 years before developing frank diabetes. (Data kindly provided by Dr G.F. Bottazzo.)

within a family, only 25% of the children will be totally HLA identical, 50% will share one of the parental haplotypes and the remaining 25% will be HLA nonidentical. In diabetic families there are marked differences from these expected figures (see Table 16.3), suggesting synergism between several (possibly allelomorphic) genes. In a review of the clinical 'risk' associated with the various allelic forms, Bodmer (1980) concluded that the greatest susceptibility was associated with the two homozygotes DR3/DR3 and DR4/DR4 and the heterozygotes DR3/DR4. A much lower probability of acquiring the disease occurs in individuals with DR3 or DR4 in association with some other allele. It is not clear whether there are different patterns of Type 1 diabetes in Caucasians and Japanese but it has been suggested that the form of disease associated with DR3 involves a more widespread and qualitatively different pattern of autoimmunity than does the disease associated with DR4.

The three complement proteins C4, C2 and Factor B are encoded by structural genes located within the MHC of chromosome 6 and allelic variants of all three (C4A, $C2^2$ and BfF1 respectively) have been associated with Type 1 diabetes. The genes for these alleles are presumably in linkage disequilibrium with the HLA-DR3 and DR4 alleles and it has not been established which association (if either) is primary.

Although autoantibody production has been unambiguously described in Type 1 diabetes, there is consider-

Table 16.3 Concordance of HLA-haplotypes in 135 pairs of siblings with Type 1 diabetes compared to 135 control pairs

Number of pairs of siblings	HLA-identical	One haplotype concordant	HLA non-identical
Expected	34 (25%)	67 (50%)	34 (25%)
Observed	74 (55%)	53 (39%)	8 (6%)

able doubt about whether this is cause or effect. Islet cell antibody has been detected in 85% of patients with Type 1 diabetes at the time of initial diagnosis, compared to 1% of a non-diabetic control population. In healthy members of diabetic families individuals with positive islet cell antibody tests tend to have the HLA phenotype characteristically associated with the disease. In contrast, the same individuals often have a raised incidence of thyroid and/or gastric parietal cell antibodies but these do not segregate with HLA phenotype.

Various conditions believed to be autoimmune occur in association with Type 1 diabetes more frequently than expected, e.g. Hashimoto's disease, pernicious anaemia, Addison's disease and coeliac disease. All are associated with HLA-DR3 or B8. For a more detailed treatment of this complex field see Gorsuch and Cudworth (1983).

Autoimmune thyroid disease

Four broad categories of AIT are recognized:
1 Autoimmune thyrotoxicosis (Graves' disease).
2 Endocrine exophthalmos.
3 Primary hypothyroidism due to atrophic autoimmune thyroiditis (AIT).
4 Goitrous AIT — Hashimoto's disease.

The aetiology of these thyroid conditions is far from clear but a role for autoimmune pathology has been recognized since the discovery of anti-thyroid antibodies in patients with Hashimoto's disease (Roitt *et al.* 1956).

Both autoimmune thyroiditis and Graves' disease appear to have a genetic component. This is supported, for example, by the increased prevalence of Graves' disease in twins and the observation that different categories of thyroid disease occur together in families more frequently than might be expected. Furthermore, both in individuals and families, there are increased frequencies of other suspected autoimmune diseases such as pernicious anaemia and myasthenia gravis.

HLA antigen associations with thyroid disease also supports the view that genetic susceptibility plays a role. There is an increased frequency of HLA-B8 in Graves' disease (relative risk 2.3) and HLA-DR3 is increased in both Graves' disease and myxoedema due to atrophic thyroiditis. Within Japanese populations Graves' disease is associated with HLA-Bw35. In the case of goitrous AIT there is a stronger association with the HLA-DR5 antigen (relative risk 7.8).

In recent years there have been a number of studies linking certain Gm alleles of IgG heavy chains with autoimmune thyroiditis. In a small study of 31 Caucasians with autoimmune thyroiditis, Farid et al. (1977) found an increase in the Gm haplotype zag, and subsequently the same authors extended their study to include 149 Caucasian patients and again found a significant increase of the Gm phenotype ag. Parallel studies in a Japanese population by Nakao et al. (1980) showed an increase in the phenotype axg when autoimmune thyroiditis was considered as a group. After subdivision into those with, and those without, goitre the phenotype ag was significantly increased in patients with atrophic thyroiditis, but not in those with goitrous thyroiditis. In a subsequent study by Nakao and colleagues there was also clear evidence of an increase in Gm (f+b+) haplotypes in Graves' disease.

Autoimmune thyroid disease is a 'typical' example of a probable multifactorial aetiology involving several genetic and environmental factors.

Miscellaneous renal disorders

The aetiological mechanisms underlying much renal disease remain obscure but autoimmune processes are thought by many investigators to play a central role. Genetic aspects of three types of nephritis are briefly summarized.

Mesangiocapillary nephritis

Friend et al. (1977) reported an association between

MCGN and a B cell alloantigen, and subsequent studies have suggested that most of such patients with nephritic factor (an antibody binding to the convertase of the alternative pathway of complement) possess the tissue type.

Anti-GBM nephritis

These patients frequently have Goodpasture's syndrome — a severe form of nephritis associated with pulmonary haemorrhage. The characteristic laboratory finding is antibody to glomerular basement membrane which appears to cross-react with alveolar capillary basement membrane. Work by Rees *et al.* (1978) has shown that almost 90% of such patients have the HLA-DR2 antigen — a highly significant association ($p = 0.0002$; relative risk 15.4). Subsequently the same group showed that patients with high levels of anti-GBM antibodies and having the haplotype HLA-DR2-B7 carried a worse prognosis.

Two animal models of anti-GBM disease have been described. Susceptibility to this type of nephritis in Brown Norway rats has been linked to the MHC locus and in guinea pigs the disease is clearly strain related.

Membranous nephropathy

Linkage between both HLA and Factor B antigens and susceptibility to this disease have been recorded. The HLA association is with B18 and DR3 whereas the Factor B association, observed in 29% of such patients, was with a rare electrophoretically fast allele (F1) occurring in only 1.7% of a healthy control population.

Myasthenia gravis (MG)

This disease, characterized by muscle weakness, occurs in about 1 in 20 000 of the population at any age and in either sex. In the last decade it has been shown that the key feature of MG is a reduction in the number of functional acetylcholine receptors. Many patients have demonstrable levels of antibody to the receptor and passive transfer of the disease to mice has been achieved with preparations of IgG from such patients. Thymomas occur in some 10% of cases and alterations in thymic architecture are the rule in most cases of non-thymomatous MG.

Myasthenia gravis was one of the first diseases to be studied for evidence of a tissue type association. Pirskanen et al. (1972) described an increased frequency of HLA-B8 in young female patients with MG. Many further studies have confirmed and refined this association. From the recent studies of Compston et al. (1980) it appears that in non-thymomatous patients in whom the age of onset of disease was less than 40 years there is a significant association with A1, B8 and DR3. In patients with a later age of onset the association is with A3, B7 and DR2. In contrast, the patients with thymomas show no strong HLA association. Some of the heterogeneous features of MG have been used by Compston et al. (1980) as the basis for a classification of three distinct disease variants (see Table 16.4).

Japanese investigators have recently focused attention on the Gm allotypes of MG patients. In Japanese patients with thymoma it was reported that the haplotype Gm axg was significantly increased compared to patients with non-thymomatous MG. In this racial group the observed HLA associations are with B12 in female patients with early onset disease and with B5 for patients with thymomas.

Animal models of MG have been described in rabbits, rats, mice, monkeys and chickens. All are generated by immunizing the animal with purified acetylcholine receptor (usually obtained from the electric organ of the electric eel). Suspicions of strain variability and possi-

Table 16.4 Characteristic features of three proposed categories of myasthenia gravis

	A	B	C
Thymus	Thymoma	No thymoma	No thymoma
Age at onset	—	<40 years	>40 years
Sex	M = F	F>M	M > F
HLA association	None	A1, B8, DR3	A3, B7, DR2
Antibodies to ACLR	High	Intermediate	Low
Anti-striated muscle	Very high	Low	Intermediate
Other auto-antibodies	Low	Intermediate	High

Adapted with permission from Compston et al. (1980).

ble H-2 linkage in mice have still to be studied systematically.

Systemic lupus erythematosus (SLE)

This immune complex mediated disease is characterized by a range of clinical features which include vasculitis, nephritis and synovitis. Patients have a wide range of circulating autoantibodies, particularly anti-DNA antibodies. The disease is markedly more frequent in females than males (9:1) and in the USA is more common in Negroes than Caucasians.

The characteristic immunological features of SLE are believed by many investigators to stem from a suppressor cell defect which results in deregulation of potentially autoreactive cells.

Genetic factors are clearly important in the aetiology of the disorder. Relatives of patients have an increased risk of contracting the disease which is between 100 and 200 times greater than the risk for the general population. Studies have shown that whereas dizygotic twins of these SLE patients have a similar frequency of the disease as the general population, monozygotic twins are 60% concordant for the disease.

Early studies showed an increased frequency of heterozygous C2 deficiency in SLE patients compared to normal blood donors. Further studies in this field have confirmed that patients with SLE have a frequency of homozygous C2 deficiency which is between 1 in 100 and 1 in 200, compared to a general population frequency of approximately 1 in 10 000. In many of the patients with C2 deficiency, including those with SLE, the HLA haplotype A10, Bw18, BfS and DRw2 is present and this has led to the suggestion that this haplotype is closely linked with a susceptibility gene predisposing to immune complex disease. In addition to the special subgroup of SLE patients with C2 deficiency there is evidence of increased frequencies of HLA-B8 and HLA-B15 and of DR2 and DR3 but in each case the relative risk is low.

A number of animal models for SLE exist — notably in the NZB mouse, but canine lupus and Aleutian disease of mink are also well studied examples.

The female NZB/WF$_1$ hybrid mouse develops a lupus-like illness with antibodies to double-stranded DNA, DNA immune complex deposition in the kidneys, haemolytic anaemia, and premature death. Murine lupus has been reviewed in depth by Talal (1978) who has presented evidence that these animals lose suppressor T cells between 1 and 2 months of age. A premature maturation of the immune system appears to occur in young animals with thymus, spleen and gonads all exerting an influence on immunoglobulin class expression. In view of the known sex bias of the disease in both man and mouse it is of interest that in the mouse androgens appear to suppress autoimmune phenomena whereas oestrogens accelerate the expression of disease. Some of the major immunological abnormalities observed in the murine model are given in Table 16.5.

Table 16.5 Immunological abnormalities observed in the NZB/WF$_1$ hybrid mouse

Premature maturation of immunocompetence
Relative resistance of immune tolerance
Decreased suppressor function
Impaired cellular immunity
Decreased thymic factor
Excessive antibody responses to a wide range of antigens

Rheumatoid arthritis (RA)

This is one of the most intensively studied immune complex diseases. Antiglobulins (rheumatoid factors) reactive with both autologous and heterologous immunoglobulin were amongst the first antibodies to be described. These antiglobulins occur in both serum and synovial fluid, and there is a significant correlation between the level of IgG antiglobulins and the formation of immune complexes involving this Ig class in the synovial space. Inflammatory processes within the joint have been demonstrated by many investigators and include complement activation, polymorph infiltration and release of damaging lysosomal enzymes such as collagenase and cathepsins. Although there is a general agreement that the disease arises from autosensitization to IgG it is by no means clear what initiates this process. No single infective organism has yet been isolated and speculation in this area has thrived for many years. The field has been reviewed by Marmion (1978) and Roitt et al. (1982).

The strong HLA-B27 association with ankylosing spondylitis acted as a major stimulus to further work on the genetics of rheumatological disease. Initially, results in patients with RA were disappointing. The frequencies of most HLA-A, -B and -C antigens were normal. More recently, however, it has become clear that the D locus antigen Dw4 is present in 40% of patients with seropositive, classical RA compared with only 9% of a control population. It has been suggested that the Dw4 gene, or a gene with which it is in linkage disequilibrium, confers susceptibility for antiglobulin production, perhaps directly through T helper cell activity or indirectly through an absence of T suppressor cells.

At the 1984 HLA workshop, data was presented on the tissue types of RA patients treated with gold, who subsequently developed nephropathy. Males appeared to be more at risk than females and all those typed as DR3 positive had nephropathy. The haplotype most at risk was B8, Cw7 and DR3. In order to minimize the possibility of renal complications (and also thrombocytopenia) assessment of DR3 status would appear to be a prudent step before instituting gold therapy.

In pauciarticular juvenile RA there is an association with HLA-B27. The identification of such children by histocompatibility testing is of some prognostic value since such individuals are likely to develop persistent spinal and sacroiliac disease and could benefit from appropriate careers advice given during adolescence. This is one area where the tissue type result may be put to the practical advantage of the patient.

Chronic active hepatitis

This form of liver disease is thought by many investigators to have an autoimmune aetiology. Patients are usually female and, in addition to hepatocellular damage, they frequently have evidence of other diseases of doubtful aetiology, including Sjogren's syndrome, ulcerative colitis, RA, thyroid disorders and myasthenia gravis.

All major immunoglobulin classes tend to be elevated and autoantibodies, especially of the IgG class, are present in many patients. Antinuclear antibodies occur in about 50% of patients with chronic active hepatitis. Anti-smooth muscle antibodies occur frequently in those patients with HBsAg-negative disease. Liver-specific antibodies reactive with lipoprotein antigens have been reported by a number of investigators but there is some controversy surrounding the exact specificity of such antibodies.

As early as 1972 Mackay and Morris reported an increased frequency of HLA-B8 in patients with chronic active hepatitis. Later studies have shown a linkage between HLA-A1, B8 and Dw3 in such patients with relative risks of 9.0 and 13.9 for B8 and Dw3 respectively. Other studies have shown a correlation between possession of the HLA-B8 antigen and titres of both autoantibodies and antibodies to viruses such as measles and rubella. Furthermore, many patients with chronic active hepatitis have relatives with detectable autoantibodies and high titres of antiviral antibodies.

Coeliac disease

Coeliac disease, or gluten-induced enteropathy, is a malabsorption syndrome associated with a range of immunological abnormalities. The dissection of primary and secondary phenomena has always been difficult in this disease but there is general agreement that, in patients on gluten-containing diets, immune complexes containing IgA and C3 accumulate in the jejunal mucosa. The damaging effects of such complexes arise from the extensive secondary events. These include the synthesis of high titre antibodies to a wide range of food proteins and, in addition, antibodies to connective tissue antigens such as reticulin. Coeliac disease may, therefore, be more properly regarded as a 'secondary' autoimmune disease.

There is a clear genetic predisposition to coeliac disease since first-degree relatives of a patient have a 1 in 10 chance of manifesting the disorder. Nevertheless, a multifactorial aetiology is strongly implied by discordant patterns in identical twins. Early studies of HLA-A and -B antigens in coeliac patients showed a high frequency of B8 (relative risk 8.3). In northern European populations there is strong linkage disequilibrium between B8 and Dw3 and between B8 and DR3. Of patients with coeliac disease approximately 75% are DR3 compared to 20% of controls. Nevertheless, there are many individuals lacking both the B8 and Dw3 antigens in whom a clinical diagnosis of coeliac disease is unequivocal. Within a coeliac family, however, it is a useful pointer to the at-risk younger sibling.

Dermatitis herpetiformis

This dermatological illness is linked to coeliac disease in a poorly understood manner. It is characterized by intensely itchy clusters of vesicles and may be associated with an abnormal jejunal biopsy. Immunofluorescent studies of skin biopsies have shown IgA deposits, sometimes in association with proteins of the alternative pathway of complement (C3, Factor B and properdin). In 17% of patients autoantibodies, including antinuclear antibodies, antireticular and antithyroid microsomes, are also detected. As in coeliac disease a gluten-free diet leads to a dramatic clinical improvement.

The 90% frequency of HLA-B8 in dermatitis herpetiformis is similar to that in coeliac disease (relative risk 8.7) but this is probably because of linkage disequilibrium between B8 and Dw3 and DR3. The latter antigens carry relative risks of 13.5 and 56.4 respectively.

References

Bodmer W.F. (1980) The HLA system and disease. *J. Roy. Coll. Phys. (Lond.)* **14**, 43.

Compston D.A.S., Vincent A., Newsom-Davis M. & Batchelor J.R. (1980) Clinical, pathological, HLA antigen and immunological evidence for disease heterogeneity in myasthenia gravis. *Brain* **103**, 579.

Farid N.R., Newton M.R., Noel E.P. & Marshall W.H. (1977) Gm phenotypes in autoimmune thyroid disease. *J. Immunogenet.* **4**, 429.

Friend P.S., Noreen H.J., Yunis E.J. & Michael A.F. (1977) B-cell alloantigen associated with chronic mesangiocapillary glomerulonephritis. *Lancet* **i**, 562.

Gorsuch A.N. & Cudworth A.G. (1983) Immunological aspects of endocrine disease in childhood. In *Paediatric Immunology*, (eds. Soothill J.F., Hayward A.R. & Wood C.B.S.), p. 427. Blackwell Scientific Publications, Oxford.

Mackay I.R. & Morris P.J. (1972) Association of autoimmune active chronic hepatitis with HL-A1, 8. *Lancet* **ii**, 793.

Marmion B.P. (1978) Infection, autoimmunity and rheumatoid arthritis. *Clin. Rheum. Dis.* **4**, 565.

Nakao Y., Matsumoto T., Miyazaki T., Nishitani H., Taratsuki K., Kasukawa R., Nakayama S., Izumi S., Fujita T. & Tsuji K. (1980) IgG heavy chain allotypes (Gm) in autoimmune diseases. *Clin. exp. Immunol.* **42**, 20.

Pirskanen R., Tilikainen A. & Hokkanen E. (1972) Myasthenia gravis and HL-A antigens. *Ann. Clin. Res.* **4**, 304.

Rees A.J., Peters D.K., Compston D.A. & Batchelor J.R. (1978) Strong association between HLA-DRw2 and antibody mediated Goodpasture's syndrome. *Lancet* **i**, 966.

Roitt I.M., Doniach D., Campbell P.N. & Hudson R.V. (1956) Autoantibody in Hashimoto's disease (lymphadenoid goitre). *Lancet* **ii**, 820.

Roitt I.M., Hay F.C., Nineham, L.J. & Male D.K. (1982) Rheumatoid arthritis. In *Clinical Aspects of Immunology*, Vol. II, 4th edn., (eds. Lachmann P.J. & Peters D.K.), p. 1161. Blackwell Scientific Publications, Oxford.

Talal N. (1978) Natural history of murine lupus. Modulation by sex hormones. *Arthrit. Rheum.* **21**, 5 (suppl.), S58.

Further reading

Gorsuch A.N. & Cudworth A.G. (1983) Immunological aspects of endocrine disease in childhood. In *Paediatric Immunology*, (eds. Soothill J.F., Hayward A.R. & Wood C.B.S.), p. 427. Blackwell Scientific Publications, Oxford.

Marmion B.P. (1978) Infection, autoimmunity and rheumatoid arthritis. *Clin. Rheum. Dis.* **4**, 565.

Roitt I.M., Hay F.C., Nineham L.J. & Male D.K. (1982) Rheumatoid arthritis. In *Clinical Aspects of Immunology*, Vol. II, 4th edn., (eds. Lachmann P.J. & Peters D.K.), p. 1161. Blackwell Scientific Publications, Oxford.

Glossary

alleles alternative versions of a gene at the same locus.

allelic exclusion mechanism preventing the expression of alternative allelic variants of immunoglobulin chains once the VDJ DNA rearrangement has occurred.

alloantigens different (allelic) forms of an antigen coded for at the same gene locus in all individuals of a species.

allogeneic see transplantation terminology.

allograft graft exchanged between two genetically dissimilar individuals of the same species, i.e. members of an outbred population, or of two different inbred strains.

allophenic (tetra-parental) chimera.

allotype genetically determined antigenic determinant encoded at a particular locus and differing in members of the same species.

antithetic antigen usually applied to serologically detected antigens (allotypes) occupying similar structural locations in different isotypic variants.

autosome any chromosome other than the sex chromosomes.

backcross (1) *laboratory animals*: mating of a heterozygote with one of the parents e.g. $(A \times B)F_1 \times A$; (2) *humans*: mating of a heterozygote with a recessive homozygote.

chimera an individual whose cells are derived from more than one zygote, e.g. irradiation chimera, an animal constructed by irradiation and reconstitution with genetically distinct stem cells.

chromosomal walking the process of isolating overlapping genomic clones by successive rounds of hybridization using end fragments of previously isolated clones.

cis on the same side, usually implies genes are inherited together, i.e. on the same chromosome.

clone a cell line derived by mitosis from a single diploid cell.

co-dominant both alleles of a pair are expressed in the heterozygote.

417

coisogenic having identical genotypes except for a single difference — only realized by mutation.

congenic approximation to coisogenic (syn) strains obtained by successively backcrossing an F_1 to a parent strain.

cosmid synthetic cloning vector which can accommodate large fragments of foreign DNA.

cross-over exchange of genetic material between homologous chromosomes during meiosis.

diploid two haploid sets of chromosomes, i.e. the chromosome number of normal somatic cells.

dominant a trait expressed in the heterozygote.

downstream towards the 3′ end of the gene.

epitope an antigenic determinant of known structure.

eukaryote any organism with cells having a nucleus and a nuclear membrane.

exon the region of a gene that codes for amino acids.

F_1 the first generation progeny of a mating.

F_2 the second generation progeny of a mating $(F_1 \times F_1)$.

gene a sequence of nucleotide bases in DNA that is transcribed into a single RNA message which then usually dictates the synthesis of a polypeptide.

genome the genetic constitution of an individual.

genotype the alleles present at one locus.

Gm allotypes (Gm groups) allotypic antigenic sites on the γ chain of human IgG.

graft rejection destruction of tissue grafted into a genetically dissimilar recipient, due to a specific immune reaction against it by the recipient.

haploid the chromosome number of gametes.

hapten small substance able to bind to antibody but only able to initiate immune responses when linked to a carrier.

heterogeneic see xenogeneic (syn).

heterograft see xenograft (syn).

heterozygote an individual with two different alleles at a given locus on a pair of homologous chromosomes.

H-2 histocompatibility system the major histocompatibility system in the mouse.

histocompatibility antigen genetically determined isoantigen carried on the surface of nucleated cells of many tissues which, when tissue is grafted onto another individual of the same species whose tissues do not carry that antigen, may incite an immune response which leads to graft rejection.

histocompatibility gene gene determining histocompatibility antigen.

HLA histocompatibility system the major histocompatibility system of man, containing the genes coding for class I, class II and class III antigens.

homozygote an individual with a pair of identical alleles at a given locus on homologous chromosomes.

hybrid a species cross.

Ia immune associated antigens. Alloantigenic differences controlled by the I region of the H-2 complex locus in mice.

idiotope an antigenic determinant expressed in the variable region of an antibody and recognized by another (anti-idiotypic) antibody.

idiotype the summation of the idiotopes characteristic of one particular antibody.

inbreeding the mating of closely related individuals.

intron a region of gene that separates two exons coding for amino acids.

InV allotypes see Km allotype.

Ir gene immune response gene.

isogeneic see transplantation terminology.

isograft see syngeneic.

isologous see transplantation terminology.

isotypic variation structural variability of antigens common to all members of the same species.

karyotype the classified chromosome complement of an individual or cell.

kindred an extended family.

Km allotype allotypic antigenic sites in the constant region of the κ chains of human immunoglobulin (previously known as InV allotypes).

linkage two gene loci on the same chromosome are linked if alleles at each locus are transmitted together more often than the 50% frequency expected by chance.

locus the precise location of a gene on a chromosome and the corresponding site on the homologous chromosome.

major histocompatibility complex (MHC) chromosomal region common to all higher vertebrates controlling synthesis of histocompatibility (transplantation) antigens.

meiosis reduction cell division which occurs in gamete production.

mitosis somatic cell division.

mosaic an individual derived from a single zygote with cells of two or more different genotypes (cf. chimera, tetra-parental, allophenic).

multifactorial inheritance due to multiple genes at different loci which summate and interact with environmental factors.

nucleotide a purine or pyrimidine base attached to a sugar and phosphate group.

oncogene a gene sequence capable of causing transformation of animal cells.

palindrome a stretch of DNA in which identical base sequences run in opposite directions.

phenotype the observable characteristics of an individual.

plasmid extrachromosomal closed circular DNA molecule found in bacteria.

polymorphism the common occurrence in a population of two or more alleles at a single gene locus (with the implication that there is no overall natural selection against these alleles).

proband the individual that draws medical attention to the family (syn: propositus).

probe(oligonucleotide) a radiolabelled DNA fragment used to identify complementary sequence(s).

prokaryote a simple unicellular organism which lacks a nuclear membrane.

propositus see proband (syn)

pseudogene a gene which cannot give rise to a functional product because of one or more mutational changes in its sequence.

race a group of historically related individuals who share a gene pool.

recessive a trait which is expressed only in homozygotes.

recombinant an individual in a linkage study in whom the marker and disease loci have assorted at parental meiosis.

recombinant DNA artificial insertion of a portion of DNA from one organism into the genome of another.

recombination the formation of new combinations of linked genes by crossing-over between their loci during meiosis.

restriction enzyme Type I an enzyme which cleaves DNA (an endonuclease) at points distant from the sequence specific recognition site.

restriction enzyme Type II an enzyme which cleaves DNA at sequence specific sites (the recognition site).

restriction fragment length polymorphism (RFLP) a

pattern of results produced by the presence or absence of a recognition site for a restriction enzyme.

reverse transcriptase an enzyme which can make complementary DNA from messenger RNA.

sex-linked inheritance of a gene carried on a sex chromosome.

sister chromatid exchange exchange of DNA by sister chromatids.

species a set of individuals who can interbreed and have fertile progeny.

syngeneic see transplantation terminology.

tetra-parental (allophenic) chimera a chimera resulting from the artificial fusion of two blastocysts at the four or eight cell stage.

trans on the other/opposite side. Usually refers to genes not inherited together, i.e. not on the same chromosome.

trans-acting refers to molecules that can act on, or with products of, the opposite chromosome.

transcription synthesis of RNA from a DNA template.

transfection transfer of DNA into an animal cell *in vitro*; a term used analogously to the term transformation (see below) in relationship to bacteria; bacterial transfection describes transfer of phage DNA.

transformation (1) uptake of DNA by bacteria; (2) conversion of an animal cell growing in culture from having a limited division potential (non-transformed state) to having an unlimited or immortal growth potential (transformed state).

translation ribosomal conversion of the mRNA message to a polypeptide chain.

translocation the transfer of chromosomal material between chromosomes.

transplantation terminology

Nomenclature	Relationship between donor and recipient of graft
Isogeneic (isologous)	The same individual
Syngeneic	Genetically identical members of the same species
Allogeneic	Genetically dissimilar members of the same species
Xenogeneic	Different species

transposon a DNA sequence able to replicate and insert itself at a new location in the genome.

upstream towards the 5′ end of the gene.

wild type the major allele present in a population.

xenogeneic see transplantation terminology.

xenograft graft from donor of different species.

xenotype structural or antigenic difference between molecules derived from different species.

Index

Page references in italic refer to tables and/or figures.

ABO blood group antigens   44, 45, 46,
    216−17
Acquired immune deficiency
    syndrome   307, 327
Actin dysfunction syndrome   366
Acute lymphoblastic leukaemia   398
Adenosine deaminase (ADA)
    deficiency   355−6
Adoptive transfer system   310
Agammaglobulinaemia
    'Swiss type'   357−8
    X-linked   345, 350
Agglutinins   45
Agranulocytosis   361
AIDS   307, 327
AIT   406−7
Allelic exclusion   201−5
Alloantibodies   217
Alloantigens   216−17
    involvement in graft rejection   45
    obsolescence   308
Allogeneic effects   320, 321
Allograft rejection   217−18, 219
    cellular basis   218−20
    genetic control   44−5
Alloimmunization   297−8
Allophenic mice see Tetraparental mice
Allotypy in immunoglobulins   97
    human   99−119
    latent   124−5
    murine   125−7
    rabbit   121−5
α-heavy chain disease   392−3, 394, 396
α$_2$-neuraminoglycoprotein see C1-INH
Altered self hypothesis   330
Alternative pathway   69−70, 71, 78−80
    activation   79
    components   71, 80−1
        deficiency   381−2
    protein interactions   79
Alu sequences   257
Am antigen allotypy
    nomenclature   102
    structural location   107, 112
Amino acid sequence analysis

experimental design   55, 63
    importance   68
Amphibian immunoglobulin structure   32,
    38
Ankylosing spondylitis, HLA
    association   251
Antibodies
    anti-idiotype   120−1
    cell cooperation in production   11, 13,
        311−13
    deficiencies   345, 346, 350−3
    experimental control   135−40, 141
    heterogeneity   134−5
    internal structure   13−20
    phenotypic markers   140−5
    response to the unexpected   129−30, 131
    specific   129
    three-dimensional structure   20−2
    see also Immunoglobulins and Monoclonal
        antibodies
Antibody combining site   21−2, 144−5
    measurement of fine specificity   140−2
    measurement of size   130−2
    multispecificity   142
    three-dimensional structure   132−3, 134
Antibody diversity   7, 129
    origins   211−13
    phenotypic   129−45
        quantitative estimates   145−50
    role of somatic mutations   208, 210−11
Antibody-fold   5, 20−2, 99
Antibody repertoires   129
    specific   146−7
Antibody-secreting cells   294
    allelic exclusion   201−5
    cloning   137−8, 139, 140
    isotype exclusion   201
    precursors   311−12, 313
Anti-GBM nephritis see Goodpasture's
    syndrome
Antigen binding site   16, 20, 21
Antigen bridging hypothesis   313−14, 315
Antigen presenting cells (APCs)   235,
    295−6, 314
    interaction with helper T cells   321−4

Antigen selection 136, *137*
Antigenic identity 46
Antigenic markers *see* Differentiation antigens
Antigens 44
  differentiation *see* Differentiation antigens
  immunoglobulin 99−102
    see also Am antigen allotypy, Gm antigen allotypy *and* Km antigen allotypy
    restricted recognition 328−30
    synthetic 316, *317*
    see also specific types
Antiglobulins 411−12
Anti-idiotype antibody 120−1, 143−4
APCs *see* Antigen presenting cells
Ataxia telangiectasia 32, 359−60
Autografts 217
Autoimmune disorders 4
  genetics 402−14
  tissue type associations *403*
Autoimmune thyroid disease 406−7
Autoimmune thyrotoxicosis 406−7
Autosomal dominant inheritance 345
Autosomal recessive inheritance 345

B cell differentiation factor 327
B cell growth factors 327
B cells *see* B lymphocytes *and* Pre-B cells
B lymphocytes
  clone expansion *295*
  cooperation with T lymphocytes 13, 311−13, 319−21
    genetic requirements 319−21
  defects 353
  differentiation re Ig gene expression 191−201
  generation 292−3
  lineage 293−4
  murine differentiation markers 302−5
  role in antibody production 11, 13, 311−13
  sites of somatic mutation 209−10
  virgin 137, 293
Bacterial killing, defects of 368−72
Bacteriophages 155−6, *157*
BCDF 327
BCGF1 327
BCGF2 327
Bence Jones proteins 15, 397
Berger's disease, HLA association *251*
β_1H globulin 85
β_2-microglobulin 5, 48−9
  amino acid sequence 50−1, *52*
  encoding gene 256, 266
β_2A globulin *see* Immunoglobulin A
Blotting 159, *164*

Bone marrow cells, role in antibody production 290, *291*, 311
Burkitt's lymphoma, translocation involved 205, *206*, 207−8, 398−400
Bursa of Fabricius 290, 292−3, *294*
  mammalian equivalent 293

c-myc gene, role in oncogenesis 207−8, 399−400
c-oncogenes *see* Cellular oncogenes
C regions 16−17
  domains 19−22
  evolution of gene diversification 39, *41*, *42*
  isotypy 97−9
  see also C_H gene cluster
C1 complex 70−5
C1 esterase inhibitor *see* C1-INH
C1-INH 84−5
  deficiency 375−7
  synthesis 86
C1q 72, *73*, *74*
  deficiency 373
  synthesis 86
C1r 72−5
  deficiency 373−4
  inhibition 84
  relationship with C1s 87−8
  synthesis 86
C1s 72−5
  deficiency 374
  inhibition 84
  relationship with C1r 87−8
  synthesis 86
C2 75, 76−7
  deficiency 377−8
  genetics 277−8
  relationship with Factor B 80, 88, 278
  synthesis 86
C3 69, 70, 77−8, 79, 83, 284−5
  deficiency 378−9
  inactivation 85−6
  inhibition 85
  polymorphism 278−80
  relationship with C4 and C5 88−91
  synthesis 86−7
C3b inactivator (C3bINA) *see* Factor I
C4 75−6
  allotype nomenclature 275−6
  deficiency 377
  inhibition 85, 86
  murine 284
  polymorphism 274−7, 284
  relationship with C3 and C5 88−91
  synthesis 86
C4b-binding protein 86
  amino acid sequence 283

C5 81−2, *83*
  deficiency 380
  inactivation 85−6
  inhibition 85
  relationship with C3 and C4 88−91
  synthesis 86
C6 82, 84
  deficiency 380−1
  polymorphism 282
  synthesis 86
C7 82, 84
  deficiency 381
  polymorphism 282
C8 82, 84
  deficiency 381
  genetics 282
  synthesis 86
C9 82, 84
  amino acid sequence 282−3
  deficiency 381
  synthesis 86
Capping 187−8
CAT box 190
CD3 (T3) antigen 306, *307*, 339−40
CD4 (T4) antigen 306−7
CD8 (T8) antigen 306−7
cDNA
  Class I 257
  Class II 267
  synthesis 152−3
  tailing 156, *158*
CDRs 16, *17*, 144−5
Cell−cell interactions 310
  B cell−T cell 311−13, 319−21
  genetic requirements 319−24
  link with Ir gene expression 315−17
  soluble factors 324−7
Cell-mediated immunity deficiency
    syndromes *348*, 354−8
Cellular differentiation 289
Cellular immunology 310
  see also Cell−cell interactions
Cellular oncogenes 205, 398
  enhanced expression 207
CGD 368−71
C$_H$ gene cluster
  human 183−4
  murine 178, 180, *184*
    allotypy 126−7
    exon-intron structure 180, *181*
    hinge region 181
    M exons 181−3
    sequential expression 189
Charon λ phages, physical maps *157*
Chediak−Higashi syndrome 362−3, *364*,
  365
Chicken, transplantation antigens 56, 60

Chromosomal markers 312−13
Chromosome translocations see
    Translocations
Chromosome walk 164, 268
Chronic granulomatous disease 368−71
Chronic myelogenous leukaemia
  translocation involved 205
CHS 362−3, *364*, *365*
Class I antigens, human 47−8
  alleles *242*
  amino acid sequences 50−1, *52*, *53*, *54*,
    56
  association with autoimmune
    diseases 402
  detection 241, *243*
  domain structure 49, 50−1
  gene frequencies in ethnic groups *243*,
    244, *245*
  inheritance 244−5, *246*
  nomenclature 243−4
  relationship with H-2 antigens 59
  structure 48-9
  transmembrane orientation 49−50, *51*
Class I antigens, murine see H-2 antigens
Class I genes 257
  expression 260−1
  mapping 261−2
  multigene family 258−60
  structure 257−8
Class II antigens, human 62−3, 245−8
  alleles *242*
  amino acid sequences 63−8
  inheritance 244−5
Class II antigens, murine see Ia antigens
Class II genes 267
  estimation of numbers 270−2
  molecular map 268−70, *271*
  structure 267
  see also Ir genes
Class III antigens 249
Class III genes 281−2, *283*
Class switching 197−201, 302−3
Classical pathway 69−70, *71*
  components 70−8
  deficiencies 372−81
Cloned DNA
  electron microscope analysis 159, *165*
  screening 156, 158−9
  sequencing 159−61, *162*
Cloning vectors 155−6, *157*
  screening for insert DNA 156, 158−9
Cloning vehicles see Cloning vectors
CMI deficiency syndromes *348*, 354−8
Coeliac disease 413
  HLA association 251
  tissue type association *403*
Cold agglutinins 120

Colony forming units *see* Stem cells
Combinatorial diversity
    antibody genes 195, 212–13
    TCR genes 338
Common variable immunodeficiency *347*,
    353
Competence 189–90
Complement 'cascade' 7–8, 70
Complement fixation 45
Complement receptors 91–2, 283
Complement system 7–8, 69–70, *71*
    allotypes *274*
      detection 273
    component biosynthesis 86–7
    component deficiencies 372–83
    component subfamilies 87–91
    control proteins 84–6
    deficiencies 375–7, 379–80
    electrophoretic polymorphism 273
    murine 284–5
    terminal components 81–4
    see also Alternative pathway, Classical
        pathway *and individual components*
Complementarity determining regions 16,
    17, 144–5
Complementary DNA *see* cDNA
Complementation 236
Complex group *b* allotypes 124
Congenic lines, development in
    mice 222–3
Constant regions *see* C regions
Convertases 69–70
Cosmid cloning 259–60
Cosmid vectors 156
Co-transcriptional splicing 197
CR1 receptor 91
    polymorphism 283, *284*
CR2 receptor 92
CR3 receptor 92
    deficiency 367–8
CR4 receptor 92
'Cross-idiotypy' 120–1
CTL *see* Cytotoxic T cells
CVID *347*, 353
Cytotoxic T cells 294
    differentiation antigens 306–7
    restriction specificity 328–30, 331–2,
      340–1

D locus 62, 246–8
D segments 175–8
δ-heavy chain disease 395
Dermatitis herpetiformis 414
    HLA association *251*
    tissue type association *403*
D_H segment 175–8, *179*
Diabetes 403–6
    HLA association *251*

Di George syndrome 360–1
Differentiation antigens 289, 294, 307–8
    detection by monoclonal antibodies 305
    functional analysis 308
    human 305–7
    murine B lymphocyte lineage 302–5
    murine T lymphocyte lineage 296–302,
      307
    rat T lymphocyte lineage *307*
Disulphide bridges 14, *15*, 19
Diversity segments 175–8
DNA sequence probes 308
    see also cDNA *and* Cloned DNA
Domains *see* Immunoglobulin domains
DP locus 248
DQ locus 248
DR antigens 63, 246–8
DR locus 62, 248
DRw antigens 248
    alleles *242*
    association with autoimmune
      diseases 402
    gene frequencies in ethnic groups *245*
Dual recognition hypothesis 313–14, *315*,
    330

Effector cells 220, 290, 294, *295*
Endocrine exophthalmos 406
Enhancer 187, 190–1
Epitope selection 136, *137*
Eu protein, amino acid sequences 22–3, *52*
Exon-intron structures
    β_2-microglobulin gene *266*
    Class I genes *258*
    Class II genes *267*
    during expression 187–9
    immunoglobulin supergene family 6

Fab 14, *15*
FACS 159, 302, *303*
Factor B 80–1
    gene structure *281*
    polymorphism 280–2
    relationship with C2 80, 88, 278
    synthesis 86
Factor D 80
    synthesis 86
Factor H 85
    deficiency 379–80
    synthesis 86
Factor I 76, 78, 85
    deficiency 379
    synthesis 86
Fc 14, *15*
First-set rejection 217–18, *219*
Fluorescence activated cell sorter 159, 302,
    303
Folding pattern 20, *21*

FR   16−17, 18
Fragment antigen binding   14, 15
Fragment crystalline   14, 15
Framework regions   16−17, 18

γ-globulin see Immunoglobulin G and
      Immunoglobulin M
γ-heavy chain disease   388−92, 396
γ-interferon   327
γA globulin see Immunoglobulin A
γG globulin see Immunoglobulin G
γM globulin see Immunoglobulin M
Gene probes   308
Gene replacement therapy   356
Genes
   conversion   263−5
   expression   187−9
   non-productive rearrangements   202−5
Genetic engineering see Recombinant DNA
      technology
Genomic DNA libraries   161, 163, 164
Germ-line genes, estimation of
      numbers   211−12
Glucose 6-phosphate dehydrogenase
      deficiency   371
Gluten-induced enteropathy see Coeliac
      disease
Gm antigen allotypy   101−2
   expression   102−5
   gene complexes
      cross-overs   113−14, 115
      deletions   114−16
      duplications   116, 117
      hybridization   116−18
      race-related   104, 105
      influence on IgG subclass
         concentration   119
      nomenclature   101−2
      occurrence in higher primates   112−13
      structural location   106−9
Gold therapy   412
Goodpasture's syndrome   408
   HLA association   251
   tissue type association   403
G6PD deficiency   371
GPLA complex   65
Graft rejection see Allograft rejection
Graft-versus-host reaction see GVH reaction
Graves' disease   406−7
   HLA association   251
   tissue type association   403
Guinea pig
   Ia antigens   65−6, 67
   transplantation antigens   56, 60
GVH reaction   57, 239
   immunological specificity   239−40

H-2 antigens   45, 217

amino acid sequences   53−7
   mutants   57, 58
   private   228−9
   public   228−9
   relationship with Class I antigens   59
   structure   52−3
   tissue distribution   48
H-2 haplotypes   229−30
   recombinant   230
H-2 locus   224−5
   antigenic specificity   227−8
   complexity   227
   genetic map   257
   location   225
   mapping   225−6
   molecular genetics   256−72
   mutation   230−3
   recombination   230
   role in T cell−B cell cooperation   319−21
   see also I region and Ir genes
H-2 restriction   328−30
HAE   375−7
Haemagglutination   45
Haemagglutination-inhibition assay   100−1
Haemochromatosis, idiopathic
   HLA association   251
Haemolysis   45
Haplotype   229
Hapten-antibody binding studies   130, 131,
      132, 134−5
Hapten-carrier protein systems, response
      specificity   313−14, 315
Haptoglobin   97
Hashimoto's disease   406−7
   HLA association   251
Heavy chain diseases   388−97
Heavy chains   14, 15
   abnormalities   395−7
   as isotypes   97
   evolution   40
   gene formation   176
   isotype switching   197−201, 302−3
   regions   16−17
   synthesis   35
   see also Heavy chain diseases
Helper T cells   13, 294
   cooperation with B lymphocytes   320−1
   differentiation antigens   306−7
   interaction with APCs   321−4
Hepatitis, chronic active   412−13
   HLA association   251
   tissue type association   403
Hereditary angio-oedema   375−7
Heteroduplex analysis   159, 165
Heterogeneity index   135
High responder strains   233, 316−17, 318
Hinge regions   20, 22−3
   deletions   389

encoding 181
evolution 41−2
susceptibility to enzymes 23, 25
Histocompatibility 215
classical genetics 215−54
Histocompatibility loci, murine
estimates of number 221
major see H-2 locus
minor 224−5
strong/weak 223−5
Histocompatibility-2 antigens see H-2
antigens
HIV 307
HLA system 241
disease associations 249−54
linkage disequilibrium 249
location 241
molecular genetics 256-72
nomenclature 243−4
polymorphism 242
products 241−9
see also Class I antigens, human and Class
I genes
HLA-A antigens see Class I antigens, human
HLA-B antigens see Class I antigens, human
HLA-C antigens see Class I antigens, human
HLA-D antigens see Class II antigens, human
Holliday model 263−4
Homology regions see Immunoglobulin
domains
Human immunodeficiency virus 307
Hybrid protein molecules 388
Hybridoma
applications 306
double 333−4
production 138−40, 141
repertoires 147−50
Hyperimmunoglobulinaemia E 366
Hypervariable regions 16, 17, 144−5
Hypogammaglobulinaemia 350, 352, 353
transient, of infancy 353
Hypothyroidism 406

I region 61, 233
involvement in GVH 239
involvement in immune response 322−4
involvement in MLR 238−9
molecular map 268−70
recombinant events 271
subregions 234
see also Ia antigens and Ir genes
Ia antigens 61, 238, 240−1
amino acid sequences 63−5
interspecies comparisons 66−8
as differentiation markers 303−4
guinea pig 65−6
human see Class II antigens, human
murine 61−2

relationship with Ir genes 241
role in antigen presentation 296, 323−4
species distribution 65−6
Idiotope 120, 143, 145
Idiotypy
in antibody specificity 142−4
in human immunoglobulins 119−21, 122
IEF 136−7, 138, 142
IgA see Immunoglobulin A
IgD see Immunoglobulin D
IgE see Immunoglobulin E
IgG see Immunoglobulin G
Igh locus, murine 174−83
allotypy 126−7
Igk locus, murine 166, 168−74, 175
Igl locus, murine 164, 166, 167
IgM 31−2, 33
IgN 38−9
IgND see Immunoglobulin E
IgRAA 38
IL-1 296, 324−5, 326
IL-2 325−7
IL-3 327
IL-4 327
IL-5 327
Immune adherence receptor 91
Immune associated antigens see Ia antigens
Immune response, cellular basis 218−20
Immune response genes see Ir genes
Immunocompetence 289−90, 291
Immunodeficiency disorders 4
classification 346−9
genetics 345, 350−83
Immunodeficiency with thymoma 352
Immunogenetic method 1
Immunoglobulin A
amino acid sequences 29
avian 39
deficiency 350−2
role 28
secretory 28, 30−1
structure 27−8, 30
subclasses 28, 97, 183−4
Immunoglobulin D 32−4
evolutionary origin 39
Immunoglobulin domains 5−7, 19−20, 108
evolutionary significance 39
Immunoglobulin E 34−5
avian 39
Immunoglobulin-fold 5, 20−2, 99
Immunoglobulin G 22
amino acid sequences 22−3, 24
chain structure 15
deficiencies see Agammaglobulinaemia
and Hypogammaglobulinaemia
homology regions 19, 108
isoallotypy 106
isotypy 97−8

order of heavy chain genes 118—19
structural mutants, murine 391, *392*
structure 25—7
subclasses 22, *25, 26,* 97—8, 119
Immunoglobulin genes
expression 187
allelic exclusion 201—5
as a function of
differentiation 190—201
molecular genetics
human 183—4
murine 164—83
rabbit 184—5
probable order 118—19
rearrangements in lymphoid
neoplasia 397—8
somatic mutations 208—11
supergene family 4—7
transcription 189—91
Immunoglobulin M 31—2, *33*
Immunoglobulin N 38—9
Immunoglobulins 11
allotypy 97
human 99—119
latent 124—5
murine 125—7
rabbit 121—5
biosynthesis 35—7
characteristics *12*
classes *11*
cleavage 23, 25—6
differentiation markers of B cell lineage
302—3
evolution 20, 37—42
heterogeneity 97—127
idiotypy 97, 119—21, *122*
internal structure 13—20
isoallotypy *103*, 105—6
isotypy 97—9
membrane-bound 36, *37, 38,* 189
molecular genetics 152—85
production 11, 13
secreted 36—7, *38,* 189
three-dimensional structure 20—2
see also *individual classes*
Immunopharmacology 8
Individual antigenic specificity 142
Intergenic conversion 264—6
Interleukin-1 296, 324—5, *326*
Interleukin-2 325—7
Interleukin-3 327
International System for Human Gene
Nomenclature 276
Intervening sequence 166
Introns 166
see also Exon-intron structures
Inv see Km antigen allotypy
Ir genes 61, 233—5

assay 235
co-dominant inheritance 237
complementation 235—7
expression linked to cell—cell
interactions 315—17
guinea pig 65
obsolete terminology 323—4
relationship with Ia products 241
role in antibody presentation 322—4
see also Class II genes
Ir-1 antigens see Ia antigens
ISGN 276
Isoallotypy in immunoglobulins *103,*
105—6
Isoelectric focusing 136—7, *138,* 142
Isotype exclusion 201
Isotype switching 197—201
Isotypy in human immunoglobulins 97—9

J chain 28, *31,* 32, *33*
Job's syndrome 366
Joining (J) segments 166, 168—9
Junctional diversity
antibody genes 212
TCR genes 338

Km antigen allotypy
nomenclature *102*
structural location *107,* 109—12

Lad antigens 238
see also Ia antigens
LAF see IL-1
λ phage, physical map *157*
λ₁ gene system, murine see *Igl* locus, murine
Lamprey, immunoglobulin structure 37
Latent allotypes 124—5
Laws of transplantation 220—1
Lazy leucocyte syndrome 366
LD antigens 238
'Lepore' hybrid 116—18
LFA-1 92
Light chain disease 397
Light chains 14, *15*
abnormalities *396,* 397
expression 193—5
isotypy 98, *99*
regions 15—16, *17*
synthesis 35
Lna antigens see Ia antigens
Local environment hypothesis 313—14, *315*
Low responder strains 233, 316—17, *318*
experimental evidence of defect site 317,
318, 320—1
Ly antigens 299
Lyb antigens 304—5
Lymphocyte activating factor see IL-1
Lymphocyte function associated 92

Lymphocyte neoplasia 397–400
Lymphocytes 220, 289–90
  classification 290
  small 290
  virgin 137, 293, 295
  see also B lymphocytes and T lymphocytes
Lymphocytotoxic tests 243
Lymphoid neoplasia 397–400
Lymphokines 326
Lyt antigens 299–302
Lyt loci 299–300

M components 386
M exons 181–3
Mac-1 92
Macrophage 1 glycoprotein 92
Macrophages 295–6
Major histocompatibility complex see MHC
Malignant disease associations
  chromosomal translocations 205–8,
    398–400
  HLA system 249–54
May-Hegglin anomaly 364
Mean survival time 219
'Mediterranean lymphoma' 393
Membrane attack sequence 81–4
Membrane exons 181–3
Membranous nephropathy 408
  tissue type associations 403
Memory cells 13, 295
  cloning 137–8, 139, 140
Mesangiocapillary nephritis 407–8
MG see Myasthenia gravis
MHC 7, 44
  human see HLA system
  murine see H-2 locus
MHC antigens see Transplantation antigens
MHC restriction 328–30
Minimum mutational distance 39
Mixed leucocyte culture 61
Mixed lymphocyte reaction (MLR) 57, 61
  immunological specificity 238–40
  I region involvement 239–9
MLC 61
MLR see Mixed lymphocyte reaction
MMD 39
'Molecular mimicry' hypothesis 253–4
Monoclonal antibodies 2
  clonotypic 334
  cross-reactivities 142
  production 138–40, 141
  use in detecting differentiation
    antigens 289, 305, 308
mRNA, conversion to DNA 152–3
MST 219
μ chain genes
  expression in pre-B cells 192–3

regulation of expression 195–6, 197
simultaneous expression with C_H
  genes 201
simultaneous expression with δ
  genes 197, 198
μ-heavy chain disease 393, 395, 396
Multigene families 4–5, 8
  see also Immunoglobulin genes
Multiple myelomatosis 387
Multiple sclerosis
  HLA association 251
Murine lupus 410–11
Murine MHC antigens see H-2 antigens
Myasthenia gravis (MG) 408–10
  HLA association 251
  tissue type association 403
Myeloma proteins 13–14, 15, 387
  repertoires 145–6, 148, 150
Myeloperoxidase deficiency 371

N regions
  antibody genes 178, 179, 212–13
  TCR genes 339
Natural killer (NK) cells
  defective activity 363
Network theory 121, 144
Neutropenia 361–2
Neutrophil defects 361–72
Neutrophil pyruvate kinase deficiency 371
19S γ-globulin 31–2, 33
NK cells see Natural killer cells
Non-correlative determinants 108–9, 110
'Non-markers' 103, 105–6
'Northern' blotting 159

OKM1 92
Oncogenes
  translocation 205–8
  see also Cellular oncogenes
One-turn–two-turn hypothesis 170, 177
Opsonization, defective 382–3

P segment 166
Panhypogammaglobulinaemia 353
Paraimmunoglobulinopathies 4, 386–97
pBR322 plasmid, physical map 157
Pelger–Hüet anomaly 364
Pemphigus, HLA association 251
Peptide segment 166
Pernicious anaemia 407
PETLES assay 235
Phage vectors 155–6, 157
Phagocytic defects 361–72
Phenotypic markers of antibody
  specificity 140–5

Plasma cells   13
   see also Antibody-secreting cells
Plasma membrane structure   46, 47
Plasmid vectors   155–6, 157
PLL see Poly-L-lysine
PNP deficiency   345–5, 356
Po group proteins   121, 122
Point mutations   213
Polyadenylation   187–8
Poly-L-lysine (PLL)
   as basis of synthetic antigens   316, 317
   response of guinea pig strains   315–6
Polypeptide chains see Heavy chains and
      Light chains
Post-salmonella arthritis
   HLA association 251
Post-shigella arthritis
   HLA association   251
Pre-B cells   291
   expression of μ chains   192–3
   leukaemias   398
   transition to early B cells   193–5
Pre-proteins   87
Pre-T cells   192
Private antigens   228–9
Pro-C3   86, 87
Pro-C4   86
Pro-C5   86
Promoter   187, 190
Promoter enhancer   189, 190–1
Properdin   81
   deficiency   381–2
   pathway see Alternative pathway
Pseudopromoter   190
Psoriasis, HLA association   251
Public antigens   228–9
Purine nucleoside phosphorylase
      deficiency   354–5, 356

R-loop mapping   159, 165
RA see Rheumatoid arthritis
Radiation chimeras   310
Ragg reagents   100–1
Reagin see Immunoglobulin E
Reciprocal-circle test   231
Recombinant DNA technology   2, 152–64
   application to immunodeficiency
      diseases   352
   application to lymphoid neoplasia
      diagnosis   397–8
Reiter's disease
   HLA association   251
Renal disorders
   role of autoimmune processes   407–8
Reptiles, immunoglobulin structure   32, 38
Responder strains see High responder strains
   and Low responder strains

Restriction endonucleases   153–5, 159
   recognition sites   155
Reticular dysgenesis   357
Retroviruses   153
'Reverse' genetics   308
Reverse transcriptase   153
Rheumatoid arthritis (RA)   411–12
   HLA association   251
   tisue type association   403
Rheumatoid factors   411–12
RNA transcript processing   187–9
Rous sarcoma virus oncogene   205

S regions   175, 180
   switch recombinations   198–201
Sacrificial receptor   30
SC   28, 30, 31
SCID see Severe combined
      immunodeficiency
Screening for cloned DNA inserts   156,
      158–9
Second-set rejection   218, 219
Secretory component   28, 30, 31
Serum carboxypeptidase B   85–6
Serum proteins, polymorphism   97
7S immunoglobulin   38
   see also Immunoglobulin G
Severe combined immunodeficiency
      (SCID)   356–7
   'Swiss-type'   357–8
   with ADA deficiency   355–6
Shark, immunoglobulin structure   37–8
Shwachman's syndrome   366–7
Single recognition hypothesis   330
Sister chromatid exchange   173–4
Skin grafts as transplantation
      models   217–18, 219
SLE see Systemic lupus erythematosus
Snagg reagents   100–1
Soluble factors involved in cell–cell
      communication   324–7
Somatic mutations in Ig genes
   contribution to antibody diversity   208,
      210–11
   mechanism   210
   sites   209–10
   timing   208–9
Southern blotting   159, 164
Specific antibody repertoire   129, 146–7
Spectrotypes   137, 138, 142
Splicing   187–9
Split genes   187
Stem cells   290–1, 292, 293
   defects   357
   determination of differentiation   192
Suppressor T cells   294
   differentiation antigens   306–7

Susceptibility genes 215−17
  estimate of number 221
Switch regions see S regions
Systemic lupus erythematosus (SLE)
  410−11

T cell antigen receptor see TCR
T cell growth factor 325−7
T cells see Cytotoxic T cells, Helper T cells,
    Suppressor T cells and T lymphocytes
T lymphocytes
  allo-reactivity 330−1, 341
  clone expansion 295
  cooperation with B lymphocytes 13,
    311−13, 319−21
    genetic requirements 319−21
  defects 358−9
  determination of MHC specificity 340−1
  differentiation markers 307
    human 306−7
    murine 296−302, 307
    rat 307
  gene rearrangement order during
    development 340
  generation 292−3
  lineage 294−5
  role in antibody production 11, 13,
    311−13
  see also Cytotoxic T cells, Helper T cells
    and Suppressor T cells
T3 (CD3) antigen 306, 307, 339−40
T4 (CD4) antigen 306−7
T8 (CD8) antigen 306−7
Tailing cDNA molecules 156, 158
TATA box 190
TCGF 325−7
TCR 332−4
  gene diversity 337−9
  gene expression 339−40
  molecular genetics 334−7
Tetraparental mice 310, 318
  experimental use 317
Theta 296−8
3rd and 4th pouch/arch syndrome 360−1
Thy-1 296−8
Thymectomy, effect on antibody
    production 311−12
Thymocytes, role in antibody
    formation 311, 312
Thymoma
  association with immunodeficiency 352
  association with MG 408−9
Thymus 290, 292−3, 294
Thymus-derived lymphocytes see T
  lymphocytes

Thymus leukaemia antigens 298−9
Ti molecule 339
Tla locus 60−1, 298−9
TL antigens 298−9
Total antibody repertoire 129
Transcription 187, 188
  Ig genes 189−91
Transcriptional competence 189−90
Transcriptional silence 189
Transferrin 97
Translocations
  association with malignant
    disease 205−8, 398−400
Transplantation
  laws of 220−1
  tumour 44−5, 215−17
Transplantation antigens 7, 44, 46, 216−17
  differentiation between major and
    minor 331
  interspecies comparisons 54, 56, 59
  mutation 57−8, 262−6
  polymorphism 262−6
  special nature 330−1
  species distribution 60
  see also Class I antigens, human and H-2
    antigens
T/t locus 60−1
Tumour graft rejection
  alloantigen theory 216−17
  susceptibility gene theory 215−16

Umbilical cord, delayed separation 367−8

V−J joining 166
V regions 16−17, 18
  amino acid sequencing 145−6, 148, 150
  as phenotypic markers 144−5
  domains 19−22
  see also V$_H$ allotypes
Variable regions see V regions
V$_H$ allotypes 123−4, 143, 184
  mechanisms 169−74
  variability in recombination 174, 175

Wa group proteins 121
Waldenström's macroglobulinaemia 32,
  387−8
WHO Scientific Group on
    Immunodeficiency 345
Wiskott−Aldrich syndrome 358−9

X chromosome inactivation mechanism 202
Xenoantisera 306
X-linked disorders 345, 350